Advances in Environment, Behavior, and Design

VOLUME 2

A Continuation Order Plan is available for this series. A continuation order will bring delivery of each new volume immediately upon publication. Volumes are billed only upon actual shipment. For further information please contact the publisher.

Advances in Environment, Behavior, and Design
VOLUME 2

Edited by

ERVIN H. ZUBE
School of Renewable Natural Resources
University of Arizona
Tucson, Arizona

and

GARY T. MOORE
School of Architecture and Urban Planning
University of Wisconsin—Milwaukee
Milwaukee, Wisconsin

Published in cooperation with the
Environmental Design Research Association

PLENUM PRESS • NEW YORK AND LONDON

Library of Congress Catalog Card Number 88-649861

ISBN-13: 978-1-4612-8047-7 e-ISBN-13: 978-1-4613-0717-4
DOI: 10.1007/978-1-4613-0717-4

Softcover reprint of the hardcover 1st edition 1989

A Division of Plenum Publishing Corporation
233 Spring Street, New York, N.Y. 10013

Contributors

Tommy Gärling, Environmental Psychology Research Unit, Department of Psychology, University of Umeå, Umeå, Sweden

Reginald G. Golledge, Department of Geography, University of California, Santa Barbara, California

Graeme John Hardie, Center for Design Research and Service, School of Design, North Carolina State University, Raleigh, North Carolina

William H. Ittelson, Department of Psychology, University of Arizona, Tucson, Arizona

Gerhard Kaminski, Department of Psychology, University of Tübingen, Tübingen, Federal Republic of Germany

Roderick J. Lawrence, Center for Human Ecology and Environmental Sciences, University of Geneva, Geneva, Switzerland

Willo Pequegnat, Environmental Design Research Association, Bethesda, Maryland

Henry Sanoff, School of Design, North Carolina State University, Raleigh, North Carolina

Herbert W. Schroeder, North Central Forest Experiment Station, U.S. Department of Agriculture Forest Service, Chicago, Illinois

Sally A. Shumaker, Behavioral Medicine Branch, National Heart, Lung and Blood Institute, National Institutes of Health, Bethesda, Maryland

Francis T. Ventre, College of Architecture and Urban Studies, Virginia Polytechnic Institute and State University, Blacksburg, Virginia

Richard Wener, Environment-Behavior Studies Program, Social Sciences Department, Polytechnic University, Brooklyn, New York

Preface

This second volume in the *Advances in Environment, Behavior, and Design* series follows the pattern of Volume 1. It is organized into six sections consisting of advances in theory, place research, user group research, sociobehavioral research, research and design methods, and research utilization. The authors of the chapters in this volume represent a range of disciplines, including architecture, geography, psychology, social ecology, and urban planning. They also offer international perspectives: Tommy Gärling from Sweden, Graeme Hardie from South Africa (recently relocated to North Carolina), Gerhard Kaminski from the Federal Republic of Germany, and Roderick Lawrence from Switzerland (formerly from Australia).

Although most chapters address topics or issues that are likely to be familiar to readers (environmental perception and cognition, facility programming, and environmental evaluation), four chapters address what the editors perceive to be new topics for environment, behavior, and design research.

Herbert Schroeder reports on advances in research on urban forestry. For most of us the term *forest* probably conjures up visions of dense woodlands in rural or wild settings. Nevertheless, in many parts of the country, urban areas have higher densities of tree coverage than can be found in surrounding rural landscapes. Schroeder reviews research that addresses the perceived and actual benefits and costs associated with these urban forests.

Graeme Hardie was invited to write a chapter discussing not only environment–behavior research *in* developing countries but also the appropriateness of environment–behavior research *for* developing countries. Having worked for many years in developing contexts, he draws on his experience and the research-and-design literature to suggest that environment–behavior issues are not the same in developed and developing contexts, nor are methods appropriate for developed contexts necessarily equally useful in developing countries. He then analyzes a number of issues, including house design, infrastructure planning, policy formulation, and fieldwork techniques relying mainly on qualitative methods.

Sally Shumaker and Willo Pequegnat started writing a chapter on advances in hospital design from a behavioral perspective. It rapidly became apparent, however, that they were addressing the failure of hospital design to address the needs of health care providers, and of nurses in particular. As a result, the chapter took a new direction when nurses were identified as a special user group within the matrix of environment, behavior, and design.

Francis Ventre's chapter on research utilization is based on the idea that research utilization involves different epistemological assumptions and different criteria from research itself, namely, those for ensuring efficacy in action and those for substantiating veracity or validity. The basic notion underlying his chapter is that many factors besides systematic research form the basis for design and policy decisions. As he has put it to us, and we agree, research is rarely the single or even the primary basis for environmental decision-making. Rather, research helps to mold the opinion-forming and managerial context within which decisions evolve. Decision-making contexts have been described by social scientists examing the policy-making apparatus of governmental agencies and business firms. These studies have been documented in a variety of places, including the journal *Knowledge*, various *Proceedings* of the Environmental Design Research Association, and earlier articles by Ventre. In this tradition, rather than reviewing the implications of environment–behavior research for policy formulation, Ventre reviews the policy environment in which environment–behavior research resides and attempts to have utility.

In Volume 1, we suggested that there are three fundamental orientations to the field: a pragmatic, social-problem, or environmental-intervention orientation (represented by sections in each volume on professional applications and research utilization); a range of substantive empirical orientations (represented by sections on place research, user group research, and sociobehavioral research); and various theoretical orientations (the lead-in section of each volume).

We have adopted a taxonomy of theories that suggests that different small-*t* theories of environment–behavior phenomena [like Campbell's (1983) theory of ambient stressors or Trist, Higgins, Murray, and Pollack's (1963) theory of sociotechnical systems applied to office buildings, and many others], as well as theories of the middle range (Merton,

1957) can be conceptualized in terms of their underlying philosophical presuppositions, unit of analysis, and locus of control of behavior, or, said more broadly, in terms of their underlying ontological, epistemological, and methodological presuppositions (Moore, 1987).

To continue this analysis, we have again invited leading theoreticians in the field, this time coming from European intellectual traditions, to bring us up to date on the latest advances in two more theoretical directions. In his chapter on ecological conceptualizations, Gerhard Kaminski from Tübingen discusses the theory deficit in the field and then analyzes Barker's ecobehavioral and Gibson's ecological optic perspectives and highlights their most recent contributions and developments—especially those of the Barker school—along with unresolved issues. He also raises many issues about theory development in the environment–behavior research-and-design field in general and hints at a set of metatheoretical principles for future theory development. Roderick Lawrence from Geneva analyzes structuralist theories, distinguishing between global and analytic structuralism and between the cognitive and Marxist structuralist perspectives, and then shows the application to analysis of the syntax and meaning of the built environment. He makes the structural analysis of space his central concern as seen from the points of view of a range of structural orientations, finally highlighting dialectical structuralism by reference to the analysis of housing in Australia and England. Both of these chapters are complex and demanding; they assume close familiarity with the basic works of Barker and Gibson in the first case and of Lévi-Strauss and perhaps Castells in the second case. William Ittelson concludes the section with a discussion of folk and scientific theories and raises provocative questions about their relationship, the role of theory in explaining versus "making contact with" the phenomena of everyday environmental experience, and the possible inappropriateness of verification and falsification as applied to theories of environment, behavior, and design.

The editors have again shared equally in the organization and content of this volume, and each has been the senior editor for half the chapters. In the preparation, we were assisted by many people. We would like to acknowledge the following people, who helped with reviews of chapter drafts: Linda Groat, Paul Maas, Ryuzo Ohno in Japan, Michael O'Neill, Harvey Rabinowitz, and Amos Rapoport, and our Editorial Advisory Board, who have continued to be a great source of support and advice on the directions for this and future volumes. The Plenum staff have again been a joy to work with, and we appreciate their assistance, especially Eliot Werner, Senior Editor and our chief adviser, and Jeanette Haglich and Alyce Hager, who have coordinated publicity and marketing.

Ervin H. Zube
Gary T. Moore

REFERENCES

Campbell, J. M. (1983). Ambient stressors. *Environment and Behavior, 15*, 355–380.

Merton, R. F. (1957). *Social theory and social structure*. New York: Free Press.

Moore, G. T. (1987). Environment and behavior research in North America. In D. Stokols & I. Altman (Eds.), *Handbook of environmental psychology*. New York: Wiley.

Trist, E., Higgins, G., Murray, H., & Pollack, A. (1963). *Organizational choice*. London: Tavistock.

Contents

I ADVANCES IN THEORY

II ADVANCES IN PLACE RESEARCH

III ADVANCES IN USER GROUP RESEARCH

IV ADVANCES IN SOCIOBEHAVIORAL RESEARCH

I

ADVANCES IN THEORY

1

The Relevance of Ecologically Oriented Conceptualizations to Theory Building in Environment and Behavior Research

GERHARD KAMINSKI

This chapter begins by defining "ecological" or "ecologically oriented" approaches and conceptualizations. There are good reasons to concentrate on two outstanding positions: that of Roger G. Barker (and some further developments by a number of his former students) and that of James J. Gibson.

The second section characterizes the problems of theory building in the environment–behavior research-and-design field and gives these problems a particular interpretation. Thus, it becomes obvious where the essential roots of the often-deplored theory deficit might be located and which kinds of contributions would therefore be needed.

The next section of the chapter focuses on the two main ecologically oriented approaches, those of Barker and Gibson, and attempts to evaluate them from metatheoretical perspectives.

The following section amplifies this analysis by taking into account

Gerhard Kaminski • Department of Psychology, University of Tübingen, D-7400 Tübingen, Federal Republic of Germany

recent extensions and modifications of the Barker approach by scholars of different backgrounds.

The last section of the chapter tries to draw conclusions in order to delineate a future perspective for research and theory development, and also for design practice in its dependence on theory and research.

WHAT IS MEANT BY *ECOLOGICAL?*

Within the wide realm formed by the union of environment and behavior research and environmental psychology, a growing number of approaches can be identified that label themselves *ecological.* Space is lacking to review all of them here. Therefore, the argument will be concentrated on the two undoubtedly most prominent ecological positions, those of Roger G. Barker and James J. Gibson. To be sure, the attribute *ecological* has in each case essentially different denotations. But they have at their core a common denominator, and this is what counts when the approaches are classified as *ecologically oriented.* Both Barker and Gibson are influenced, while appropriating the term *ecology,* by certain—although partially different—biological concepts and perspectives. Their common denominator is the intent to describe, analyze, and explain the behaving and the functioning of (especially) human organisms (and aggregations) in natural living conditions. For both of them, the term *ecological* signifies an ambitious, encompassing orientation and research program.

The Barker approach and its further development will be given more extensive treatment here because it appears—at least up to this time—to have been more highly regarded and referred to in the environment–behavior field. However, this situation may (and probably will) change sooner or later. Anyway, to explicate the Gibsonian approach thoroughly in all its imaginable consequences for the environment–behavior research-and-design field would require a chapter of its own.

How do these two approaches differ with respect to their ecological perspective? Barker is convinced that psychology needs an ecological perspective as an analogue to ecology in the biological sciences (Barker & Wright, 1955) and as a complement to experimental approaches within psychology. He has been concerned about the representativeness and the relevance of his own prior experimental research with children (Barker, 1987). Thus, he initiated a "psychological ecology" that aims to describe and analyze naturally occurring behavior in its natural surroundings (Willems, 1965). This approach necessitated the development of a "naturalistic" methodology (Willems & Alexander, 1982; Willems &

Raush, 1969), which should help in answering questions like: What is happening in everyday life? How is it structured? What are the natural units in this stream of happenings? The individual stream of behavior appeared to be segmented in hierarchically organized units of molar, goal-directed actions ("episodes"; Barker, 1978b). But most of these happenings appeared also to be embedded in and determined by supraindividual systemic units called *behavior settings* (Barker, 1968, 1978a), which ought, therefore, to become the focal subject of a supradisciplinary "ecobehavioral science." Barker's classical ecobehavioral approach has been further developed and partially modified by some of his former students and co-workers, especially by Wicker (1979a, 1981, 1987), but also by Bechtel (1977, 1984) and Gump (1987).

Gibson's "ecological approach" (1979) developed in a fundamentally different way (Reed, 1987b). It arose from his many years' commitment to research on perceptual processes and their theoretical articulation. Gibson rejects having perception—or the psychology of perception—begin with the "stimulus," as is urged by traditional experimentation. Instead he holds that the perceptual systems of animals, and also of people, have evolved through a very long process. Their functioning must be sufficient to secure the organism's survival. This requirement implies that perceptual systems support behavioral adaptation to the environment. The functioning of perceptual systems can be understood adequately only if the surrounding conditions, which are relevant to the organism's survival, are conceptualized and described in such a way that it becomes evident what it really is that perception has to recognize. These considerations result in Gibson's "ecological approach to perception" (1979), comprising, among other components, an "ecological optics" and an "ecological theory of perception." In this context, Gibson developed not only categories for the description of the environment on an "ecological level," but also an innovative conceptualization of the functional interrelatedness of environment, perception, and action.

PROBLEMS OF THEORY DEVELOPMENT IN THE ENVIRONMENT–BEHAVIOR RESEARCH-AND-DESIGN FIELD

If "relevance to theory building" is at issue, then what is sensed to be problematic should be obvious. But what is meant by *theory?*

Classes and Functions of Theoretical Structures

A "theory deficit" in the environment–behavior research field (as well as in environmental psychology) is increasingly and unanimously deplored (Moore, 1987; Moore, Tuttle, & Howell, 1985; Stokols, 1983, 1987). The word *theory* has been used in a rather broad and encompassing way. It includes concepts, conceptualizations, and conceptual frameworks, as well as explanatory "theories" in a narrower sense of the word (Moore, 1987; Stokols, 1983). Common to all of them is that they are meant to be projected into reality and also to be used in our coping with reality. Reality appears, after the application of conceptualizations of any kind, to be specifically organized.

Ensuring the precision of scientific communication as well as transparence and reliability in the handling of those structuring instruments necessitates some differentiation, not only in the logical format of the "theoretical" structures, but also in the purposes they have to serve in the production and the application of scientific knowledge (Moore, 1987; Wicker, 1987):

1. Some theoretical structures, like the concepts of *personal space, cognitive map,* and *synomorphy,* have an orienting and descriptive function insofar as they help to delimit sections of reality and to accentuate certain aspects of it.

2. Certain theoretical structures have a systematizing function. Barker's (1968) list of "action patterns" mediates an ordered overview of all kinds of happenings within all kinds of behavior settings. The interior of behavior settings can also be systematized by defining a set of "penetration zones" (resembling roles), or a set of factors may taxonomize the variety of environmental attitudes (Craik & Zube, 1976; McKechnie, 1974).

3. Conceptual or theoretical structures may fulfill heuristic functions. They can help to generate new questions and hypotheses. Once you know what is meant by *social distance,* you may become curious about how it covariates with culture, with age, or with gender.

4. Conceptual or theoretical structures fulfill integrative functions. They may connect separate or loosely tied phenomena in a plausible and scientifically fruitful manner. The concept *behavior setting* connects a variety of events into a systemic unit (Gump, 1987). Another typical example is the concept of *goal-directed action.*

5. In a narrower sense, theories that meet specific formal standards form the basis of hypothesis testing. They contribute to the production of consistent, relatively precise, and valid knowledge. Part of that knowledge may also be applicable to the solution of practical problems.

All of these categories of theoretical structures are needed for different purposes: integrative middle-range theories, as well as verifiable explanatory theories, and descriptive-taxonomic conceptualizations, as well as innovative new concepts.

However, in making use of those structuring instruments, one obviously does not evade a certain kind of trade-off: the more a theory tries to meet the standards of verifiability, the less it is suitable to integrating a broader range of heterogeneous phenomena, and vice versa (McGrath, 1981).

A practitioner may, in principle, make use of all those kinds of theoretical structures (Zeisel, 1981). Every architect is probably guided by more-or-less implicit sociobehavioral concepts, models, and hypotheses. Encountering a new sociobehavioral theoretical structure, the architect can react to it in different ways (Seidel, 1985); by assimilating the new structure to familiar conceptual systems and making use of it in a more-or-less arbitrary way, or by accommodating to it and using it strictly in accordance with its genuine operational meaning and in connection with the proper empirical knowledge, which has been produced and accumulated on its basis and around it.

Tracing the Background of Theory Deficit

What appears to be lacking in all these considerations of theory deficits and theoretical structures are reflections on the particular set of circumstances that might, after all, be responsible for theory deficit and for the particular difficulties of theory development in the environment–behavior research-and-design field. And on the other hand, it seems not to be at all clear which status of theoretical articulation should be striven for in the long run. Would it make sense to conceive of a "grand theory" covering the entire realm (Wapner, 1987)? Or should we envisage a conglomeration or an ordered system of coordinated theoretical structures of different types?

But what should be apprehended as the domain to be organized conceptually by theoretical structures in the environment–behavior field? Hardly anyone who is in any way committed to what is called *environment–behavior studies* or *environmental psychology* would disagree that the reality under discussion encompasses at least two, so to speak, overlapping subdomains: (1) organism–environment interrelationships of different kinds and (2) "environment" in a very broad sense. And probably no one would oppose the view that the "interrelationships" subdomain (i.e., all kinds of interfaces between organism and environment) is not very distinctive of environment–behavior research or of environmental psychology. Rather, the distinctiveness of these fields is located within the "environment" component of the total domain (Proshansky, Ittelson, & Rivlin, 1976; Stokols, 1982). However, with respect to this crucial subdomain, there is less agreement on its essence than one might expect (cf. Russell & Ward, 1982). Yet, these many divergencies seem not to bear the primary responsibility for the theory deficit in environment–behavior research and related fields.

It seems reasonable to hypothesize that it is the "vastness" of the "environment" subdomain that challenges and taxes theory building and that also makes it so exceptionally difficult. *Environment* is a very ambiguous term. The mere appropriation of this label was a sufficient condition for sociobehavioral approaches or investigations to be, so to speak, automatically subsumed under "environment–behavior studies" or "environmental psychology." Thus, a fairly large bulk of diverse contributions aggregated and resulted in the well-known, bewildering status quo. This overwhelming diversity becomes manifest if one realizes, in each of those contributions, what *environment* stands for and compares them in this respect with one another. "Personal space" is an "environment," as is a city; noise, as well as outer space; an object for daily use, as well as a hospital. Thus, the most outstanding characteristic of the most distinctive subdomain of environment–behavior research and environmental psychology seems to be what may be called its vastness. Even at an intuitive level, it is evident that this characteristic will obstruct theory development. On the other hand, a strong need for theory exists. No wonder that complaints about the theory deficit in environment–behavior research and in environmental psychology most often refer to the integrative function of the theories to be developed.

The vastness of "environment" appears in three manifestations, each of which has already given rise to serious attempts to cope with its complexity (see Kaminski, 1987b): a bewildering diversity of pertinent research approaches (e.g., Fisher, Bell & Baum, 1984; Russell & Ward, 1982; Stokols, 1978); a likewise confusing diversity of pertinent theoretical positions (Altman & Rogoff, 1987; Moore, 1987); and an even more embarrassing plethora and diversity of reality components—i.e., elements and constellations of surrounding conditions—(Moore, 1987; Moore *et al.*, 1985). The descriptive-taxonomic task that emanates from the last-named, but primordial, "diversity" or "vastness" manifestation could and should also be part of the first two kinds of taxonomic endeavors. There, however, it has so far been neglected, although this appears to be exactly what is urgently needed to overcome the stagnation in theory development: a descriptive-taxonomic articulation of the "environment" subdomain. As long as this is lacking, the various kinds of theoretical structures that have accumulated in the environment–behavior field can not be related precisely, reliably, and specifically enough to reality, especially to segments of the "environment" subdomain. The definition of mutual interrelationships between theoretical structures suffers accordingly. All this vagueness has many detrimental consequences, not the least of which is difficulty in the further development of theories themselves.

This is where ecologically oriented conceptualizations come in, with what is probably their most fundamental and important contribution to theory development. So that their specific merits will become evident, they will be characterized and evaluated in relation to the hitherto best known descriptive-taxonomic system attempting to articulate the domain of environment and behavior research in its entirety, which has been developed by Moore (1979, 1987) and adopted by the Environmental Design Research Association (Moore *et al.*, 1985).

A Descriptive Taxonomy of Domain Components

The Moore–EDRA taxonomy referred to here is the first module of a (four-component) "multidimensional framework" (Moore, 1987, pp. 1383ff.), which attempts to articulate the research–design circular interplay in longitudinal perspective and in sociocultural context. Within this first descriptive-taxonomic module, the entirety of the domain is broken down into descriptive subdomains and their elements.

The "environment" subdomain, in particular, is partitioned into "places": elements at different scales (room, apartment, house, housing estate, neighborhood, city, region). What can be conceived of as the "interrelationships" subdomain appears here further subdivided in two second-order subdomains: types of people defined by certain enduring characteristics (age, kind of handicap, gender, etc.) and types of sociobehavioral phenomena produced by all these people (internal physiological responses, internal psychological responses, external social responses, etc.).

Elements of different subdomains can be combined in each case to a specific set of, one could say, "reality markers." Each set frames a particular, roughly traced subfield of reality, which may suggest specific inquiries and investigatory operations. Thus, choosing the set "room, infant, internal physiological response" could suggest questions like: "How do an infant's (internal) physiological responses change if the infant is transferred from a familiar to an unfamiliar room"? The extension of these subdomains in time and especially their being embedded in historical sociocultural change can be noticed additionally; the result is a further enrichment of the questioning.

This structuring strategy stimulates and facilitates analyzing and locating empirical approaches, and it helps the researcher to detect gaps and blanks in the research landscape and to envisage and frame new empirical approaches. Furthermore, it conveys to the researcher an impression of the "vastness" of the environment subdomain in its entirety, and it explicitly makes an attempt to articulate this domain by hierarchical differentiation under spatiotemporal aspects. Thus, it pleads for theory development and assists it by collecting and displaying a great variety of domain elements that ought to be taken into account by theoreticians. But it does not give suggestions on how to progress to theories. In an ecological perspective, the most problematic implication of the Moore–EDRA taxonomic system is its *a priori* separation of "places" (understood, in the first instance, as spatial units)—as plausible as it may appear—from the "happening"-like components of the domain. Why this separation is problematic will soon be made clear.

SPECIFIC CONTRIBUTIONS OF THE TWO MAIN ECOLOGICAL APPROACHES TO CONCEPTUALIZING THE ENVIRONMENT–BEHAVIOR FIELD

Paradoxically, the first contribution of both these ecological approaches results in aggravating the situation that they really should help to master: the vastness of the environment subdomain is expanded still further by an ecological perspective. This follows from three fundamental aims and aspirations of psychological ecology:

1. *The width of its claim.* The ecological perspective is an integral component of basic research and therefore aims at fundamental findings and general knowledge. It cannot confine itself to specific types of environments; on the contrary, it should bring forward ecological viewpoints of all varieties and extend them even to phylogenetic and historical change.

2. *Its nearness to reality.* The primary tool of an ecological perspective is observation practiced at the point where the organism encounters reality behaviorally; and there the ecologist is almost forced to take note of an abundance of procedural details. Psychological ecology cannot content itself with environment-related cognitions or verbal data, as they merely contain, on the whole, selective and relatively coarse-grained representations of environments and environment–behavior relations.

3. *Its dependence on psychological progress.* Psychological ecology not only describes but also interprets and finds explanations for what is being observed. Explanatory principles and schemata, which also help to refine observation, are provided by different psychobiological disciplines. They undergo steady development and revision, which lead, on the whole, to progressive complication rather than to simplification.

These three aims hold, in general, for Barker's and Gibson's approaches, although these implications must be differentiated when looked at in detail, as will very soon become obvious.

Barker's Ecobehavioral Approach

It is not possible to review the Barker approach here in its entirety. Rather, the argument will be confined to issues of conceptualization and theory.

Principles of Partitioning Reality. Organizing the environment–behavior realm is not a primary aim of "psychological ecology," as it has been conceived of by Barker and his associates. But as psychological ecology, because of its self-definition, claims, in principle, to deal with all possible surrounding conditions, it should also have an answer to the question of how "environment" in its entirety can be structured. Its answer ought to be of particular interest because psychological ecology has as its goal maintaining exceptionally intimate contact just with the "environment" subdomain. Therefore, a comparison of Barker's ecobehavioral approach and the Moore–EDRA taxonomy promises to become especially informative.

Barker's ecobehavioral perspective cannot be adequately incorporated into the Moore–EDRA taxonomy. The incompatibility follows from the discrepancies in their fundamental presuppositions. The Moore–EDRA taxonomy partitions the "environment" subdomain hierarchically in "geographically" defined, spatial subunits ("places") of different scale. The "behavior setting" unit, on the other hand, is not delimited primarily by geographic criteria, but by "functional" criteria that are defined with reference to "happening" complexes where fixed spatial boundaries and demarcations have only subordinate, not constitutive, relevance. Moreover, the second-order subdomain, "responses," of the Moore–EDRA taxonomy—with its rather elementaristic and constructivistic implications—is not suited to assimilating and classifying the "behavior setting" in its essence as a systemic happening complex.

So this comparison makes it evident that Barker's ecobehavioral perspective organizes reality by means of a completely different fundamental principle. The Moore–EDRA taxonomy starts partitioning the "environment" subdomain within its solid, spatiomaterial substance; it can delimit "place" units without any people being present. The psychological ecologist, such as Barker, on the other hand, could not start partitioning the environment subdomain without people being actually present because the primary partitioning units are constituted by people, by their natural, spontaneous spatiotemporal gatherings and the natural functional boundaries of the happening complexes occurring within them. The partitioning of the spatiomaterial substance comes second, as "milieu" units, which depend for their delimitation on the functional boundaries of the happening complexes. The particular, temporally limited, concrete happening complex (behavior setting) "weekly market," for example, extends as far as this type of happening ("standing patterns of behavior," Barker, 1968) is occurring. This spatial extension and delimitation may not at all coincide with the marketplace, defined as a "geographic" unit.

"Natural Units." Barker's preference (1968) for "natural" units, the most eminent example being the "behavior setting," implies several distinguishable denotative facets. A fundamental aspect is methodological. From a "naturalistic" viewpoint (Willems & Raush, 1969), which is characteristic of the ecological perspective, reality is left untouched as much as possible. Because any kind of invasion or intervention runs the risk of adulterating reality, "unobtrusive" methods (Webb, Campbell, Schwartz, & Sechrest, 1966) have to be applied above all observation. All knowledge gained by obtrusive, interfering, or "reactive" methods must be questioned from a naturalistic viewpoint. The naturalistic stance deserves to be given fundamental priority, although subsequent and more systematic research admittedly requires also the application of intrusive and constructive methodology.

The naturalistic viewpoint is usually associated with the assumption that happenings in reality are in themselves structured in a "natural" way (Barker, 1968), so that naturally existing entities—like behavior settings or their "natural" subunits—can be searched for, observed, and analyzed in their genuine being. These entities—and preserving them in observation and scientific analysis—have priority over construing relatively artificial happening complexes of dubious ecological representativeness and over breaking up natural units and isolating parts of them ("tesserae," i.e., artificial fragments; Barker, 1978b) for the purpose of enforced, "inorganic" data aggregation.

These assumptions imply another one: that natural units have "natural" boundaries, by which they delimit themselves spatiotemporally against their surroundings. Respecting these boundaries has priority over violating them, that is, over imposing more-or-less artificial boundaries on reality, an imposition that would result in cutting out "tesserae" from naturally occurring, organic happening units.

All these assumptions and prerequisites imply once more that reality in the very format of happenings is given priority, especially in the format of behavior-setting units, because these are the most complex kinds of fundamental units. *Most complex* here means that they include social-actional happenings and their physical surroundings ("milieu") in an inseparable interrelatedness. Components or special aspects—like spatiomaterial components or aspects of social interactions alone—could be abstracted, but they could never attain the status of completeness, of originality, or of authenticity that the complex "mother" unit possesses (Gump, 1987). Thus, the "natural" behavior-setting-happening units deserve to be given priority.

One further facet of Barker's preference amounts to interpreting these natural units as systems—that is, as systemic units with a qualified dynamic internal organization—as Barker (1968) held for the "behavior setting." Respecting the systems character of natural units in scientific analysis then takes priority over approaches neglecting or ignoring it. Thus, Stokols's plea (1987) in favor of "composite" in relation to "partitive" analysis seems to tie in with this kind of priority setting, as well as other scholars' committing themselves to "transactional" units and/or their analysis (e.g., Altman & Rogoff, 1987; Wapner, Kaplan, & Cohen, 1973).

The conceptual unit *behavior setting* can always serve as a central orienting model for what constitutes a "natural unit" in environment–behavior research and in environmental psychology. A marketplace, a filling station, or a restaurant, looked upon merely as geographic entities or as pure "settings," is only the milieu part of a behavior setting. None of these is *per se* a "natural unit" for the environmental psychologist, as they are not systems of happenings, each with its own dynamics. Terms like *restaurant* and *filling station* are ambiguous, unfortunately. They can designate either proper "behavior settings" or merely the physical substratum of them. Geographically delimited units are, of course, not irrelevant to environment–behavior research or environmental psychology, nor are they altogether "unnatural." But their being functionally related to ecologically "natural units" remains questionable.

Merits and Weaknesses in Comparison. Obviously, the Moore–EDRA taxonomy and Barker's ecobehavioral approach are rivals, at least insofar as they are alternative attempts to structure the "environment" subdomain. Neither of them could claim to be the right one or to be superior to the other one. Each has its advantages and its weak points, compared with the other one. The Moore–EDRA taxonomy excels through constituting environmental subunits by means of a "hard," objective criterion, which makes the "place" units appear palpable, definite, and highly reliable. This criterion secures a high degree of consensuality—also among different disciplines—with respect to those spatiomaterial environment components. Moreover, this principle of unit formation ought to be most congenial to the mentality of architects, urban planners, geographers, usance managers, and others who are used to handling or even constructing systems with rather clear-cut spatiomaterial boundaries. Furthermore, this principle is guaranteed to be applicable all over the environment subdomain, articulating it in its entirety.

On the other hand, the behavior setting, as the fundamental unit of Barker's ecobehavioral approach, may appear rather "spongy," impalpable, and not very trustworthy. The criteria by which a behavior setting is delimited and constituted (the "k 21-test" in which the functional independence of a particular happening complex is judged with a set of seven rating scales; Barker, 1968) are usually perceived as very complicated and "soft" (Kaminski, 1986a). Probably, it is much easier for social scientists than for experts trained in the physical sciences to appreciate this kind of unit. However, Barker's principle of unit formation has several important merits. It has its roots in the postulate that social (co)determination of behavior is, on the whole, more powerful than behavior's being (co)determined by the surrounding physical conditions. Everyday experience lends this postulate some plausibility: Whether one actually teaches or writes an essay in a classroom, or whether one really plays on a playground is not, after all, determined by the surrounding physical conditions (by the "place") so much as by the "behavior settings" that are occurring there just at that moment. In the classroom case, the behavior setting could be a worship service or a carnival or a geography lesson. And in the playground case, the behavior setting could also be (exceptionally) a worship service or a carnival or a geography lesson, or it could be just an ordinary playground-behavior setting (Barker, 1987). The highest order regulation of the the activity, which occurs within particular physical surroundings, originates in the system of social rules ("program"; Barker, 1968) that is "valid" at the time when the behaving individual is present. The program is, so to speak, the largest actually present and effective social context endowed with regulating power. It also regulates the usage of the surrounding physical conditions, that is, the milieu.

Therefore, from the viewpoint of psychological ecology, it is not only justified but even obligatory in partitioning the environment subdomain to search for all those program-guided, activity-regulating actual happening systems called *behavior settings* as an absolute priority, as difficult as their delimitation may be. For although the geographically defined environmental units may, at first sight, appear to be the "harder" or more substantial ones, nevertheless, viewed in an ecological perspective, they turn out not to stay invariant against variability in social usage.

Thus, the specific virtue of the ecobehavioral principle of partitioning reality is that it submits first and foremost to observable social reality, to manifestations of naturally occurring, spatiotemporal, concrete collective organization. In this characteristic lies, however, also the very reason for the limitations and the weaknesses of the ecobehavioral approach. For this partitioning principle cannot be expanded arbitrarily like the principle of spatiomaterial hierarchization, on which the *place* concept is based.

To be sure, a behavior-setting unit may be subdivided into "subsettings" or "synomorphs," the smallest subunits sometimes being called *cytosettings* (Barker & Wright, 1955), and the fundamental principle of partitioning can then be retained: A "weekly market" behavior setting may be broken down into "stall" behavior settings, which, in turn—at least in those cases where several salesmen or saleswomen are selling relatively independently—could be broken down into subsettings, which can again be further broken down into still smaller "happening" complexes called *synomorphs* (Gump, 1987). All these subunits are delimited by "functional" boundaries in the sense that whatever happens within certain spatiotemporal limits possesses a relatively higher degree of functional homogeneity, of interconnectedness, compared with the connections to the "outside." Or still better: The interconnectedness in certain more-or-less encompassing happening complexes delimits where the spatiotemporal boundaries extend.

If one expects reality articulation "above" the level of behavior-setting units, then Barker's approach is a disappointment. However, the ecobehavioral approach does not ignore this problem completely. Barker proposed some "supra-behavior-setting" structures: certain functional bonds between single behavior settings, "multisetting synomorphs," behavior-setting "genotypes," "authority systems," "town," or "community" (Barker, 1968). But all this kind of superstructure is far less elaborated than the behavior-setting unit itself as the main focus of the partitioning system. Thus, questioned about what this system really achieves in partitioning the environment subdomain in its entirety, Barker would have to admit that it starts articulating the "middle ground" and then expands from there in opposite directions: to smaller scale units and, with comparatively less success, to a few larger scale units. It has a strong bias for the "mid field." In principle, it would be possible, of course, to cultivate the realm of larger scale organization in accordance with Barker's approach. But doing this would necessarily imply developing and applying different principles for delimiting and constituting environmental units because larger scale units of social organization are no longer observable and directly palpable as concrete entities, as behavior settings are.

Another merit of Barker's ecobehavioral perspective becomes salient when it is compared with approaches in the social sciences that attempt to structure reality merely with respect to social organization. In this case, the contact of social organization with concrete spatiotemporal surrounding conditions is in danger of being lost. The behavior-setting concept, on the other hand, maintains consistently and universally the integration of social organization and socially regulated individual activity with concrete spatiotemporal conditions. Regrettably, however, the spatiomaterial components undergo only relatively coarse-grained consideration in Barker's approach, which may give rise to disappointment in anyone especially interested in the physical environment.

How, then, should the researcher and the practitioner be influenced by Barker's ecobehavioral approach? In which ways can they draw on it? They both can gain from it, in important respects, certainly much more than from the Moore–EDRA taxonomy, because Barker's concepts help them identify a reality that is much more richly and concretely preorganized as concerns happenings in everyday life. Both could make rich "conceptual use" (Seidel, 1985) of this conceptual articulation. The researcher could borrow from it suggestions for a great variety of research questions, prearticulated in more detail than is directly attainable from the Moore-EDRA construction-set principle. Making use of Barker's ecobehavioral approach presupposes, of course, that a researcher is interested in just those types of happenings to which the behavior-setting concept is calibrated.

The potential profits for the practitioner also depend on specific tasks, interests, and professional attitudes. Someone who, as an architect, is about to design a particular place—say, a school building—is invited by the ecobehavioral approach to start by imagining the "happening complexes" that are expected to occur there later on, as well as the demands they will impose on the milieu via Barker's "synomorphy" axiom (Gump, 1987). This procedure is obviously akin to the programming period in architectural planning. It amounts to Bechtel's "enclosing behavior" conception (1977). The specific contribution of the Barker approach in this context would be the principle of first taking into consideration all kinds of supraindividual happening systems before considering "synomorphy" for individual activities. (Therefore, *enclosing happening systems* would have been a more specific, although less elegant title.) There is, in detail, much kinship between this kind of exploiting of the ecobehavioral approach conceptually for the benefit of design practice and Alexander's "pattern language" design philosophy (Alexander, Ishikawa, & Silverstein, 1977). In correspondence with programming strategies, the ecobehavioral approach should also be useful in the conceptual preparation and articulation of postoccupancy evaluation.

Someone who wants to make "instrumental use" of the classical ecobehavioral approach would, however, not find much that could substantiate these hopeful expectations (Kaminski, 1986a; Winett, 1987).

What could a theoretician, concerned with theory development in the environment–behavior relations realm, gain from the ecobehavioral approach? Whatever one's predilections might be, one could not evade the behavior setting as a reality of sufficiently verifiable factualness. Moreover, one encounters Barker's rather coarse-grained theory of the internal dynamics of behavior settings, which interprets them as (supraindividual) self-stabilizing, quasi-stationary systems (Barker, 1968). Undoubtedly, this approach contains many more suggestions for theory development. But what might challenge a theoretician most in this context would be the need to elaborate further on the interrelations existing between one behavior setting and the coordinate as well as the superordinate units and on the interrelationships between the behavior setting and the individual as a systemic whole. Besides, the interrelationships between a behavior setting and its supraindividual subunits of different scales also demand further elaboration.

Gibson's Ecological Approach

Gibson developed a descriptive articulation of the "environment" subdomain, which can be viewed, in some sense, as a mixture of the Moore–EDRA taxonomy and Barker's principles of partitioning reality. Gibson's "ecological" conceptualization of "environment" arose, so to speak, as a by-product of the generation of a new theory of perception. This approach can be described here only roughly and selectively in its most pertinent characteristics.

Similarities to the reality articulation by the Moore–EDRA taxonomy exist in that Gibson's approach also reflects physical reality in its entirety, extending from the macrocosmic level of the universe to the microcosmic level of atoms and their components. Thus, he also makes use of the hierarchization ("nesting") principle, applying it to "places" as well as to temporal units. He, too, can partition environment and delimit environmental units without people being actually present because he apprehends persisting units of the surrounding conditions as existing in themselves.

On the other hand, there are many fundamental differences: From the vast totality of surrounding conditions, Gibson has selected a middle-ranged, partial reality, the world of the "ecological level." It is constituted by the principle that everything in it is "animal-scale" or, especially, "human-scale." The term *environment* is reserved for the surrounding conditions of this calibration. In this ecological world, animals, especially people, and "environment" exist in complementarity. Environment is perceived and acted upon. It is organized in nested "natural units": event units and more-or-less persistent units. Terrestrial surrounding conditions are fundamentally subdivided into "medium" (air), "substances," and "surfaces," which are characterized above all by their "layout." "Places" are natural subunits of "habitats"; they are delimited not geographically, but functionally. Therefore, they have no sharp boundaries. Because Gibson has discussed environment–behavior relatedness primarily from an evolutionary point of view, he is not interested in taxonomizing environmental units in terms of living conditions in civilized, urbanized countries, as is suggested by the perspective of environmental design. Instead, he has established more fundamental "ecological" categories (like "terrain," "shelters," "detached objects," "other animals," "written language," and "progressive pictures"). All environmental units possess (positive or negative) "affordances"; that is, they offer some particular kind of usage, which may lead to more-or-less beneficial or detrimental consequences.

Likewise, Gibson has not provided a categorization of people, as the Moore–EDRA taxonomy has, because he is interested only in very fundamental, general theoretical viewpoints. So, with few exceptions, he has merely distinguished between people and different kinds of animals. Gibson's counterpart to the socio-behavioral response categories of the Moore–EDRA system is a scattered, partially systematized enumeration of activity ("behavior") types like "grasping," "opening," "carrying," "sitting," "throwing," "cutting," and "drawing," which are sometimes subsumed under broader categories like "locomotion" and "manipulation."

In contrast to the Moore–EDRA taxonomy, Gibson's descriptive articulation of reality is embedded in a consistent theoretical framework. It is mainly focused on a new, an "ecological" theory of perception.

There are also similarities to Barker's ecobehavioral approach: Gibson, also drawing on concepts of biological ecology, has likewise applied functional principles in partitioning the "environment" subdomain. Like Barker's approach (with its "behavior setting"), Gibson's approach has a particular focal calibration, namely, the perceiving and behaving individual (animal or person), guided by the affordances in the ecological, molar environment. And Gibson has also proposed "natural units," criticizing and devaluing "unecological" ones. The most complex natural unit ("event" in Gibson's case, as compared to "behavior setting") has priority. It cannot be composed of, nor be reconstructed from, simpler ones. On the contrary, simpler units, being more-or-less artificial, can be derived from complex ones by selection and abstraction. The types of "behavior" that Gibson considers remind one sometimes of Barker's "action pattern" and "behavior mechanism" categories, and the concept of *affordance* resembles in some respects Barker's *synomorphy* axiom. The affordance of a "place" corresponds to a certain degree—and on a higher level—to the characteristic, behavior-relevant features of the "milieu" of a behavior setting. This correspondence is even closer at the level of "synomorphs," understood as the smallest natural subunits of behavior settings.

On the other hand, however, these two approaches are separated by deep-rooted differences. Barker's gives clear priority to a supraindividual "natural" unit, and to a dynamic social-systemic unit ("behavior setting"), which executes control over the individual inhabitants, whereas Gibson's approach gives priority to units of individual perceiving and behaving. Barker utilized a rather coarse-grained and atheoretical description by means of extensive "naturalistic" observation of behavior. Gibson analyzed environment–behavior relatedness phenomenologically (and, in part, experimentally) within a thoroughgoing theoretical framework that has its roots in an evolutionary orientation. This framework necessitates relatively fine-grained descriptions of individual transactional processes. The huge diversity of social-cultural phenomena remains at the periphery of his considerations, although Gibson, more than Barker, allows for humans' potential to change their world.

How, then, can research, design practice, and theory building profit from Gibson's approach? Its eminent heuristic value need not be further shown (McLeod & Pick, 1974; Noble, 1987; Reed, 1987a; Reed & Jones, 1982). However, the main focus of research activity inspired by Gibson's ideas is located in basic research. Applied research that might also be stimulated by these ideas would probably deal, in the first instance, with problems of the "dynamic" environment, where surrounding conditions change rather rapidly, especially in connection with locomotion or with the production and reception of "progressive pictures" like films or telecasts.

A practitioner of environmental design could, in principle, make ample "conceptual use" of Gibson's approach, as it is always possible to interpret oneself as a creator of "affordances" (Scheu, 1985). Taking notice of Gibson's stimulating descriptive conceptualizations may sensitize designers to many details of perceptual and behavioral person–environment interrelatedness, and it would help them in differentiating Barker's "synomorphy" concept and increasing its precision and concreteness. This potential, however, still waits to be disclosed and conveyed. It is again Alexander (Alexander *et al.*, 1977) who, with the reasoning in his more elementary "patterns," comes relatively close to Gibson's "affordance" concept.

Gibson's contribution to theory building is obvious yet problematic. He created an almost revolutionary theory of perception, which will probably be under debate for a long time to come. But he seems also to have believed that every aspect of environment–behavior relations that needs explanation can be explained by a theory of perception (especially his)—to express it in a pointed manner. However, he has admitted that many problems are still left unresolved in his theorizing. So attempts to cope theoretically with the "vastness" of environment will continue.

FURTHER DEVELOPMENTS OF BARKER'S ECOBEHAVIORAL PERSPECTIVE AND THEIR THEORETICAL IMPLICATIONS

Until now the attributes *ecological* or *ecobehavioral* referred mostly to Barker's classical approach. *Ecological* so far encompasses several denotative facets (Fuhrer, 1983; Kaminski, 1978). In analogy to biological ecology, it signifies a particular methodological orientation, which favors observing natural behavior in natural surroundings. *Ecological* also characterizes the kinds of reality evidenced by this very methodology, one focal and most essential type of component being the behavior setting. Moreover, *ecological* means "objective" with respect to the surrounding conditions, as opposed to a "subjective," cognitively represented environment. Furthermore, *ecological* or *ecobehavioral*, as Barker uses the terms, implies certain descriptive-taxonomic categories applied primarily to behavior settings (like "action patterns," "behavior mechanisms," and "penetration zones"; Barker, 1968). Finally, *ecological* stands for favoring certain process-theoretical conceptualizations (especially action theory), which refer to the systemic functioning of a behavior setting and to the interaction between a behavior setting and an individual inhabitant.

In the meantime, the term *ecological* has undergone some change in psychology and its vicinity. Partly, it has been expanded and further enriched, and partly, it has been simplified, often without any specific connection with Barker (Fuhrer, 1983, 1986a; Kaminski & Bellows, 1982; Smith & Connolly, 1980; Willems & Alexander, 1982). Gibson anchored this term in an evolutionary perspective of perception and animal–environment relatedness in general. On the other hand, some approaches have emerged that are evidently based in important respects on Barker's perspective, without making use of the term *ecological*. Because it is not possible to report here on all these contributions with adequate thoroughness, only a short overview of the main direction of recent developments can be given, with a main emphasis on problems of theory building and theory modification.

Among those more recent approaches that explicitly relate themselves in some way to Barker's ecobehavioral perspective, six trends toward theoretically modifying or expanding his classical position can be distinguished; and in addition, (point 7) a few solitary attempts to test some of its assumptions in experimental arrangements or in field experiments must be mentioned.

These trends, as diverse as they may appear at first glance, are not independent of each other; they have, on the whole, some general characteristics in common: The description and the functional analysis of the spatiomaterial details of behavior settings are paid relatively little attention (exceptions: Gump, 1987; Molt, 1986; Moore, 1986). Interest is attracted mainly by the social-behavioral components of behavior settings, and cognitive aspects come increasingly to the foreground. This can probably be understood as a reaction to the forbidding effort required by the study of behavior settings in the classical way, relying on observational methodology. This assumption harmonizes with the fact that many of these developments are still merely conceptual, empirical exemplifications or corroborations still being needed (but cf. Gump, 1987; Kirmeyer, 1985; Moore, 1986; Wicker, 1979b, 1987; Wicker & King, 1988).

Modifications of the Conceptual Unit "Behavior Setting"

Amazingly, most of the proposals for a conceptual modification of the behavior-setting unit seem to revive earlier stages in the development of this concept. Thus, Barker and Wright (1949) at first admitted physical characteristics and boundaries as primary delimiting and defining properties of a behavior setting (e.g., a "chair," p. 136). Besides, being generally perceived as an appropriate location for particular kinds of behavior was also a defining characteristic for a behavior setting at that time (pp. 136ff.), a stance that evidenced the still-existing relationship to Lewin (1936, 1951). Recently, a more-or-less pronounced tendency to subordinate the *behavior setting* concept to the *place* concept, with its primarily geographic mode of delimitation, has been observed (Barker, 1987, pp. 1418ff.; Bronfenbrenner, 1977; Fuhrer, 1985, 1986b; Gump, 1987; Moore, 1986; Stokols, 1981; Willems, 1976). Surely, this approach would eliminate the enormous difficulties of a functional delimitation by the "k 21-test"; but at what a price! Another tendency amounts to characterizing and perhaps even delimiting a behavior setting by "perceptual" or cognitive features or boundaries (Kaminski, 1986a), for example, by something like a "social image" or a "perceived social field" (Kruse, 1986; Stokols, 1981). A third tendency involves extending the original behavior-setting definition by applying it also to certain kinds of "multisetting synomorphs" (Barker, 1968) like "firms" or smaller organizations (Wicker, 1987, p. 617; Wicker & King, 1988). This tendency eventually leads to "structuralizing" the behavior setting itself; to giving "structural" features and conditions priority in its definition and delimitation in lieu of the "happening-system" character; and to enlarging the set of constituent "facets" of behavior settings.

Changes in the Embedding of the Behavior-Setting Unit

Originally, the behavior-setting unit was "naturally" and fundamentally embedded in the superordinate unit *town* or *community* (Barker & Schoggen, 1973; Barker & Wright, 1955). Within this context, it is characterized also by certain bonds to other behavior settings or to particular superordinate systems ("authority systems"; Barker, 1968). This kind of embedding is often—at least, implicitly—abandoned, if the concept is modified (Bronfenbrenner, 1977; Stokols, 1981). If the behavior setting is viewed in isolation, freed from these concrete ties, it can be used conceptually in a more flexible manner. On the other hand, the original kinds of embedding can be strengthened or supplemented. Bronfenbrenner provided a new concept ("mesosystem") for designating certain kinds of intersetting ("microsystem") bonds. However, the conceptual interrelations between his terms "setting" and "microsystem" ("school," "family," "workplace," "peer group") and the behavior-setting concept are not precisely established.

Furthermore, Bronfenbrenner pointed to an entirely different kind of social organization ("social networks"), which can, in part, be viewed as being superimposed on the organizational level of behavior settings and also as constituting certain bonds between them (see also Kaminski, 1987a; Roehrle, 1985; Wicker, 1987, p. 639ff.). Moreover, Bronfenbrenner elaborated on the embedding of "settings" in larger sociocultural contexts (1979; also Moore, 1986; Wicker, 1987, p. 637ff.).

Wicker's shift of interest to firms or organizations caused him to enumerate and to systematize several bonds between behavior settings and their surroundings by pointing to "resources" that they need for their coming into existence and their functioning (Wicker, 1987). He introduced "information" as a particular component of intersetting exchanges and of intrasetting functioning, an introduction that has far-reaching consequences for the behavior setting notion itself. Fuhrer (1986a) held that Lewin's "gatekeeper" concept might help in articulating certain external relations of behavior settings (see also Barker, 1987, p. 1419). Barker (1987) pointed to the importance of temporal schemata of a different scale that are imposed on behavior settings "from the remote ecological environment" (p. 1425).

Development of Classifications and Taxonomic Dimensions for Behavior Settings

It was Barker himself who first pointed out that behavior settings could, in principle, be categorized in a multiplicity of ways (1968; 1987, p. 1427). This certainly holds true even where this concept has undergone extensive modification. The criteria and principles for categorization or taxonomization can be rather simple features or rather complex relational properties. Most creative in this respect have been Stokols and associates (Stokols, 1981; Stokols & Jacobi, 1984; Stokols & Shumaker, 1981). They have combined different taxonomic perspectives. On the one hand, they have focused their categorizations primarily on "places" as such (e.g., whether they are "oriented to single individuals," "to coacting aggregates," or "to organized groups," or whether they are "same-occupant-places" vs. "variant-occupant-places," or "single-function-places" vs. "multiple-function-places"). On the other hand, further categorizations are based primarily on characteristics of the sociobehavioral happenings at "places": the number of people typically or actually present and the actual level of systemic organization of the sociobehavioral processes (independently acting individuals, "aggregates," or group processes). Still further categories refer to certain kinds of relationships ("associations") between individuals or groups and particular "places" or "settings" (e.g., place specifity and place dependence). It transpires, finally, that "settings," defined similarly to Barker's "behavior settings," are only a rather special type of unit, definable by a conjunction of some of those taxonomic features. Thus, it also becomes possible to distinguish "settings" from "nonsettings." This rather complicated taxonomic system seemingly amounts to weakening the unique position of Barker's "behavior setting." Wicker (1987, 629ff.) proposed several developmental and functional differentiations of behavior settings and behavior-setting-like systems, which are also applicable as taxonomic categories and dimensions (see also Gump, 1987).

Consideration of the Individual's Relatedness to Behavior Settings

When Barker's team (1978b) observed children's behavior over entire days, most of the children's behavior seemed to be determined by contextual conditions like behavior settings. It may have been the impressiveness of this observation that led to setting aside the question of what is left over for the individual in his or her specificity (Wicker, 1987, p. 616) and his or her unique personal entity. Although this problem has not been ignored entirely by the Barker school, less attention has been devoted to it than it deserves. However, it has attracted the interest of other scholars. The problem has several distinct facets: Disharmony between an individual and a particular behavior setting may arise if the behavior setting does not contribute sufficiently to fulfilling the individual's wishes and goals (Barker, 1978a, pp. 199ff.; 1987, p. 1524, 1428ff.; Boesch, 1986; Wicker, 1987, pp. 625ff.), be they short-lived action goals or more complex and enduring "personal projects" (Little, 1983). A second facet refers to the question of how the individual's partakings in several distinct behavior settings may be related to each other (Bechtel, 1984): In which respects and to what extent does an individual (or do different individuals) remain consistent in the course of participating in different behavior settings, in addition to adapting herself or himself to the different behavior-setting programs (Pawlik & Buse, 1982)? Or in which ways and to what extent does a child's or a youth's excessive participation in special kinds of behavior settings (e.g., training or competition in top sports) have an impact on his or her participational competence and performance in other types of behavior settings, for example, at school or at home (Kaminski, Mayer, & Ruoff, 1984)? In a third facet of the problem, attention is directed to the contribution of behavior settings to the individual's socialization (Barker & Schoggen, 1973, 1978; Bronfenbrenner, 1979; Silbereisen, Eyferth, & Rudinger, 1986; Silbereisen, Noack, & Eyferth, 1985).

"Psychologization" of the Behavior Setting and Its Occupants

It is one of the great merits of Barker's classical ecobehavioral approach that it was rigorously based on the naturalistic observation of behavior. This methodology brought a rich harvest of novel ecopsychological data. But this kind of data alone does not yield access to the deeper structure of actions and interactive happenings, unless by interpretive inferences. This may explain why Barker and his associates have refrained from dealing systematically with the psychological background of the observed behavior. In this way, they have tolerated serious alienation from the mainstream of psychology and its further development (Kaminski, 1983, 1986a). *Psychologization* here stands for endeavors toward a reconciliation between the ecobehavioral approach and traditional fields of psychology. This begins with interpreting happenings within behavior settings by virtue of pertinent psychological models (Gump, 1987, pp. 714ff; Kaminski, 1986b; Wicker, 1987, pp. 624ff.). The complex unit *behavior setting* and its behaving inhabitants afford many diverse possibilities for "psychologization":

A first focus is on social structures and processes. The "manning" concept of the Barker school (Wicker, 1979b) can be apprehended as taking first steps in this direction. Wicker (1987, pp. 627ff.) further elaborated on the "social dynamics" in behavior settings (see also von Cranach, Ochsenbein, & Tschan, 1987; Fuhrer, 1986b).

A second focus is on the cognitive implications of participating in a behavior setting. One fundamental aspect of the inhabitant's information processing is knowledge, especially knowledge of the "program." (Gump, 1987, p. 714, introduced the term "action structure.") Barker (1968) gave a few hints on his own ideas about the acquisition of "program" knowledge. Kruse (1986) drew on the "script" concept of Schank and Abelson (1977; Abelson, 1981; Schank, 1982), which seems apt, in principle, to model the action sequences of a behavior setting's inhabitant. Fuhrer (1985, 1986b) investigated the acquisition of program knowledge empirically. After tracing the process of getting acquainted with a new behavior setting longitudinally by diary methods, Kaminski (1987) held that knowledge about a particular behavior setting obviously encompasses much more than program knowledge. Stokols's notion (1981) of the "perceived social field" of a "place" can—with all its denotational facets and differential dimensions—also be applied to behavior settings and can be interpreted as a special aspect of behavior-setting knowledge. Bronfenbrenner (1979) also stressed the enduring cognitive aspects of settings.

An entirely different perspective focuses on the inhabitant's actual functioning. The person is continually in need of being oriented in the course of the behavior-setting happenings. In this perspective, the program or the rule structure of a behavior setting can be apprehended as a means of facilitating orientation and of relieving capacity load by the reduction of complexity (Kaminski, 1982). However, this interpretation refers only to one special aspect of processing behavior-setting participation. This view gives rise to the question of how this processing in its entirety can be modeled and investigated psychologically (Kaminski, 1987a; see also Neisser's approach, 1976). Gibson's "ecological approach" (1979) is in some respects close to this perspective, although he did not include forms of sociocultural organization at higher levels of complexity, like behavior settings.

Longitudinal Perspectives on Behavior-Setting Changes

The behavior setting can be viewed as a quasi-stationary system that stabilizes itself by "maintenance circuits" (Barker, 1968). However, some behavior settings cease, while new ones come into being (Barker & Schoggen, 1973). But how does this happen in detail? Interest in the "life cycle" of behavior settings has emerged (Stokols, 1982; Wicker, 1981, 1987). The cycle can be segmented into phases of different functional types. The life cycle metaphor refers to the behavior-setting system as a whole. But the longitudinal perspective can also be applied to the inhabitant of a behavior setting, who can be viewed as passing several stages during repeated participation (cf. also Wapner, 1981). After being familiarized with the behavior setting, the person may be associated with it progressively, so that a "dependence" (Stokols & Shumaker, 1981) or even an "identification" may develop.

Empirical Exemplification, Application, and Validation of Behavior-Setting-Related Hypotheses

The global and vague term *theoretical structures* has been differentiated in the second section of this chapter into five subcategories, primarily with respect to the different purposes that they can serve. It may have been noted that, in all the preceding considerations of ecologically oriented conceptualizations, at least one of these subcategories was missing, namely, hypotheses or testable theories. This should not be a cause for surprise, for psychological ecology, because of its very nature, implies a tremendous expansion of the field of scientifically relevant phenomena and their interrelations (the "vastness" mentioned above). So the primary and most pressing task is to find and develop appropriate descriptive units, categories, dimensions, differentiating aspects, and integrating frameworks, and to make descriptive use of them. The generation of specific hypotheses stating more-or-less regularities and/or making circumscribed predictions would imply a narrowing and specializing of the perspective that must appear to be premature and therefore inappropriate. Thus, Barker (1987) separated "discovery research" from "verification research" (p. 1414), and Wicker has defined and tried out something like a new "third way," besides description and hypotheses testing (1986, 1987, pp. 616, 642ff.; Wicker & King, 1988).

Nevertheless, there is a certain developmental dynamic among the ecologically oriented descriptive conceptualizations, based on the illustrative exposure of critical new evidence, on the application of plausibility criteria, and on a consensual acknowledgment of conceptual modifications or extensions. To give only one simplified example, finding the occurrence of "maintenance" operations within behavior settings corroborates the notion that behavior settings are self-stabilizing, quasi-stationary systems (Barker, 1968). It suffices to recall illustrative examples (in which "deviating" actions within behavior settings have the function of "improving" a behavior setting) to legitimize a fundamental revision of the behavior-setting concept. It has thereby undergone, through a consensual shortcut, invalidation and modification, in that, from now on, this kind of systemic conceptual unit will imply the denotative component of a "life cycle," and it will admit and raise new kinds of questions (Wicker, 1987, referring to the "negotiated-order" concept, pp. 627ff.).

But there also exist—at present, still rather few—attempts to apply stricter, more traditional forms of validation or testing in connection with typical ecobehavioral conceptualizations (Fuhrer, 1985, 1986a; Gump, 1987; Kaminski, 1987a; Kirmeyer & Diamond, 1985; Moore, 1986; Wicker, 1979b, 1987). However, closer inspection reveals that what is being investigated and tested are not hypotheses inherent in the fundamental and original ecobehavioral conceptualizations as such. Rather, the descriptive conceptual classics have been extended, enriched, and combined with concepts of different derivation in order to make at all possible the formation of sufficiently precise, testable hypotheses. This extension already holds true for the near-classical example of "undermanning theory" (or "staffing" theory; see Wicker, 1987), where frequency and "intensity" of "experiences of satisfaction and responsibility" come in (Barker, 1987, p. 1424). This extension also holds true for "postclassical" variables such as the (good or poor) "spatial definition" of behavior settings (or "activity pockets"), acoustic and visual separation (in child care centers), and developmentally appropriate activities (Moore, 1986), and for "work overload," "stress," and Type A/Type B behavior patterns (Kirmeyer & Diamond, 1985). Yet, once these extensions have been created and applied in empirical investigations, they "react" to the classical descriptive conceptualizations by introducing specifications and operationalizations, thus raising the theoretical status of these conceptualizations (cf. Moore, 1987, p. 1394), and also perhaps shifting and narrowing their denotation. Anyway, we have the chance to learn that, under special circumstances, the interaction between spatiomaterial, social, and personal components in determining different types of activities is a rather complicated and peculiar one, far more than could be predicted from everyday experience or from simple psychological models (cf. Moore, 1986).

IMPLICATIONS FOR FUTURE RESEARCH AND APPLICATION

Because this chapter has two distinguishable thematic foci (namely, general problems of theory building and the appraisal of ecologically oriented conceptualizations), an attempted look into the future must go in at least two different directions, although there is substantial overlap.

Let us begin with the second point: The future of ecologically oriented approaches ought to begin with terminological clarifications. To hope for—and to strive for—a terminological unification in the usage of the term *ecological* would certainly be unrealistic in view of the ambiguity that the term has adopted. But what seems possible and necessary is to reflect on the interrelations between the fundamental intentions standing behind the use of this term and the approaches designated as "ecological." It appears that there are only two fundamental approaches in the environment–behavior field that lay claim to the term *ecological:* that of Barker and that of Gibson. Thus, the question arises whether—or how—these two approaches can be related to each other, or perhaps even combined and coordinated. But what would be a reasonable long-range, superordinate goal for such an attempt? Two conceivable general aims come into view if one takes biological ecology as an orienting model (as both Barker and Gibson, in fact, did, although in rather different ways and with very different consequences).

The first general aim is description. This kind of ecology would attempt to attain, in principle, a complete inventory of all sorts of environments and all types of organism–environment interrelationships. Taken literally, this is, of course, a utopian goal. A more realistic one would entail a mixed strategy, that, on the one hand, would develop taxonomic dimensions and principles, which should, however, be universally applicable. On the other hand, prototypical examples of environments and environment-behavior relations could be selected, according to criteria like representativeness, importance, and heuristic fruitfulness and would then be described in detail (cf. Wicker, 1986). All this would need multidisciplinary effort and interdisciplinary coordination. Barker's ecobehavioral approach evidently leans primarily toward this descriptive strain of ecology, without having sought contact with its scientific neighborhood (Smith, 1977). What is needed are expanded descriptions of many kinds of terrain that have been left undiscovered and undescribed by the classical ecobehavioral approach (e.g., happenings beyond publicity in homes, happenings in connection with mass-media consumption, happenings in cities and metropolitan areas, happenings in modern work in all kinds of industry and organizations, and happenings in all kinds of modern traffic and in tourist undertakings).

In a second strain of ecology, the general aim is the systematic analysis of functional organism–environment interrelations as they are effective and traceable in everyday life under natural living conditions. These kinds of endeavors surpass the descriptive-taxonomic ones in that they penetrate much more deeply into the details of complicated processes. Obviously, Gibson, more than Barker, is closely connected to this kind of ecology. It may focus on the individual (e.g., Gibson, 1979), on supraindividual social systems (e.g., on behavior settings), or on interrelationships between an individual and social systems (discussed above, in this chapter). Functional-ecological analysis is differentiated still further; it can and it must specialize according to the width of time perspective. At one extreme of the dimension, functional analysis is specialized in organism–environment transactional processes that happen within single seconds or even within fractions of a second, for example, perceptual processes or certain perceptual-motor coordinations in reaction to rapid changes in the (natural) surrounding conditions. At the other extreme of the time dimension stands the attempt to reconstruct the entire phylogenetic development of a certain type of transactional process. Gibson's primary interst was located at the first pole, being focused mainly on perceptual transactions, because he seemed to be convinced that not much more was necessary. He was also interested in the other extreme, the phylogenetic perspective. But there are many stages in between, even as concerns perception, for example, the ontogenetic scale or variations in the analytic perspective corresponding to transactional processes of intermediate scale (perhaps emcompassing several seconds or minutes or hours, etc.). Many possibilities have been left open for future work.

The same is true if the ecologist looks at supraindividual systems (instead of at individual organisms). Barker's functional conceptions (1987) with respect to behavior settings are specialized on procedural units in the (time perspective) scale of human goal-directed actions (p. 1425). This calibration has already been complemented, in part, by the "life cycle" perspective, on one side, and by fine-grained "psychologizations," on the other side of that time perspective dimension. But here also, far more possibilities are open for future ecological research, theoretical as well as empirical, for example, the perspective of historical change or the phylogenetic scale. Of course, there are already approaches that specialize in these other perspectives in neighboring disciplines, but they are apparently not yet integrated into the conceptual framework of a "human ecology," in which the Barker school, as well as the Gibsonians, would each cover special subfields.

Attempts to imagine future research, especially on the behavior-setting unit, encounter a complicated situation: If the term *behavior setting*, which has become ambiguous, will have been disambiguated by conceptual analysis and if the classical "behavior setting" will then stand out clearly again, it could well be the subject of further empirical research (see, e.g., Kirmeyer, 1985). This research would not necessitate applying the "k 21-test" for delimiting behavior settings because the typical examples described by Barker (1968) would probably serve as models for unit delimitation with sufficient reliability. A fruitful general aim would be to refine the descriptive and the theory-related functional analysis of the "social dynamics" of typical behavior settings (Wicker, 1987). In contrast to traditional kinds of (mainly experimental) analyses of group processes, this functional analysis would systematically include also the spatiomaterial components "milieu" and "behavior objects" (Moore, 1986; von Cranach *et al.*, 1987). Refinement could mean attempting to delimit systematically the functional subunits of behavior settings ("synomorphs" or "cytosettings") and to develop theoretical models that would describe the dynamic interplay between them ("transitions" between "segments" in their temporal sequence; Gump, 1987, p. 721). This refinement could also make the behavior-setting concept more easily applicable for the design professions.

Of course, taking up these suggestions, which are primarily derived from current trends in conceptual develoment, demands some inventiveness with respect to the creation of adequate methods for investigation (Kirmeyer, 1985; Moore, 1986).

Some further considerations are devoted here to future perspectives for application. They are confined to the ecological approach of the Barker school and its proximate vicinity, whose implications for design have already been characterized sporadically in earlier sections of the chapter (cf. Barker, 1987, p. 1427; Wicker, 1987, p. 636). The essence of the following statements can be summarized in the assertion that this approach lends itself merely to "conceptual use" (Seidel, 1985) and not—at least in its classical, mainly descriptive structure—to "instrumental use." Comparing the Barker approach with the operant paradigm of behavior modification, Winett (1987) pointed out that the Barker approach never developed a proper technology. Even if all the suggestions for further development of the ecobehavioral approach should be taken up, it would nevertheless remain unlikely that instrumental use could be made of the outcomes, because analyzing the functioning of behavior settings would, in any case, hardly result in something like simple, generally applicable empirical laws (cf. Moore, 1986; Wicker, 1987).

However, this drawback would not preclude employing this approach for the production of quite specific empirical knowledge, gained in the very context where a specific design problem has to be solved. Bechtel (1977) demonstrated which kinds of design-relevant information could be extracted from applying the Behavior Setting Survey method. He introduced incisive methodological modifications in replacing the classical survey with interviews and questionnaires (see also Wicker, 1981). Bechtel seems to have held that empirical data from behavior-setting surveys have, in themselves, a normative character, so that recommendations for environmental design could be read out of them almost immediately. This assumption, however, is not justified. Data from systematic observation are undoubtedly revealing in many respects, but this information always needs to be interpreted, weighed, and evaluated by someone who is qualified. This state of affairs would not be altered by any further development of the ecobehavioral approach, in principle. However, some of the recent developments and part of what has been recommended for future research promise to increase the relevance of this approach to environmental design.

This approach could assist, in principle, in the conceptual structuring of programming and evaluation by allowing for many kinds of supraindividual happening systems, as well as for all forms of individual transactional processes. By further refining the conceptual and empirical analysis of happenings in everyday life, this approach could help to differentiate—and also to complicate—the "synomorphy" principle, giving it a more realistic and concrete shape (Gump, 1987; Molt, 1986; Moore, 1986). This differentiation could facilitate detecting and defining potential or factual deficiencies of "synomorphy." The concept of *synomorphy* therefore needs to be essentially extended and enriched, so that it also includes all kinds of semiotic aspects of person–environment transactions or longitudinal changes in person–environment interrelatedness viewed in different time scales. Taxonomies of behavior settings or behavior-setting-like systems could assist the practitioner's orientation by facilitating the primordial structuring of programming or evaluation tasks.

Concerning the future of theory development in the environment–behavior research-and-design field in general, a few final remarks must suffice here. If the theory deficit is indeed due substantially to the "vastness" of the "environment" subdomain, then this vastness is what needs to be coped with. The Moore–EDRA taxonomy (Moore *et al.*, 1985) made an important attempt to articulate the environment subdomain in its entirety, but this taxonomy is rather void of heuristics and impulses for theory development, largely because it partitions reality primarily into spatiomaterial, "geographic" subunits. Barker's ecobehavioral approach is an alternative that is endowed with some remarkable strengths. But on one crucial point, the very vastness of the environment subdomain, it falls short of what is needed because its articulation of reality is rather specialized and is limited to a middle range. The same is true—although in a different way—of Gibson's approach. Thus, one is left with the impression that these kinds of problems have not yet been analyzed metatheoretically with adequate thoroughness. Where should we go from here?

Notably, one property is common to all three approaches in their attempt to articulate the "environment" subdomain descriptively: They all make use—although in a different manner—of the hierarchization principle, which can obviously be viewed as a basic strategy for complexity reduction.

This hierarchization principle implies that theoretical structures can and must be provided for on all levels of the hierarchies. This necessity again makes it evident that, in no attempts at theory building can we evade a "breadth–depth dilemma" (Kaminski, 1986b; cf. McGrath, 1981). Either we expect from theoretical structures that they will cover a wide range, or we expect them to articulate minute details; these positions, of course, are not mutually exclusive. Apparently, we can place expectations that are not mutually compatible on theoretical structures. Therefore, when we consider theory development and attempt to pursue it, the first and foremost fundamental "given" that we are confronted with seems to be a trade-off. What we need most are means of handling this type of dilemma: perspectives that reflect our interests and aims, criteria for appraising different types of theoretical structures with respect to different kinds of aims, and strategies for setting priorities and for decision making in these dilemmas. In other words, what we need first are certain metatheoretical principles that would guide theory development in the environment–behavior research-and-design field in general and also in specific cases. Such principles have been proposed by the author elsewhere (Kaminski, 1987b); it is not possible to elaborate on them here.

ACKNOWLEDGMENTS

I thank Walter Nilson and Sandra Kirmeyer for their assistance in raising the linguistic qualities of the text. Thanks also to Gary Moore and

Ervin Zube for their valuable hints and proposals concerning the subject matter of this chapter.

REFERENCES

Abelson, R. P. (1981). Psychological status of the script concept. *American Psychologist, 36,* 715–729.

Alexander, C., Ishikawa, S., & Silverstein, M. (1977). *A pattern language.* New York: Oxford University Press.

Altman, I., & Rogoff, B. (1987). World views in psychology: Trait, interactional, organismic, and transactional perspectives. In D. Stokols & I. Altman (Eds.), *Handbook of environmental psychology.* New York: Wiley.

Barker, R. G. (1968). *Ecological psychology.* Stanford, CA: Stanford University Press.

Barker, R. G. (1978a). Behavior settings: Human habitats and behavior machines. In R. G. Barker & Associates, *Habitats, environments, and human behavior.* San Francisco: Jossey-Bass.

Barker, R. G. (1978b). Stream of individual behavior. In R. G. Barker & Associates, *Habitats, environments, and human behavior.* San Francisco: Jossey-Bass.

Barker, R. G. (1987). Prospecting in environmental psychology: Oskaloosa revisited. In D. Stokols & I. Altman (Eds.), *Handbook of environmental psychology.* New York: Wiley.

Barker, R. G., & Schoggen, P. (1973). *Qualities of community life.* San Francisco: Jossey-Bass.

Barker, R. G., & Schoggen, P. (1978). Behavior-generating machines: Models Midwest and Yoredale. In R. G. Barker & Associates, *Habitats, environments, and human behavior.* San Francisco: Jossey-Bass.

Barker, R. G., & Wright, H. F. (1949). Psychological ecology and the problem of psychosocial development. *Child Development, 20,* 131–143.

Barker, R. G., & Wright, H. F. (1955). *Midwest and its children.* New York: Harper & Row.

Bechtel, R. B. (1977). *Enclosing behavior.* Stroudsburg, PA: Dowden, Hutchinson & Ross.

Bechtel, R. B. (1984). Patient and community. The ecological bond. In W. O'Connor & B. Lubin (Eds.), *Ecological approaches to clinical psychology.* New York: Wiley-Interscience.

Boesch, E. E. (1986). Verhaltensort und Handlungsbereich. In G. Kaminski (Ed.), *Ordnung und Variabilitaet im Alltagsgeschehen.* Goettingen: Hogrefe.

Bronfenbrenner, U. (1977). Toward an experimental ecology of human development. *American Psychologist, 32,* 513–531.

Bronfenbrenner, U. (1979). *The ecology of human development.* Cambridge: Harvard University Press.

Craik, K. H., & Zube, E. H. (1976). *Perceiving environmental quality.* New York: Plenum.

Fisher, J. D., Bell, P. A., & Baum, A. (Eds.). (1984). *Environmental psychology* (2nd ed.). New York: Holt, Rinehart & Winston.

Fox, K. A. (1985). *Social system accounts.* Dordrecht: Reidel.

Fuhrer, U. (1983). Oekopsychologie: Some general implications from a particular literature. *Journal of Environmental Psychology, 3,* 239–252.

Fuhrer, U. (1985). Handlungsanalyse durch Setting-Programm-Analyse. *Zeitschrift fuer Experimentelle und Angewandte Psychologie, 32,* 194–216.

Fuhrer, U. (1986a). Beyond the behavior setting. (Book review.) *Journal of Environmental Psychology, 6,* 359–369.

Fuhrer, U. (1986b). Der Erwerb von "Programm-Wissen" in Abhaengigkeit vom Vorwissen sowie von der Komplexitaet des settingbezogenen Kontextes. *Zeitschrift fuer Experimentelle und Angewandte Psychologie, 33,* 11–42.

Gibson, J. J. (1979). *The ecological approach to visual perception*. Boston: Houghton Mifflin.
Gump, P. V. (1987). School and classroom environments. In D. Stokols & I. Altman (Eds.), *Handbook of environmental psychology*. New York: Wiley.
Kaminski, G. (1978). Environmental psychology. *German Journal of Psychology, 2*, 225–239.
Kaminski, G. (1982). *What follows from looking at behavior settings in a naturalistic way*. Paper read at the Second Annual Symposium in Environmental Psychology. University of California, Irvine, April 1982.
Kaminski, G. (1983). The enigma of ecological psychology. (Book review.) *Journal of Environmental Psychology, 3*, 85–94.
Kaminski, G. (1986a). Das Behavior Setting-Konzept im Rueckspiegel. In G. Kaminski (Ed.), *Ordnung und Variabilitaet im Alltagsgeschehen*. Goettingen: Hogrefe.
Kaminski, G. (1986b). Paradigmengebundene Behavior Setting-Analyse. In G. Kaminski (Ed.), *Ordnung und Variabilitaet im Alltagsgeschehen*. Goettingen: Hogrefe.
Kaminski, G. (1987a). Cognitive bases of situation processing and behaviour-setting participation. In G. R. Semin & B. Krahé (Eds.), *Issues in contemporary German social psychology*. Beverly Hills, CA: Sage.
Kaminski, G. (1987b). *Ecologically oriented conceptualizations in environment-behavior-design-research*. Reports from the Psychological Institute, University of Tuebingen, No. 23.
Kaminski, G., & Bellows, S. (1982). Feldforschung in der Oekologischen Psychologie. In J.-L. Patry (Ed.), *Feldforschung*. Bern: Huber.
Kaminski, G., Mayer, R., & Ruoff, B. A. (1984). *Kinder und Jugendliche im Hochleistungssport*. Schorndorf: Hofmann.
Kirmeyer, S. L. (1985). Employee reactivity and adaptation to observation on the job. *Journal of Environmental Psychology, 5*, 355–372.
Kirmeyer, S. L., & Diamond, A. (1985). Coping by police officers: A study of role stress and type A and type B behavior patterns. *Journal of Occupational Behaviour, 6*, 183–195.
Kruse, L. (1986). Drehbuecher fuer Verhaltensschauplaetze oder: Scripts fuer Settings. In G. Kaminski (Ed.), *Ordnung und Variabilitaet im Alltagsgeschehen*. Goettingen: Hogrefe.
Lewin, K. (1936). *Principles of topological psychology*. New York: McGraw-Hill.
Lewin, K. (1951). *Field theory in the social sciences*. New York: Harper.
Little, B. R. (1983). Personal projects: A rationale and method for investigation. *Environment and Behavior, 15*, 273–309.
MacLeod, R. B., & Pick, H. L., Jr. (Eds.) (1974). *Perception*. Ithaca, NY: Cornell University Press.
McGrath, J. E. (1981). Dilemmatics. *American Behavioral Scientist, 25*, 179–210.
McKechnie, G. E. (1974). *Manual for the environmental response inventory*. Palo Alto, CA: Consulting Psychologists Press.
Molt, W. (1986). Das Bahavior Setting Strassen. In G. Kaminski (Ed.), *Ordnung und Variabilitaet im Alltagsgeschehen*. Goettingen: Hogrefe.
Moore, G. T. (1979). Environment-behavior studies. In J. C. Snyder & A. J. Catanese (Eds.), *Introduction to architecture*. New York: McGraw-Hill.
Moore, G. T. (1986). Effects of the spatial definition of behavior settings on children's behavior: A quasi-experimental field study. *Journal of Environmental Psychology, 6*, 205–231.
Moore, G. T. (1987). Environment and behavior research in North America: History, developments, and unresolved issues. In D. Stokols & I. Altman (Eds.), *Handbook of environmental psychology*. New York: Wiley.
Moore, G. T., Tuttle, D. P., & Howell, S. C. (1985). *Environmental design research directions*. New York: Praeger.
Neisser, U. (1976). *Cognition and reality*. San Francisco: Freeman.

Noble, W. (1987). Perception and language: Towards a complete ecological psychology. In A. Costall & A. Still (Eds.), *Cognitive psychology in question*. Brighton: Harvester Press.
Pawlik, K., & Buse, L. (1982). Rechnergestuetzte Verhaltensregistrierung im Feld. Beschreibung und erste psychometrische Ueberpruefung einer neuen Erhebungsmethode. *Zeitschrift fuer Differentielle und Diagnostische Psychologie, 3,* 101–118.
Proshansky, H. M., Ittelson, W. H., & Rivlin, L. G. (Eds.). (1976). *Environmental psychology* (2nd ed.). New York: Holt, Rinehart & Winston.
Reed, E. S. (1987a). James Gibson's ecological approach to cognition. In A. Costall & A. Still (Eds.), *Cognitive psychology in question*. Brighton: Harvester Press.
Reed, E. S. (1987b). Why do things look as they do? The implications of J. J. Gibson's the ecological approach to visual perception. In A. Costall & A. Still (Eds.), *Cognitive psychology in question*. Brighton: Harvester Press.
Reed, E. S., & Jones, R. (Eds.). (1982). *Reasons for realism*. Hillsdale, NJ: Erlbaum.
Roehrle, B. (1985). Zur Oekologie sozialer Stuetzsysteme. In P. Day, U. Fuhrer, & U. Laucken (Eds.), *Umwelt und Handeln*. Tuebingen: Attempto.
Russell, J. A., & Ward, L. M. (1982). Environmental psychology. *Annual Review of Psychology, 33,* 651–688.
Schank, R. C. (1982). *Dynamic memory*. Cambridge: Cambridge University Press.
Schank, R. C., & Abelson, R. P. (1977). *Scripts. plans, goals, and understanding*. Hillsdale, NJ: Erlbaum.
Scheu, R. (1985). Ueberlegungen zur oekopsychologischen Konzeptualisierung der Mensch-Umwelt-Beziehung. In P. Day, U. Fuhrer, & U. Laucken (Eds.), *Umwelt und Handeln*. Tuebingen: Attempto.
Seidel, A. D. (1985). What is success in E&B research utilization? Environment and Behavior, 17, 47–70.
Silbereisen, R. K., Noack, P., & Eyferth, K. (1985). Untersuchungen zu Jugendtreffpunkten. In P. Day, U. Fuhrer, & U. Laucken (Eds.), *Umwelt und Handeln*. Tuebingen: Attempto.
Silbereisen, R. K., Eyferth, K., & Rudinger, G. (Eds.). (1986). *Development as action in context*. Berlin: Springer.
Smith, M. B. (1977). Some problems of strategy in environmental psychology. In D. Stokols (Ed.), *Perspectives on environment and behavior*. New York: Plenum.
Smith, P. K., & Connolly, K. J. (1980). *The ecology of preschool behaviour*. Cambridge: Cambridge University Press.
Stokols, D. (1978). Environmental psychology. *Annual Review of Psychology, 29,* 253–295.
Stokols, D. (1981). Group × place transactions: Some neglected issues in psychological research on settings. In D. Magnusson (Ed.), *Toward a psychology of situations*. Hillsdale, NJ: Erlbaum.
Stokols, D. (1982). Environmental psychology. In A. Kraut (Ed.), *G. Stanley Hall Lecture Series*, Vol. 2. Washington, DC: American Psychological Association.
Stokols, D. (1983). Theoretical directions of environment and behavior research. *Environment and Behavior, 15,* 259–272.
Stokols, D. (1987). Conceptual strategies for environmental psychology. In D. Stokols & I. Altman (Eds.), *Handbook of environmental psychology*, New York: Wiley.
Stokols, D., & Jacobi, M. (1984). Traditional, present oriented, and futuristic modes of group-environment relations. In K. J. Gergen & M. M. Gergen (Eds.), *Historical social psychology*. Hillsdale, NJ: Erlbaum.
Stokols, D., & Shumaker, S. A. (1981). People in places: A transactional view of settings. In J. H. Harvey (Ed.), *Cognition, social behavior, and the environment*. Hillsdale, NJ: Erlbaum.

von Cranach, M., Ochsenbein, G., & Tschan, F. (1987). Actions of social systems: theoretical and empirical investigations. In G. R. Semin & B. Krahé (Eds.), *Issues in contemporary German social psychology*. Beverly Hills, CA: Sage.

Wapner, S. (1981). Transactions of persons-in-environments: Some critical transitions. *Journal of Environmental Psychology, 1*, 223–239.

Wapner, S. (1987). A holistic developmental, system-oriented environmental psychology: Some beginnings. In D. Stokols & I. Altman (Eds.), *Handbook of environmental psychology*. New York: Wiley.

Wapner, S., Kaplan, B., & Cohen, S. B. (1973). An organismic-developmental perspective for understanding transactions of men and environments. *Environment and Behavior, 5*, 255–289.

Webb, E. J., Campbell, D. T., Schwartz, R. D., & Sechrest, L. (1966). *Unobstrusive measures*. Chicago: Rand McNally.

Wicker, A. W. (1979a). Ecological psychology: Some recent and prospective developments. *American Psychologist, 34*, 755–765.

Wicker, A. W. (1979b). *An introduction to ecological psychology*. Monterey, CA: Brooks/Cole.

Wicker, A. W. (1981). Nature and assessment of behavior settings: Recent contributions from the ecological perspective. In P. McReynolds (Ed.), *Advances in psychological assessment*, Vol. 5. San Francisco: Jossey-Bass.

Wicker, A. W. (1986, August). *Substantive theorizing: Clothing for a naked emperor?* Division 34 presidential address, presented at the meeting of the American Psychological Association. Washington, DC.

Wicker, A. W. (1987). Behavior settings reconsidered: Temporal stages, resources, internal dynamics, context. In D. Stokols & I. Altman (Eds.), *Handbook of environmental psychology*. New York: Wiley.

Wicker, A. W., & King, J. C. (1988). Life cycles of behavior settings. In J. E. McGrath (Ed.), *Research toward a social psychology of time*. Beverly Hills, CA: Sage.

Willems, E. P. (1965). An ecological orientation in psychology. *Merrill-Palmer Quarterly, 11*, 317–343.

Willems, E. P. (1976). Behavioral ecology, health status, and health care: Applications to the rehabilitation setting. In I. Altman & J. F. Wohlwill (Eds.), *Human behavior and environment*, Vol. 1. New York: Plenum.

Willems, E. P., & Alexander, J. L. (1982). The naturalistic perspective in research. In B. B. Wolman (Ed.), *Handbook of developmental psychology*. Englewood Cliffs, NJ: Prentice-Hall.

Willems, E. P., & Raush, H. L. (1969). *Naturalistic viewpoints in psychological research*. New York: Holt, Rinehart & Winston.

Winett, R. A. (1987). Empiristic-positivist theories of environment and behavior: New directions for multilevel frameworks. In E. H. Zube & G. T. Moore (Eds.), *Advances in environment, behavior, and design*, Vol. 1. New York: Plenum.

Zeisel, J. (1981). *Inquiry by design*. Monterey, CA: Brooks/Cole.

2

Structuralist Theories in Environment–Behavior–Design Research

APPLICATIONS FOR ANALYSES OF PEOPLE AND THE BUILT ENVIRONMENT

RODERICK J. LAWRENCE

Structuralist theories and methods have gained momentum since the late 19th century, largely owing to the exchange theory of the French anthropologist Marcel Mauss, the linguistic theories of Roman Jakobson at the Prague school of structural linguistics, and the contribution of Ferdinand de Saussure. Nonetheless, *structural* explanation became an integral part of analytical approaches in the social sciences only in the second half of this century, largely as a result of the contributions of the French anthropologist Claude Lévi-Strauss, the Swiss psychologist Jean Piaget, and the Marxism of Louis Althusser. Today, there are published studies in fields including archaeology (Clarke, 1977; Hodder, 1982), architecture (Eco, 1972, 1973, 1976; Glassie, 1975; Lüchinger, 1981), linguistics (Macksey & Donato, 1972; Sturrock, 1979), social anthropology (Leach, 1976; Lévi-Strauss, 1968), and urban sociology (Castells, 1977, 1978). This complex range of contributions cannot be considered a unified set that suggests a shared perspective and a linear development of structure theory and method. For example, although the interpretation of Lévi-Strauss owes much to Mauss and Jakobson (Badcock, 1975;

Roderick J. Lawrence • Center for Human Ecology and Environmental Sciences, University of Geneva, 1211 Geneva 4, Switzerland.

Chiari, 1975), the advocates of structural Marxism reject the overriding importance which his cognitive structuralism attributes to symbolic dimensions and a synchronic perspective. Beyond such divergences, which are discussed in this chapter, the debate about orthodox structural explanations has led to *poststructuralist* contributions, led by Jacques Derrida, and including the works of Pierre Bourdieu (1977) and Anthony Giddens (1976, 1979), so that the notions of *structure* and *structuralism* need to be defined within the history of ideas and contributions in this field (Chiari, 1975; Sturrock, 1979). Therefore, I do not intend to provide a definition that will be unanimously acceptable. Rather, the aim of this chapter is to advance the search for a conceptual definition and research methods for structuralist studies of people and the built environment.

This chapter begins with a discussion of diverse interpretations of structural theories and structuralist methods. It shows that at least two distinctly different interpretations of structure and structuralism have been applied to the study of people and their physical surroundings. Beyond the two main classes of interpretations, a range of theories and approaches is presented in order to show that there are divergences not only between scholars in diverse disciplines, but also within disciplines, such as structural archaeology, as Hodder (1982) and Tilley (1982) ably showed. This chapter first focuses on several interpretations and applications that are specifically linked to comprehending the reciprocal relations between people and the built environment. Second, it presents a critique of these contributions in order to identify their strengths and weaknesses. Finally, this chapter suggests and illustrates how the redefinition of orthodox structuralist theories and methods can enhance our understanding of the reciprocal relations between people and the built environment. The concluding paragraphs present an outline of dialectical structuralism and indicate some implications for future interdisciplinary research and design practice, by referring to a cross-cultural and historical analysis of domestic space in Australia and England.

WHAT IS STRUCTURALISM?

The advocates of structuralist theories and methods share a perspective that maintains that scientific studies of human societies and culture cannot be restricted to observable phenomena. A corollary of this stance is that human knowledge and understanding depend fundamentally on theoretical analysis that goes beyond manifest empirical phenomena and encompasses the tacit ideas, processes, or structures responsible for this material culture. However, beyond this shared perspective, there is much ambiguity about the notions of *structure* and *structuralism* in the social sciences, which authors such as Barthes (1967), Boudon (1971), Lane (1970), Macksey and Donato (1972), and Robey (1973) have discussed at length. These contributions suggest that this ambiguity is reflected, at least in part, in the association of these notions with concepts such as *system* and *holism,* and also in two different interpretations of structuralism. In the first interpretation, the term *structure* refers to the systematic nature of an object or event (or of a set of objects and events). In the second interpretation, the term *structure* is commonly used to describe and *explain* the systematic composition of objects and events—including language, texts, and music—in terms of the social signification that specific elements or component parts acquire as signs. Both these interpretations have been adopted by scholars from diverse disciplines with different, sometimes contradictory, theoretical perspectives, for example, cognitive psychology and Marxism. Nonetheless, these two interpretations have rarely been differentiated in architectural discourse or by authors such as Doxtater (1984), Glassie (1975), and Lüchinger (1981) and, in behavioral studies of people and the built environment, including Rapoport (1982). In order to clarify the theoretical and methodological contributions of structuralist approaches, this chapter compares and contrasts two dominant interpretations, which are not considered mutually exclusive.

Global Structuralism

In the first interpretation, the term *structure* refers to a system of observable interrelationships between a set of variables that are examined according to their reciprocal relations. This notion is what Piaget (1971) termed "global" structuralism (p. 98). From this perspective, Boudon (1971) stated that "a structure is always the theory of a system—and it is nothing else" (p. 133). An example of this interpretation can be given from the field of structural linguistics. In this context, the term *structure* commonly refers to "the manner in which the individual elements of a specific language are arranged in relations of mutual independence for the purpose of communication" (Robey, 1973, p. 1). Moreover, Robey noted that, because the relations between elements differ from one language to another:

> It follows that the separate components of a system can only be understood in the light of the system as a whole, and therefore, that the primary object of

> linguistic study must be the structure of the system itself rather than the individual linguistic fact. (p. 1)

Likewise, Lyons (1973) stated that:

> Linguistic units derive both their existence and their essence from their interrelationships. [Furthermore] every distinct language is a unique relational structure; and the units which we identify in describing a particular language—sounds, words, meanings etc.—are but points in the structure, or networks, of relations. (p. 6)

This kind of interpretation led Robey (1973) to define *structuralism* "as that approach—to any subject—which has as its object . . . the laws of solidarity, the reciprocal relations of the different facts under observation, rather than considering these facts in isolation" (pp. 1–2).

Within the field of architectural research, Glassie (1975) used this interpretation of structure and structuralism to decipher the mathematical rules that can account for the size, shape, and composition of rooms of vernacular houses built in one region of the State of Virginia. Glassie developed a typology of 17 basic types of houses, founded on the combinations of room size, room shape, and the locations of doors, windows, and chimneys as illustrated in Figure 1. Glassie maintained that the house builders consciously or unconsciously used these geometrical and compositional rules—which he called an "architectural competence"—when these dwellings were constructed. Although Glassie's interpretation does have strong empirical grounds for acceptance, there is no understanding of the reasoning for, or the meaning of, the architectural competence because he did not relate his analysis to other

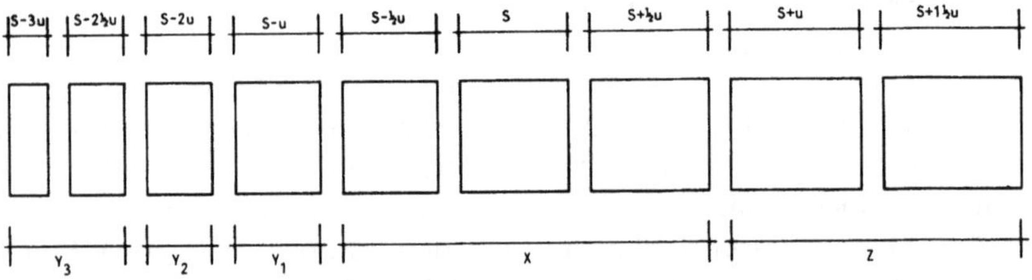

FIGURE 1. The geometrical and compositional rules, developed by Glassie, that are the basis of an architectural competence. Above the shapes are the designations for the transformations of the square. Below them are set out the classes of shapes. This competence was developed on the basis of s = 15 feet, u = 3 feet. (Adapted from Glassie, 1975.)

nonphysical aspects of the design and use of the built environment, notably the nomenclature and use of rooms. In this respect, Glassie upheld the innate capacity of the human mind, as the advocates of cognitive structuralism are wont to do. However, unlike many structuralist scholars, who do not limit their research to material culture, Glassie was unable to show that the mathematical rules he presented are signs of, or homologous with, other aspects of human culture in Virginia. In sum, this example of global structuralism includes a very limited number of variables, and there is no understanding of the meaning of the mathematical rules they present.

Analytical Structuralism

The second interpretation of *structure* and *structuralism* not only encompasses the first but also represents a more distinctive body of theoretical and methodological concepts, derived primarily from studies in the fields of human cognition, structural linguistics, and Marxism. The second interpretation of *structuralism* endeavors to *explain* the interrelationships between different sets of variables that are postulated to be linked by unobservable logicomathematical models, or what structuralists commonly term *transformational interactions*. This interpretation is what the Swiss psychologist Jean Piaget (1971) termed "analytical" structuralism (p. 98). According to Piaget (1971), "the notion of structure is comprised of three key ideas: the idea of wholeness, the idea of transformation, and the idea of self-regulation" (p. 5). Although this kind of structuralism has been adopted by cognitive structuralists and structural Marxists, the interpretation of these two groups is quite different and warrants our attention.

Cognitive Structuralism. Within the field of cognitive structuralism, it has been commonplace to assume that the theory and methods of structural linguistics are directly or indirectly applicable to the analysis of all aspects of human culture, insofar as all of these, like language, may be interpreted as systems of signs: "Structuralism in this sense is more or less coterminous . . . with *semiology*, or *semiotics*, the science of signs" (Robey, 1973, p. 2). In general, the aim of semiotic studies is, as Necipoğlu (1981), said, "to study sign systems instituted by society within given contexts, in order to understand the culturally codified conventions on which the process of signification is based" (p. 260).

The linguistic analogy, and its application to all aspects of human culture, has been the subject of a heated debate in a range of disciplines, including social anthropology and architecture (cf. Boudon, 1971; Broadbent, Bunt, & Jencks, 1980a; Broadbent, Bunt, & Llorens, 1980b; Colquhoun, 1981; Eco, 1973; Jackson & Smith, 1984; Macksey & Donato, 1970; Markus, 1987; Robey, 1973). Culler (1973) explained the reasoning for the use of this analogy in the following way:

> A language is not simply a collection of sentences, for there are well-formed sentences of English that have never been uttered, and an adequate description of English would have to account for these potential grammatical and ungrammatical sentences. The language must therefore be considered as a system of items and rules that have been assimilated by native speakers; a description of the language is an explicit representation of their implicit knowledge. The facts which linguistics must explain are various, but they are all facts about this implicit knowledge: that any sentence on this page, read backwards, is not a well-formed English sentence, that *Flying planes can be dangerous* is ambiguous, that *The enemy destroyed the city* and *The city was destroyed by the enemy* are synonymous, and so on. One might say, by analogy, that the sociologist or anthropologist attempts to explicate the implicit knowledge which enables people to function as members of a particular group or society. (p. 24)

From this perspective, the French anthropologist Claude Lévi-Strauss (1968) maintained not only that kinship systems are examples of algebraic structures, but that marriage rules are signs of and homologous with other sets of rules and meanings embedded in the everyday life of people. In one ethnography, for example, Lévi-Strauss (1968) examined the concentric layout of the village plans in the Trobriand Islands. Lévi-Strauss criticized functionalist interpretations of built environments, including that by Malinowski (1929) of the villages on the Trobriand Islands which is limited to an analysis of manifest culture.

The plan of an Omarakana village, presented in Figure 2, is arranged in two concentric rings. A central, public space is surrounded by yam storehouses, which are sacred and to which taboos are attributed. A circular street is located around these storehouses, and the dwelling units of the married couples are constructed around the outer edge. Malinowski (1929) noted that, in contrast to the central and sacred space of the village, the peripheral zone is profane. Beyond the oppositions between *center* and *periphery* and between *sacred* and *profane*, the opposing functions of storing and cooking food are also evident, because, as Lévi-Strauss (1968) noted, "in the storehouses of the inner ring *raw* food is stored and cooking is not allowed," and simultaneously, "food can be *cooked* and *consumed* only in or around the family dwellings of the outer ring" (p. 139). Moreover, only unmarried men can live in the inner ring, and married couples must live on the periphery. Finally, the two concentric rings in an Omarakana village are opposed with respect to gender, so that the central space is considered a male domain, whereas the peripheral zone is a female domain.

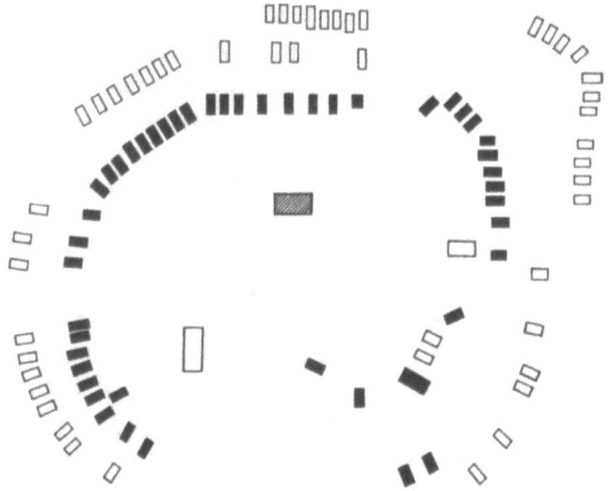

FIGURE 2. Plan of an Omarakana village (Adapted from Lévi-Strauss, 1968, after Malinowski, 1929.)

Lévi-Strauss (1968) extended his study by examining those rules prohibiting and permitting the marriage of the residents in the village. This analysis concludes that the layout of an Omarakana is a complex metaphor of the interpersonal relationships and daily activities of the inhabitants. In sum, this structural analysis illustrates that, although the spatial organization of a village can be described according to the orientation, the relative position, and the demarcation of sets of objects, activities, and people, these manifest, interdependent variables alone cannot account for the social meaning of the design and use of spaces in the village unless tacit ideas and values, including the categorization of people, food, and activities, are also considered as an integral component of the unit of analysis. Lévi-Strauss showed that the spatial metaphors are related not only to a north–south and an east–west axis, but also to the bipolar oppositions between male and female, bachelorhood and marriage, center and periphery, and sacred and profane. Moreover, even when some villages were moved, Lévi-Strauss noted that the same layout, geometry, and relative position of buildings and spaces were reconstructed.

Structural Marxism. The structural analysis presented above and, in general, all analytic structuralist studies enable us to simultaneously describe and explain a human-made object or event in an holistic way, by considering the interrelations between both tacit and manifest components. This ability is particularly useful given that many cultural products, especially the built environment, are too complex to be represented by a single interpretation or model. Nonetheless, the context of society examined by many structuralist archaeologists or anthropologists, including Lévi-Strauss, is very different from the context of many contemporary societies, so that analogies between the two are questionable. Therefore, if one wishes to compare the contributions of structuralist scholars, it is important to recognize that, in contrast to many cognitive structuralists, the majority of structural Marxists have endeavored to explain products and processes in contemporary socialist and capitalist societies. Unfortunately, the implications of comparing structuralist studies in such different societies have rarely been discussed, even by those who have presented a so-called critical or radical perspective (Gregory, 1978). Nonetheless, as Lawrence (1987), for example, showed, differences between the design and use of the built environment in past and present nonindustrialized societies and contemporary industrialized countries are not only, nor completely, reflected by differences between their built environments. In principle, it is also necessary to consider the implicit and explicit contextual contingencies that relate to the production, the use, and the subsequent transformation of all material culture, especially buildings, during the course of time. In this respect, structural Marxism has offered important insights at the level of urbanization, more than at the level of the design and layout of buildings (cf. Lebas, 1982). In general, the nature of tacit structural determinants—economic, political, sociodemographic, and ideological variables—are examined in order to explain production and reproduction processes, including those that create and nurture the fundamental foundations of urbanism. Diverse interpretations of this kind have provoked the "structuralist debate" among social scientists (cf. Giddens, 1976; Gregory, 1978; Gregory & Urry, 1985; Jackson & Smith, 1984). This debate has led to the acceptance or rejection of structuralist theories and methods by scholars from diverse disciplines and, simultaneously, to the development of different interpretations of structuralism, notably those grounded in psychological theories of human cognition in contrast to those developed from Marxist theories of human societies. The debate concerning these interpretations now warrants our attention.

STRUCTURALIST INTERPRETATIONS

Structuralism and Human Cognition: The Contribution of Lévi-Strauss

For those analytical structuralists who have grounded their approach in human cognition, the world exists as it is apprehended by the human mind. Therefore, by analyzing how people apprehended specific objects or events, and especially by studying the way people classify and use space, objects, activities, and time, it is possible to infer fundamental principles about human culture. Leach (1970) suggested that this kind of structuralism, as interpreted and applied by Lévi-Strauss and other social anthropologists, adopts the following approach:

> What we know about the external world we apprehend through our senses. The phenomena which we perceive have the characteristics which we attribute to them because of the way our senses operate and the way the human brain is designed to order and interpret the stimuli which are fed into it. One very important feature of this ordering principle is that we cut up the continua of space and time . . . into segments so that we are predisposed to think of the environment as consisting of vast numbers of separate things belonging to named classes, and to think of the passage of time as consisting of sequences of separate events. (p. 21)

From this perspective, Lévi-Strauss (1968) suggested that the structuralist method includes the definition of study "as a relation between two or more terms, real or supposed"; the formulation of a matrix of possible permutations between these terms; and the use of the matrix as the general object of study, which can yield necessary connections only at this general level. The empirical phenomenon considered are "only one possible combination among others, the complete system of which must be constructed beforehand" (Leach, 1970, p. 25).

As Leach (1976) suggested, a structuralist analysis of traffic signals illustrates this approach. The color spectrum is a natural continuum bounded by violet and red wavelengths. There is no natural point at which green changes to yellow or yellow changes to red. Nonetheless, by a process of discrimination, the human brain subdivides this continuum into segments, so that green, yellow, and red are "perceived" as three different colors, as shown in Figure 3. This ordering permits anyone who is not color-blind to "perceive" that green is the "opposite" of red, as black is the "opposite" of white. In many human cultures, this discrimination between colors is associated with positive, safe (green) and negative, dangerous (red) values. The meanings attributed to these colors may well derive from the fact that the "danger" ascribed to red derives from its "natural" association with blood. By establishing paired oppositions of this kind, each color is consistently given the same value.

The example of traffic signals is an interesting one because the addition of a third color, yellow, is frequently included to denote caution. It so happens that yellow is midway in the color spectrum between red and green. In this case, the structure of the color spectrum and the ordering of traffic movement correspond. Yet, *there is no inherent reason in the color spectrum itself that red should indicate "stop" rather than "go."*

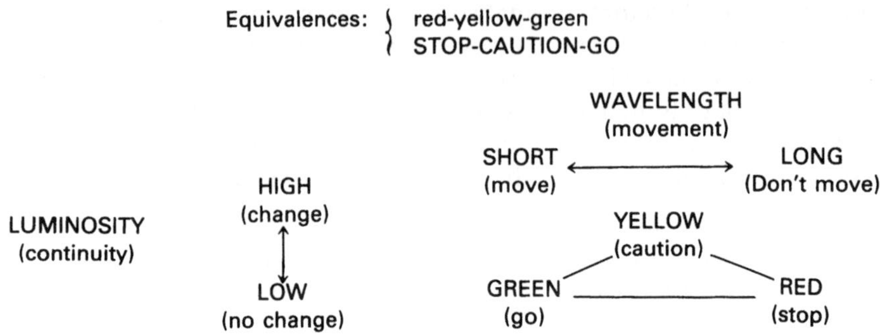

FIGURE 3. Example of a traffic signal color triangle. (Adapted from Leach, 1976.)

Indeed if the values assigned to red and green traffic signals were reversed, the ordering of neither system would not change.

For the advocates of cognitive structuralism, the example of traffic signals shows that the classification of cultural products (i.e., traffic signals) reproduces, and is therefore homologous to, the structural interrelationships between the constituent parts of natural phenomena (i.e., the color spectrum) as these are apprehended by the human brain. From this perspective, but at a more general level, Leach (1970) stated:

> Lévi-Strauss' thesis is that by noticing *how* we apprehend Nature, by observing the qualities of the classifications which we use and the way we manipulate the resulting categories, we shall be able to infer crucial facts about the mechanism of thinking. (p. 26)

In sum, for Lévi-Strauss and other advocates of cognitive structuralism, the role of the human mind is crucial, in contrast to the role of the social rules and norms of human groups or societies. Although the products of different human cultures are varied, this structural interpretation upholds the idea that, "since all cultures are the product of human brains, there must be, somewhere, beneath the surface, features that are common to all" (Leach, 1970, p. 26). Unlike functionalist or positivist interpretations, which focus only on the manifest characteristics of human culture, this structuralist interpretation upholds the idea that "the universals of human culture exist only at the level of structure never at the level of manifest fact" (Leach, 1970, p. 27). Hence, in the example of traffic signals discussed earlier, it is the contrast between colors and the change from one color to another that provide the information; each color is attributed meaning and acquires relevance only in relation to the others, and in the specific context in which they occur.

In general, this structuralist interpretation relies on arguments originally developed by linguists, notably in the seminal contribution of Ferdinand de Saussure (1916, 1974), who illustrated the fundamental similarities between human languages when compared at the level of their grammatical rules. Saussure also applied the classical structuralist method of binary oppositions by distinguishing between *substance* and *form*, *language* and *parole*, *signifier* and *signified*, and *synchronic* and *diachronic*. This structural approach led to the analysis of phonemes, the most elementary constituent parts of language.

Leach (1970) discussed the influence of Saussure and Jakobson, another linguist and a colleague of Lévi-Strauss, on the latter's approach:

> Lévi-Strauss repeatedly makes an assumption that other modes of structural expression, such as kinship systems and folk taxonomies, are organized like human language. This culture/language analogy has been developed out of Jakobson's distinctive feature theory and Lévi-Strauss has not exploited the additional insights which might have been derived from Chomsky's thinking about generative grammars. Incidentally, Chomsky himself has expressly declared that Lévi-Strauss' use of linguistic analogies is unjustified . . . , though he agrees that Jakobson's argument must constitute a basic part of any general linguistic theory, including his own. (pp. 27–28)

Beyond his recourse to a linguistic analogy, Lévi-Strauss mentioned three other important influences on the development of his structuralist approach, including a personal fascination with geology, psychoanalysis, and Marxism. Jackson and Smith (1984, p. 127) suggested that, for Lévi-Strauss, these three subjects had a shared commitment to developing different levels of understanding by examining the underlying, implicit forms and processes of geological structure, human behavior, and the economic dimensions of human society.

An example of Lévi-Strauss's approach was presented earlier in this chapter. It clearly illustrates his concern with deciphering and explaining the tacit homologous interrelationships between kinship rules, food categories, and the layout and relative position of houses. What is not apparent in this example, despite Lévi-Strauss's respect for Marxism, is the role of sociopolitical factors (such as the regulation of rules related to marriage, eating, and house construction) in the social and spatial organization of the village. These sociopolitical factors are also implicit dimensions, but they are ignored. In principle, for Lévi-Strauss and other advocates of cognitive structuralism, the innate capacity of the human mind to classify and structure natural and human-made objects and events overrides other factors and is determinant, so much so that they ignore those social variables that may play an intervening or secondary role. This criticism is addressed in the following section of this chapter.

Structuralism and Marxism: The Contribution of Castells

Within the social sciences, there has been a persistently strong movement against cognitive structuralism. This movement has derived its arguments and momentum from humanism, on the one hand, and Marxism, on the other. Nonetheless, as Jackson and Smith (1984) noted with respect to their analysis of theoretical and methodological issues in geography, Marxism is not "an undifferentiated and amorphous whole, but a complex set of ideas spanning the whole spectrum of humanist and structuralist analyses" (p. 5). The advocates of humanism—whether they adopt existentialism, phenomenology, or idealism as their perspective—are opposed to both cognitive structuralism and structural Marxism, because both adopt a passive view of human beings. In this respect, Jackson and Smith (1984) stated:

> An exclusive use of *a priori* theoretical categories effectively reduces man to an object: relations between people become relationships between things; individuals are totally manipulable; and it is the overarching system (for many purposes irreducible) which provides individuals with their definitions of situations, their purposes and their bases for interaction. (p. 9)

Lebas (1982, pp. 9–10) noted that, although there is no explicit theory of personality in the works of Marx, there is a theory of the economic and political formulation of social needs, which she maintained can be related to human motives that express social (class) interests. Nonetheless, having reviewed a range of Marxist perspectives, Lebas (1982) stated that all "focused on the mechanisms whereby the modern capitalist or industrial city denied the satisfaction of social needs" (p. 70). In short, the meaning of the urban environment was politicized.

The earlier sections of this chapter discuss and illustrate how one kind of analytical structuralism founded on human cognition seeks to decipher the common features of the human mind, irrespective of individual variations or particular cultural settings. In like manner, a second kind of analytical structuralism—namely, structural Marxism—endeavors to identify the nature of those tacit, structural determinants of the social and spatial organization of societies, irrespective of their manifestations in the observable features of any particular society. In general, economic, political, sociodemographic, and ideological variables are examined.

During the last three decades, there has been a significant growth of interest in Marxist theory and its application to studies of the built environment, especially the social and spatial organization of cities. This interest has been reflected in numerous publications stemming from the fields of urban geography, urban history, and urban sociology (cf. Castells, 1977, 1978; Harloe, 1976; Harvey, 1973; Hindess & Hirst, 1975, 1977; Lefebvre, 1972; Soja, 1980, 1985). On the one hand, Harvey, a geographer, has emphasized economic dimensions by examining the distribution of surplus value, one example of which is land rent. From this perspective, "the advanced capitalist city is essentially the built environment" (Lebas, 1982, p. 30). On the other hand, Castells, a sociologist, has emphasized sociopolitical dimensions. Consequently, "the actual physical form of the city is important only to the extent that it can be understood in terms of its involvement in the process of social reproduction of labour power—the producer of surplus value" (Lebas, 1982, p. 30). A review of some of these contributions shows that there are three distinctive characteristics of studies in this field grounded on structural Marxism, namely, their lack of concern for empirical evidence, their discard of history, and their rejection of humanism. According to Jackson and Smith (1984):

> It is anti-empiricist because reality is not held to be present in appearance. Experience cannot be trusted, and observable data are merely an outcome of the hidden realities which must be divined by rational thought. Structural Marxism's anti-historicism is expounded most vociferously by Hindess and Hirst (1975, 1977) who contend that all attempts to formulate historical explanations must take the form of teleological doctrines rather than scientific theories. Finally, the movement's anti-humanism is augured by the denial of a separate human or social science. It is confirmed by a rejection of the possibility for conscious, purposive social action, in favour of explanations of behaviour couched in terms of "structural causality." (p. 142)

These criticisms are examined here by reference to the contribution of Manuel Castells. In his first book, Castells (1977) maintained that studies of the urban environment should not be limited to deciphering patterns of interaction between places and human activities but should focus on establishing those structural laws of the production and functioning of urban settings. Therefore, it is necessary to examine how the organization of society defines and is defined by "the economic system, the political system and the ideological system" (Castells, 1977 p. 126). Castells adopted this approach; his analysis of housing and segregation is only one illustration of his causal interpretation that the social and spatial organization of a society is a product determined by economic factors. Castell maintained that housing, like other consumer goods, follows the general laws of the distribution of products. Nonetheless, unlike many consumer goods, housing has a fixed location, so that the inhabitants reside in groups according to their socioeconomic capacity. Hence, the spatial stratification of residences in an urban agglomeration is determined by the distribution of housing units among the population. In Castells's own words (1977):

> . . . although one can observe a social differentiation of space, there is no possible general law in terms of geographical regularities, but always particular expressions of the articulation of class relations (economic, political, ideological) with the distribution of a product (housing) which includes among its qualities those of its spatial environment. (p. 177)

Castells discussed the role of political, ideological, and institutional structures, which support and nurture this fundamental economic basis of urbanism. These ideas are derived from Marxist theory, which had no established tradition of dealing with urban development. Subsequently, in *City, Class and Power* (1978), Castells acknowledged that the application of *a priori* theories are untenable, because "the theoretical coding [had] been too rapid, too formal; the reality analysed was more complex than the models used" (p. 12). Castells admitted that realistic theories need to be derived from, or supported by, empirical examples. Consequently, Castells examined the concept of *collective consumption* with respect to processes of urban development. For him, collective consumption involves not only sets of objects (e.g., housing units) but also a matrix of social relations that define and are defined by sets of structural processes grounded in, and emerging from, production and consumption. This theory of collective consumption barely incorporates the spatial dimensions of the built environment, owing to Castells's theoretical perspective (1978): "Space in itself has no meaning because it is socially constructed." Therefore, it is necessary to reject spatial empiricism "precisely in order to carry out empirical research into the spatial expression of social relations" (p. 181). In other terms, it is not the spatial organization of society that defines social relations, but the latter that engenders spatial organization. Hence, Castells appears to have replaced one form of determinism with another while ignoring the contributions of other scholars. Harvey (1973), for example, considered any deterministic interpretation with caution:

> The conception of society as a totality of internally related parts with inner laws of transformation cannot be assumed *a priori* to be a superior conception of things. Marx appealed to history as proof and affirmation of his ontology, and sought to demonstrate its superiority through the practice of the method it defined. (p. 296)

Although extremist interpretations of structural Marxism have become less common, as illustrated above, this interpretation has continued to be criticized by scholars from diverse disciplines, for the lack of supporting empirical evidence and for the predominant role ascribed to economic, political, and social variables that predetermine the subordinate role of the individual (cf. Giddens, 1976; Gregory & Urry, 1985; Jackson & Smith, 1984). Some of these criticisms, particularly the latter, are specific to structural Marxism, whereas others are also applicable to structuralism founded on human cognition. Both specific and general criticisms are reexamined in the following paragraphs, which present a range of structural studies of the interrelations between people and the built environment.

APPLICATIONS TO STUDIES OF THE BUILT ENVIRONMENT AND HUMAN BEHAVIOR

There are diverse interpretations of the interrelationships between human behavior and the built environment, which are ably discussed at length in the chapters of this and the accompanying volumes of this series. Within the vast amount of published research, relatively few studies have used either of the two dominant interpretations of analytical structuralism presented earlier in this chapter. Nonetheless, in recent decades, the built environment and human behavior have been considered from structuralist perspectives, largely owing to the application of anthropological and semiotic theories. Some of these contributions are now reviewed.

THE SYNTAX OF THE BUILT ENVIRONMENT

In a seminal paper, Hillier and Leaman (1973) challenged those interpretations of people and their physical surroundings that envisage individuals as existing autonomously within the built environment. This interpretation, which these authors termed "the man-environment paradigm," is founded on Kantian dualism, and it leads to mechanistic interpretations of human behavior in terms of "stimuli" from the "environment" into the "person." Hillier and Leaman maintained that the chief shortcoming of this paradigm is that it ignores the important role of the symbolic and semiotic systems that are an inherent part of human culture. Broadbent *et al.* (1981a,b) concurred with this point of view, whereas Rapoport (1982) refuted the pertinence of a semiotic approach. However, Rapoport was apparently interested only in observable phenomena!

In other academic disciplines, several authors, including Gregory and Urry (1985), have maintained that, during the development of social theory by human geographers, "the 'social' was separated from the 'spatial'" (p. 2). Consequently it has been common practice for the built environment to be described and explained solely in terms of spatial processes and geometrical or formal languages. This approach implies that the spatial characteristics of the built environment can be interpreted only as physical elements that do not encompass other symbol systems that exist in logical space. Hillier and Leaman (1973, p. 511) proposed a logical space model, based on their interpretation of cognitive structuralism, rather than on structural Marxism, in order to explain the tacit knowledge that defines and is defined by the spatial organization of the built environment. In a subsequent publication, Hillier, Leaman, Stanstall, and Bedford (1976) endeavored to show that the "spatial language" of settlement patterns and building designs is expressed in nonverbal communication systems, and that, therefore, they belong to the domain of the unconscious. The authors admire and interpret the structuralist perspective of Lévi-Strauss, which they have applied to an analysis of selected examples of the built environment in a range of cultural and geographical contexts. Hillier, Leaman, and their coauthors have developed a typology of morphological archetypes, (some of which are shown in Figure 4) that they have suggested includes the generative rules, or what they have termed the "immanent laws" that constitute the spatial form of the built environment. Moreover, unlike Glassie (1975), Stiny and Mitchell (1978a,b), and Flemming (1981), who have failed to develop an explanation or understanding of the meanings of the inventory of geometrical forms that structure the built environments they have examined, Hillier *et al.* (1976) maintained that different types of settlement patterns and building designs are capable of a systematic and symmetrical classification of "syntactical structures." Beyond this typological analysis, they also suggested that the spatial patterns of settlement correlate with the social organization of the inhabitants, to produce what they termed "morphic language." These authors defined the *syntax* of a morphic language as:

> A set of related rule structures formed out of elementary combinations of the elementary objects, relations and operations. These can be introduced, independently or cojointly, in a *minimum set up* for the morphic language to produce recognisable patterns. (p. 150)

They also stated that, in order to understand the social production of architecture, it is necessary to know a "a particular set of historical determinants and contingencies" and "the 'immanent laws' of the constitution of spatial structure itself" (p. 179). In essence, this approach uses similar concepts and methods to describe society and the built environment. From this perspective, "a clear, if somewhat complex, 'inverse law' began to suggest itself as the relation between social and spatial structure" (p. 179). This interpretation, supported by ethnographic evidence from several human settlements, led the authors to conclude that:

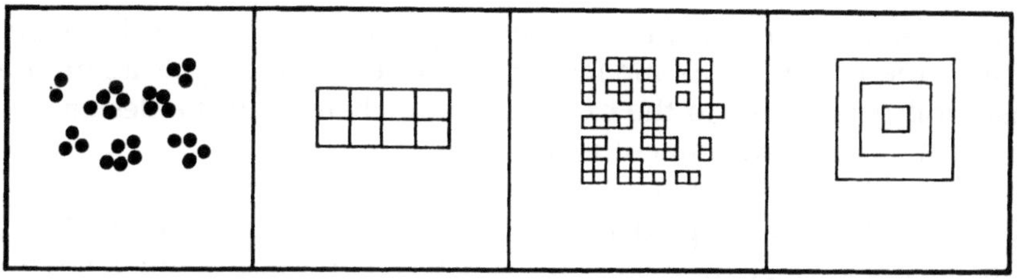

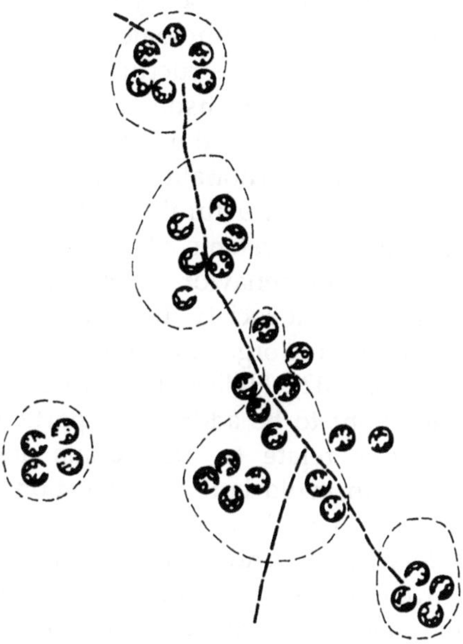

FIGURE 4. The first four archetypes of the space syntax developed by Hillier, Leaman *et al.* (1976), showing (1) a "cloud of mosquitoes"; (2) a "pure block"; (3) a universal neighbor"; and (4) the "concentric rings." Beneath these archetypes is a Yao compound village that the authors presented as one example of one-syntax. (Reprinted here by the kind permission of Bill Hillier.)

> *Space is not a reflection of society, but a set of strategies in relation to social form, as often as not offering an alternative basis for encounters, other than those dictated by social structure.* The streets of the city, for example, do not always reflect the social structure; they can be the means by which the pattern of social differences is forgotten, and the inhomogeneous is assembled. (p. 1980)

Leach (1978) challenged the space syntax model proposed by Hillier and his colleagues. By reference to several ethnographic studies that describe the social and spatial organization of a range of human settlements, Leach (1978) confirmed, in contrast to the space syntax model, that one cannot infer the generative syntax of the built environment simply by looking at the layout of settlement "patterns on the gound"; furthermore, "one cannot infer anything at all about the society that makes use of the resultant settlement" (p. 397). Apart from these criticisms, the space syntax model is not verified by the application of a historical perspective to the examination of changes in the design and use of the built environment in specific contexts.

In a subsequent publication founded on earlier studies, Hillier and Hanson (1984) also adopted a structural interpretation in order to define and illustrate what they termed "a new theory and method for the investigation of the society-space relationship" (p. x), including a description of the morphological features of the built environment that can be related to social structures and processes in a systematic way. Initially, Hillier and Hanson (1984) presented several criticisms of various theoretical interpretations of the built environment and human behavior, including territoriality, cognition, functional studies of building use, and orthodox structuralism. They noted that cognitive structuralists maintain that systems of rules (i.e., structures) generate but do not determine the behavior of individuals; yet, it is unclear how these structures represent society. In other terms, "structuralism pays attention to how structures organize society but not to how society organizes structures" (p. 202). Hillier and Hanson suggested that both of these questions ought to be addressed, particularly how the origin and locus of structures relate to individuals and society, whereas structuralism "seems to avoid both the question of the *origin* of structure, and the question of its locus." (p. 202) These authors relied heavily on their understanding and critique structuralism and apparently ignored the contribution of structural Marxism (and criticisms of it), which do provide cues for answers to these questions. These authors maintained that the construction of a social theory of the built environment is a question of understanding the relations between the principles of generating patterns. Hillier and Hanson (1984, p. 50) upheld the idea that the built environment *and* social encounter patterns can be interpreted as a morphic language. Their interpretation of morphic languages challenges the orthodox structuralist interpretation of a *rule* or set of rules. Whereas cognitive structuralists claim that such rules are *anterior* to events and are located in the human brain, Hillier and Hanson rejected this biological model and proposed a sociological one. They argued that the space syntax model they were presenting overcomes the shortcomings of orthodox structuralism, because it is not a deterministic set of anterior prescriptive rules (or cognitive structures) but an analytical tool for describing extant social and spatial patterns.

Hillier and Hanson (1984) applied their interpretation to an analysis of settlement layouts and building designs by developing a series of analytical techniques stemming from the theory of graphs, as shown in Figure 5. A principal axiom of this approach is that "spatial organisation

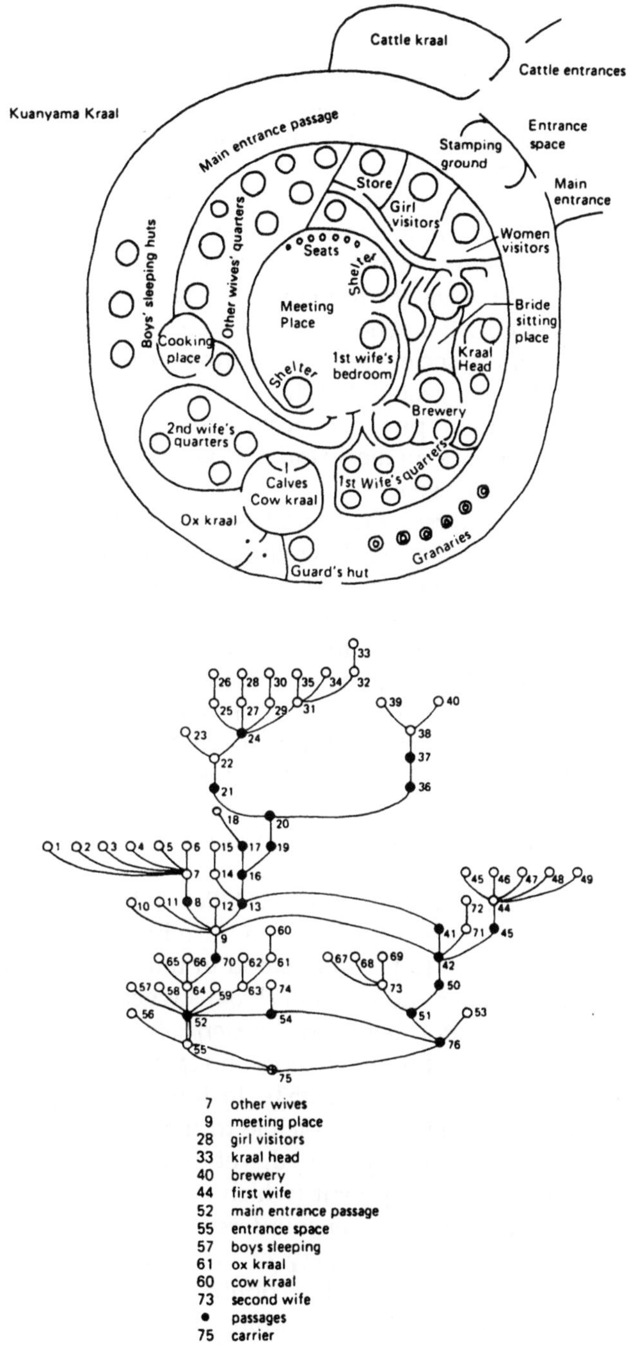

FIGURE 5. Plan of a "chiefly" kraal of the Ambo people with a "justified permeability map" developed by Hillier and Hanson (1984). (Reprinted with kind permission of the authors.)

is a function of the form of social solidarity" (p. 142). The examples presented, from diverse nonindustrialized cultures (all having strong internal rules and shared values rather than polyvalence), show that, despite variety in form and function, they all have an elementary relation between the inhabitant and the visitor in the sense that the inhabitant is in the deeper, often nondistributed parts of the building and interacts with the visitor through the shallower, often distributed parts of the building that form its principal circulation system. Then, the authors discussed how this schema does *not* apply to institutional buildings in contemporary societies (such as hospitals, prisons, and museums). In fact, they suggested that this schema is reversed, in the sense that hospital patients and prisoners occupy the primary cells for limited periods of time, whereas medical staff and wardens move freely in the circulation and ancillary spaces.

These examples express the dialectical relationship of the interface between, and/or segregation of, different groups of people and of people and artefacts. Here, the authors adopted the notion of social class—not merely in socioeconomic terms, but also with respect to the hierarchy of professionals and bureaucracies—to explain the spatial organization of public buildings, institutional quarters, and the offices of corporations. This notion implies that architectural and urban forms respond more directly to societal rules and customs than to sociopsychological variables (such as territoriality). Although this is an important difference from the approach of cognitive structuralism, these societal parameters are not examined in terms of group or generational differences within societies in order to confirm their approach.

At this point in the development of Hillier and Hanson's argument, the fundamental notion of a boundary is introduced to enable societies and spatial organization to be understood in terms of spatial and transpatial categories. Hillier and Hanson (1984) suggested that "it is useful to make a clear distinction between the *morphological*, or *arrangemental, principles* of a society and the actual *social* and *spatial mechanisms* by which these principles are realized" (p. 253). In the final chapter and postscript, the authors applied this principle to the study of contemporary housing estates (which are segregated, yet densely populated communities) and suburban development (which is a dispersed form of community with few interpersonal encounters). These types of residential quarters are related to the ideology of industrial production and reproduction in capitalist and socialist countries. Apart from overlooking group and individual differences within particular societies (or even large housing estates), the authors limited their study to a synchronic investigation, although they maintained that their revised "space syntax model" is generative rather than static. The lack of a historical or temporal perspective in much contemporary research of this kind has been challenged by Colquhoun (1981) and rejoins our earlier discussion of some criticisms of structuralist theories and methods. Apart from this important issue, it is noteworthy that Hillier and Hanson (1984) included a range of cultural, spatial, and sociopolitical parameters in their structural interpretation. This is highly commendable.

Architectural Semiotics

Semiotic analyses of the built environment uphold the principle that architecture, like language, is composed of a system of signs, but for the purpose of nonverbal communication (cf. Abel, 1980; Broadbent *et al.*, 1980a,b; Jencks & Baird, 1969; Preziosi, 1979). Meaning is conveyed by means of a culturally specific set of conventions or codes. Two decades ago, Jencks and Baird (1969) published a collection of essays that show that buildings can be understood if the coded meanings attributed to them by architects and builders (and sometimes their clients) can be interpreted by laypeople. The immediate inference from this principle is that human behavior and related responses to the design of the built environment are inseparable from these encoded meanings. In any specific context, the meaning attributed to the built environment is not solely dependent on its shape, size, and pragmatic functions; rather, the design and layout of buildings may comprise a set of signs that are culturally specific, as some theorists, including Eco (1972, 1976), Markus (1987), and Preziosi (1979), have shown.

Those scholars in architecture and the social sciences who have adopted and directly transferred notions from structural linguistics apparently assume that the structure of discursive language also functions at the much broader level of nondiscursive expression in daily life. Doxtater (1984, p. 2) noted that this assumption implies that the spatial components of the built environment are devoid of any *unique* elements that contribute to its symbolic meaning. During the last decade, it has been common for scholars in architecture and the social sciences to adopt and transfer notions from structural linguistics, in order to study the meaning of the built environment. The pertinence of applying any linguistic analogy has been challenged by Knox (1984) and Rapoport (1982), whereas Colquhoun's essay (1981) not only predates these contributions but also provides a more constructive discussion of the limits of semiotic analysis in architectural theory and research. Yet, it was Chomsky (1957, 1968), the renowned linguist, who presented the seminal discussion. He argued that behavioral systems, such as gestures, lack some of the fundamental characteristic of language, which includes a fixed set of constituents and precise rules. Chomsky illustrated the uniquely reflexive properties of language, which neither nonverbal communication nor artifacts possess. In essence, some linguistic information is required for the interpretation and comprehension of human activities and the built environment, or their constituents. Hence, language and other sets of symbols can be compared only metaphorically. Similarly, Doxtater (1984), Lavin (1981), and Sanders (1986) have stressed that the inherent differences among the built environment, human behavior, and language ought to be acknowledged, whereas Markus (1987) extended this point of view by examining the reciprocal relations among ideas, language, and the design, the classification, and the use of buildings. Markus astutely showed that the classification of objects, people, and activities ultimately becomes "an embodiment of these classification systems" (p. 467).

Many semiotic interpretations of the built environment have been subject of a complex debate about the nature of the meanings attributed to the built environment (cf. Bonta, 1979; Broadbent *et al.*, 1980a, Eco, 1972, 1973, 1976; Krampen, 1979; Rapoport 1977, 1982). A review of some of these publications indicates the need for a comprehensive and coherent analysis, which Groat (1981, 1987) maintained has not been forthcoming, despite the rapid growth of studies in this field.

Research on this subject usually presents three classes of meaning in architecture, which duplicate the three levels of the semiotic elaborated by Morris (cf. Broadbent, 1980). Gutman 1972 summarized these meanings as:

> *syntactical* meaning, or the meaning that an element of form or style acquires by virtue of its location in a chain of form or style elements; *semantic* meaning, or the meaning it acquires because of the norm, idea or attitude that it represents or designates; and *pragmatic* meaning, or the meaning that is understood in relation to the architect, client or social group that invents or interprets the building's form or style. (p. 299)

Despite reservations about any linguistic analogy, studies of these three kinds of meaning have commonly adopted elements of linguistic theories, based on the assumption that there are innate conventions that enable buildings, or their constituent parts, to convey meanings in the same way that language does (e.g., Jencks & Baird 1969; Broadbent *et al.* 1980a,b). According to Krampen (1979), this interpretation induces "logical errors," whereas Knox (1984) states that it "is highly codified and mechanistic, deliberately and systematically abstracting symbols from their historical and social context" (p. 113). In this respect, many of these studies are positivistic, focusing attention on buildings, rather than on

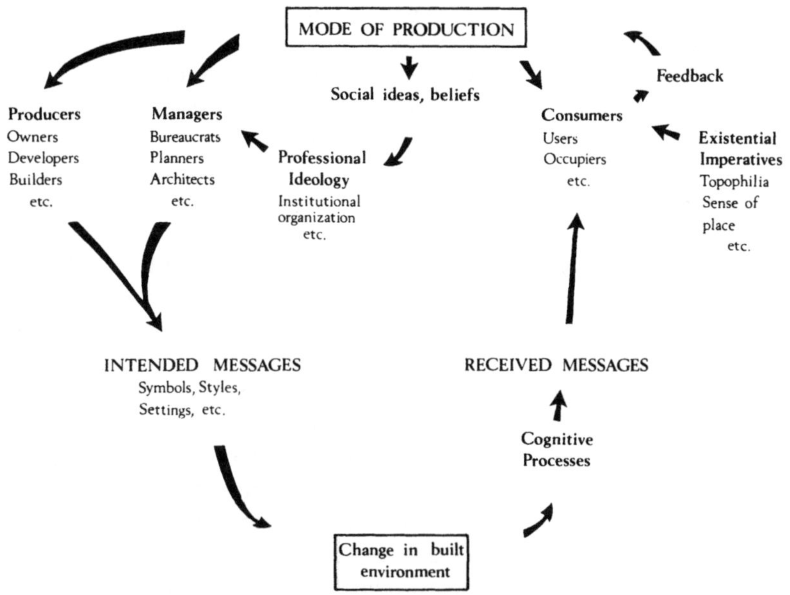

FIGURE 6. The interrelationships between signs, symbols, and settings were formulated in terms of an analytical framework by Knox (1984). (Reprinted here by the kind permission of the author.)

the interrelationships between people and buildings, which cannot be considered separately from the production and meaning of the built environment. In contrast to this recurrent approach, Bonta (1979) presented an analysis of texts that shows that the interpretation and evaluations of specific buildings by architectural critics and historians change during the course of time. Therefore, their judgments cannot be delimited from social and professional values at specific times. Hence, apart from Bonta's contribution, as Knox (1984) noted, "none of these approaches to architectural semiology directly confronts the fundamental question of communications *by whom*, to what audience, to *what purpose* and to *what effect?*" (p. 114)

The design and use of the built environment ought to be related to the societal context of building construction policy in order to overcome the "myopic approach" that Knox (1984) attributed to many studies in this field. Although Knox admitted that the built environment can inhibit, facilitate, encourage, or modify individual and group behavior, he outlined an approach that examines how the built environment not only sustains but reproduces "social divisions, ideological forces and political fantasies" (p. 109). Knox suggested that an integrative perspective, which incorporates Marxian theory and also the role of the individual, can develop a new sociology of the built environment. One organizing framework for this kind of perspective was presented by Knox (1984) as shown in Figure 6. Although this framework does incorporate change, it should also account for a long-term historical perspective for the following reasons.

Although buildings are functional objects, they serve a range of purposes and they are attributed a range of values. Lawrence (1987) discussed and illustrated how a building is commonly attributed an economic value, an exchange value, an aesthetic value, and a use value, whereas, in addition to these, a house is usually attributed a sentimental and a symbolic value. Although the first four values are tangible and can be quantified, the latter two cannot. All these values are not simply expressed by individuals but are acquired, nurtured, transmitted, reinforced, or modified by interpersonal communication (e.g., between parents and children, or between members of the same social or professional group). Moreover, a distinction can be made between the *intended meaning* of buildings by individuals or specific groups (e.g., architects or public administrators) and the *attributed meaning* (by people of different ages, socioeconomic status, or cultural backgrounds). Hence, the communication between different groups of people clarifies one or more systems of values in a society at precise points in time.

Since World War II, the systems of values have multiplied and diversified to such an extent that it has become extremely difficult to account for all those values associated with town and country planning, urban design, and house planning. Moreover, the values and meaning attributed to buildings change; consequently, it is necessary to examine the meanings of the built environment in terms of a temporal perspective. This principle is ably illustrated by the transformation of the design, meaning, and use of the first-floor level and the attic of residential buildings in Swiss towns since the beginning of this century (Lawrence, 1986). Until that time, the first floor contained the most prestigious apartment(s), and the attic was a menial area for clothes drying or servants' quarters, whereas today, the attic contains the most prestigious and expensive apartment(s), and the first floor is considered humble.

The preceding discussion and examples illustrate that the meaning and use of the built environment are not intrinsic in a set of physical characteristics, nor in the nomenclature applied to rooms and their facilities. Moreover, it is now clear that it is necessary to analyze a range of economic, political, and social variables, including building regulations and tenancy agreements, in order to comprehend the nature of changes both in the design and use of the built environment and in the values attributed to the built environment by different people. The following and last section of this chapter presents and illustrates how a redefinition and then a reapplication of structuralist theories and methods can overcome recurrent shortcomings and can also enhance our current understanding of the reciprocal relations between people and the built environment.

REDEFINITION OF STRUCTURALIST THEORY AND METHODS

Several structuralist interpretations of people and the built environment have been presented. Three recurrent shortcomings of these interpretations can now be briefly summarized as:

1. The determinant influence of human cognition is assumed at the expense of a range of cultural, social, and, therefore, economic and political variables; or conversely, the determinant influence of these latter sets of variables presupposes the impact of individual human factors that often are ignored.
2. There is a concern with theory but a lack of supporting empirical evidence.
3. A historical perspective is frequently omitted.

Although some of the studies presented earlier in this chapter have addressed one or another of these criticisms, each of them ought to be considered simultaneously if structuralist theories and methods are to improve our comprehension of the design, meaning, and use of the built environment. The following paragraphs briefly show that the recurrent criticisms of structuralist interpretations of people and buildings are not inherent in structuralist theories and methods.

Toward a Definition of Dialectical Structuralism

This chapter has illustrated the polarized positions in structuralist interpretations of the interrelations between individuals and society, but it would be misleading to suggest that these opposing interpretations are limited to structuralist theories and methods. In fact, the roles of the individual and society have been the subject of an enduring debate in many academic disciplines, in the philosophy of science, and in recent discourses on social theory (cf. Giddens, 1976, 1979; Hawthorn, 1976). Yet, one of these two interpretations is not necessarily exclusive of the other if a third perspective, which explores the dialectical relations between them, is applied. From this perspective, the goal-oriented behavior of individuals is not divorced from the contextual contingencies—societal codes, norms, and rules—that not only regulate the design and use of the built environment and human activities but are also the medium of interpersonal communication. Given that the built environment, human behavior, and contextual contingencies change during the course of time, an adequate explanation of the interrelations between an individual and society, at one level, and people and the built environment, at another level, should account for the dialectical nature of these interrelations. Whereas orthodox structuralist interpretations—both cognitive-structuralist and structural-Marxist—do not usually account for these dialectics, some poststructuralist scholars, including Bourdieu (1977), Giddens (1976, 1979), and Soja (1980, 1985), explicitly do. As Soja (1985) stated:

> Spatiality is simultaneously the product of a transformation process and transformable itself. As a social product, spatiality can be continuously reinforced or reproduced over time, presenting an appearance of stability and persistence. But it can also be substantially restructured and radically reconstituted, invoking again its origins and grounding in social practice and the labour process. The production of space is thus not simply a mechanical extrusion of a frozen matrix which acts passively to contain society. Spatiality and temporality, human geography and human history, intersect in a complex social process which creates a constantly evolving historical sequence of spatialities, a spatio-temporal structuration of social life which gives form not only as the grand movements of societal development but also to the recursive practices of day-to-day activity. (p. 94)

This account of the built environment as both the medium and the product of the nonstatic interrelations between people and their physical surroundings is far removed from the space syntax model and some semiotic interpretations presented above. Moreover, the grand merit of the axioms presented by Soja is that they can be illustrated by the concept of *habitus* formulated by Bourdieu (1977) and the concept of *structuration* presented by Giddens (1976, 1979). The position advocated here is founded on the notion of *dialectical structuralism*, a term used to prevent any confusion with those orthodox interpretations of structuralism presented above. The perspective is structuralist in the sense that, at one level, it identifies and analyzes the reciprocal relations between a range of spatial, cultural, social, political, and economic variables and, at another level, it examines the interrelations among all these variables and the activities of individuals and groups, both in specific contexts at a precise time, and also during the course of time. From this perspective, implicit structures can be interpreted in a dynamic rather than a static way, or in Giddens's own terms (1979), as being in a continual "form of *structuration*" (p. 210). In sum, Tilley (1982) stated:

> The social formation is a totality of human experience and action, the entire *ensemble* of the relations between individuals and groups and of their relationships with their natural and social environment. It is a dynamic whole,

always in the process of structuration; there can be no elements outside it, nor can the elements within it be understood without dialectically referring them to it. The possibility of transformation embedded within the social totality can now be discussed. (p. 34)

This principle enables us to reexamine, from a new perspective, the proposition of orthodox structuralism that the built environment and human behavior may be understood as a communicative code. A *code* is a rule-governed system that comprises a general set of possibilities for understanding and communicating physical, social, and behavioral characteristics of human culture. Contextual contingencies establish a range of acceptable spatial and behavioral configurations, and *acceptable* may well imply different sets of meanings for different groups of people at the same point in time.

After this brief outline of the nature of dialectical structuralism, it remains to be shown how such a structuralist perspective can enhance our understanding of the reciprocal relations between people and the built environment in contemporary settings.

A Structural, Historical, and Cross-Cultural Analysis of Homes

A detailed historical and ethnographic study of domestic space in Australia and England has indicated that the design and use of domestic space and facilities are related to a matrix of factors that change during the course of time (Lawrence, 1982). Considered independently, however, these factors tell very little to the architect or the social scientist about household life. For example, it has become evident that two societies stemming from the same culture have different meanings and uses for the room labeled the *kitchen*, which can be explained only by ethnographic observation and then detailed historical research. Therefore, and in contrast to Rapoport's claim (1982), neither the meaning of the physical form of the house nor the activities that occur in it can be understood simply by observation. In the context of this study, it has become apparent that the meaning and use of houses in Australia and England are independent of the label commonly given to rooms, or of their size, shape, and furnishing, because it is related to a host of factors. A structuralist interpretation of the interrelations between these factors can explain why a woman in Australia washes her children's clothes in a room demarcated from the kitchen (the laundry), whereas her counterpart in England commonly undertakes the same activity in the kitchen and usually in the kitchen sink (Lawrence, 1982).

In colonial Australia, for example, despite the transformation of the detached kitchen, the development and use of domestic appliances, the

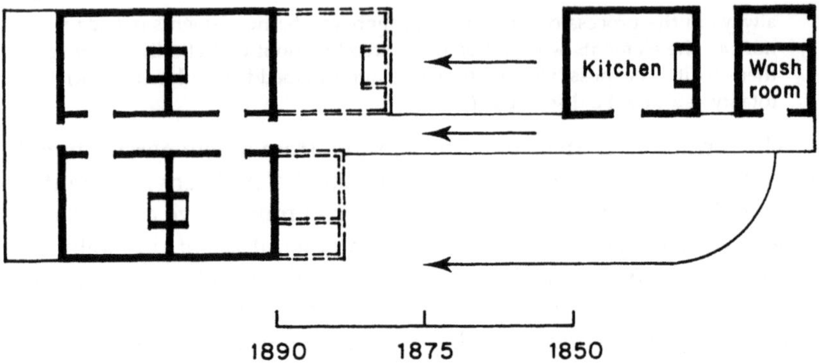

FIGURE 7. The transformation of the Australian detached kitchen, washroom and toilet into service rooms integrated under the main roof of the house, via the back veranda, occurred between 1850 and 1890.

growing availability of a range and construction materials, the enactment of building regulations, and the increase in the size of houses, the structural organization of domestic space has remained the same. At first sight, this proposition appears to be untenable owing to the gradual integration of detached rooms and services—the kitchen, the washhouse, and the toilet—with the other living rooms of the house. Nonetheless, this spatial transformation, during several decades, which is represented diagrammatically in Figure 7, did not change the explicit demarcation between the space and activities associated with the preparation and eating of food, the laundering of clothes, bathing, and ablutions (Lawrence, 1982). This ordering of domestic space and activities in Australian houses can be contrasted with the design, the meaning, and the use of houses in England during the same period of time.

In England, dishwashing, laundering, and (sometimes) bathing have commonly been considered one set of activities that remove dirt from objects—crockery, clothes, and the body—and these objects were commonly brought into conjunction and cleansed in one room, as illustrated in Figure 8. There were no building regulations or health acts that discouraged this custom. The scullery often served this practice until the advent of gas and electrical heating appliances. In Australia, however, the kitchen (whether detached from or integrated into the house) has commonly been reserved for the preparation of food and for dishwashing. A separate laundry was usually provided, and a separate bathroom has been taken for granted. This organization of domestic space and activities is also shown in Figure 8. It suggests that dishwashing, laundering, and bathing have traditionally been considered three different

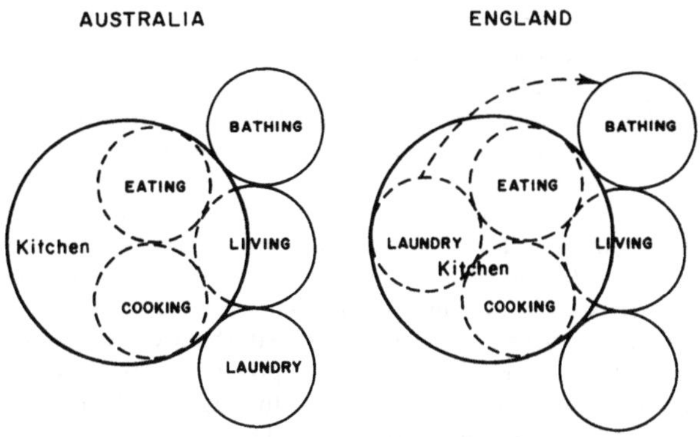

FIGURE 8. Diagrammatic representation of activities and facilities that are commonly associated or demarcated in the domestic kitchen in Australia and England.

classes of activities. It was not the removal of dirt but the objects that were conceptually important: therefore, three demarcated rooms were commonly provided for the removal of dirt from these sets of objects. Since the 1920s, health and housing regulations in Australia have prohibited the use of the kitchen for cleaning clothes!

This cross-cultural study of domestic space has illustrated that it is not the intrinsic qualities of house interiors but their differential features that are the bearers of social meaning. Therefore, in any approach to the study of the built environment, a distinction ought to be made between the extant sets of distinctive or differential features, according to the following principles. Initially, it would seem important to consider human-made space, first, by distinguishing the connotation and use of architectural types from the connotation and use of specific buildings; then, by considering the connotation of architectural types (and that of specific buildings) with respect to their morphology or configuration, in relation to their structural or constructional elements, and also with respect to their decorative or applied elements; finally, by considering what distinctions should be made between interior architectural space and the objects it contains when personalized by people.

At the level of "analytical structures," this study also illustrates that the reciprocal relations between the design, the meaning, and the use of domestic space (and, indeed, the built environment in general) can be interpreted in terms of one or more of the following analytical codes, elaborated by Lawrence (1987):

1. *A code for the classification of activities and space.* The label which is given to an activity encodes that event with a meaning. Similarly, the label

given to a room in a house implies a social meaning and use. These significations vary between different cultures, and between different groups of people in the same culture.

2. *A code for the relative position of activities and space.* There is a set of social ideas and images which suggest the location of activities and spaces in the dwelling. Such a code distinguishes between the front and back of the dwelling. This ordering of activities and spaces is extended outside the dwelling to the public realm, particularly the street.
3. *A code for the association of demarcation of activities and space.* There are customary social systems for the classification of sets of activities and spaces in non-functional terms. Hence the binary oppositions of clean and dirty, day and night, public and private *may* be used to associate or demarcate domestic activities, and to suggest their position relative to other activities and spaces in dwelling.
4. *A code for domestic activities which indicates the meaning of one activity* (a food event) *in the total range of domestic activities* (the food system). The significance of a food event suggests not only what food is eaten, how it is embellished, when it is served and who is present at the table, but also where it is served. Hence the greater significance of a food event suggests more embellishment of the food, and more people being present, at a relatively formalized setting. (pp. 158–159)

Unlike laws, or other deterministic interpretations of the interrelations between people and the built environment, these codes enable us to account for and comprehend the polyvalent meanings and functions attributed to the design of the built environment and human activities within the framework of extant architectural, cultural, societal, and behavioral parameters. In this way, recurrent shortcomings of orthodox structuralist theories and methods, particularly those applied to the interpretation of people and the built environment, have been addressed.

IMPLICATIONS FOR FUTURE RESEARCH AND APPLICATION

This chapter has discussed and illustrated a range of structuralist interpretations of the relations between people and the built environment. It has shown that, to different degrees, these interpretations enable the design, the meaning, and the use of the built environment to be considered in terms of the reciprocal relations between a range of material and nonphysical variables. Thus, whereas some contributions emphasize the structure of spatial relations and the autonomous spatial laws of building construction and transformation (Flemming, 1981; Hillier *et al.*, 1976; Hillier & Hanson, 1984), other contributions focus on the innate capacity of the human mind to conceptualize and comprehend the social and spatial organization of human culture (Glassie, 1975; Lévi-

Strauss, 1968). Finally, a third set of contributions emphasizes how the social processes of production and consumption define and are defined by the spatial organization of the human-made environment (Castells, 1977, 1978; Gregory & Urry, 1985; Jackson & Smith, 1984). Despite theoretical and methodological differences between these interpretations (and their shortcomings, which have been presented here) they all confirm that future research should not only focus on the reciprocal relations between manifest objects and events but also analyze those implicit ideas, rules, and symbols associated with a range of cultural, societal, and individual human factors, which are not static. Therefore, this chapter requests and illustrates how future studies founded on the principles of dialectical structuralism can examine the *interrelationships* between the material characteristics of the built environment and the nonphysical world of ideas, images, rules, regulations, and symbols and, concurrently, how these interrelations define and are defined by a range of cultural, social, political, economic, and psychological variables during the passage of time.

ACKNOWLEDGMENTS

The author expresses his gratitude to Gerhard Kaminski, Gary Moore, Joan Ockman, Donald Sanders, and Erwin Zube for their helpful comments in the writing of this chapter, and to Bich Do for her skills in the preparation of the manuscript for publication.

REFERENCES

Abel, C. (1980). The language analogy. *Architectural Association Quarterly, 12*(3), 39–47.

Badcock, C. (1975). *Lévi-Strauss: Structuralism and sociological theory*. London: Hutchinson.

Barthes, R. (1967). *Elements of semiology*. London: Jonathan Cape. (Translated from *Eléments de sémiologie* by A. Lavers & C. Smith. Paris: Éditions Gonthier, 1965.)

Bonta, J. (1979). *Architecture and its interpretation*. London: Lund Humphries.

Boudon, R. (1971). *The uses of structuralism*. London: Heinemann. (Translated from *À quoi sert la notion de structure* by M. Vaughan. Paris: Gallimard, 1968.)

Bourdieu, P. (1977). *Outline of a theory of practice*. Cambridge: Cambridge University Press. (Translated from *Esquisse d'une théorie de la pratique* by R. Nice. Genève: Libraire Droz, 1977.)

Broadbent, G. (1980). Architectural objects and their design as a subject for semiotic studies. *Design Studies, 1*(4), 207–216.

Broadbent, G., Bunt, R., & Jencks, C. (Eds.) (1980a). *Signs, symbols and architecture*. Chichester: Wiley.

Broadbent, G., Bunt, R., & Llorens, T. (Eds.). (1980b). *Meaning and behaviour in the built environment*. Chichester: Wiley.

Castells, M. (1977). *The urban question: A marxist approach*. London: Edward Arnold. (Translated from *La question urbaine* by A. Sheridan. Paris: Maspero, 1972.)

Castells, M. (1978). *City, class and power*. London: Macmillan.

Chiari, J. (1975). *Twentieth-century French thought: From Bergson to Lévi-Strauss*. London: Paul Elek.

Chomsky, N. (1957). *Syntactic structures*. The Hague: Mouton.

Chomsky, N. (1968). *Language and mind*. New York: Harcourt, Brace & World.

Clarke, D. (Ed.) (1977). *Spatial archaeology*. London: Academic.

Colquhoun, A. (1981). *Essays in architectural criticism: Modern architecture and historical change*. Cambridge, MA: MIT Press.

Culler, J. (1973). The linguistic basis of structuralism. In D. Robey (Ed.), *Structuralism: An introduction*. Oxford: Clarendon.

Doxtater, D. (1984). Spacial opposition in non-discursive expression: architecture as ritual process. *Canadian Journal of Anthropology*, *4*(1), 1–17.

Eco, U. (1972). The componential analysis of the architectural sign/column. *Semiotica*, *5*(2), 97–117.

Eco, U. (1973). Function and sign: Semiotics of architecture. In J. Bryan & R. Sauer (Eds.), *Structures implicit and explicit*. Philadelphia: University of Pennsylvania Press.

Eco, U. (1976). *A theory of semiotics*. Bloomington: Indiana University Press.

Flemming, U. (1981). The secret of the case Giuliani Frigerio. *Environment and Planing B*, *8*(1), 87–96.

Giddens, A. (1976). *New rules for sociological method: A positive critique of interpretative sociologies*. London: Hutchinson.

Giddens, A. (1979). *Central problems in social theory: Action, structure and contradiction in social analysis*. London: Macmillan.

Glassie, H. (1975). *Folk housing in middle Virginia*. Knoxville: University of Tennessee Press.

Gregory, D. (1978). *Ideology, science and human geography*. London: Hutchinson.

Gregory, D., & Urry, J. (Eds.). (1985). *Social relations and spatial structures*. London: Macmillan.

Groat, L. (1981). Meaning in architecture: New directions and sources. *Journal of Environmental Psychology*, *1*(1), 73–85.

Groat, L. (1987). Recent developments in architectural theory: Implications for empirical research. *Journal of Environmental Psychology*, *7*(1), 65–76.

Gutman, R. (Ed.). (1972). *People and buildings*. New York: Basic.

Harloe, M. (Ed.). (1976). *Captive cities*. London: Wiley.

Harvey, D. (1973). *Social justice and the city*. London: Edward Arnold.

Hawthorn, G. (1976). *Enlightenment and despair: A history of sociology*. Cambridge: Cambridge University Press.

Hillier, B., & Hanson, J. (1984). *The social logic of space*. Cambridge: Cambridge University Press.

Hillier, B., & Leaman, A. (1973). The man-environment paradigm and its paradoxes. *Architectural Design*, *63*(8), 507–511.

Hillier, B., Leaman, A. Stanstall, P., & Bedford, M. (1976). Space syntax. *Environment and Planning B*, *3*(2), 147–185.

Hindess, B., & Hirst, P. (1975). *Pre-capitalist modes of production*. London: Routledge & Kegan Paul.

Hindess, B., & Hirst, P. (1977). *Mode of production and social formation*. London: Macmillan.

Hodder, I. (Ed.). (1982). *Symbolic and structural anthropology*. Cambridge: Cambridge University Press.

Jackson, P., & Smith, S. (1984). *Exploring social geography*. London: George Allen & Unwin.

Jencks, C., & Baird, G. (Eds.). (1969). *Meaning in architecture*. London: Barrie & Rockcliff.
Knox, P. (1984). Symbolism, styles, and settings: The built environment and the imperatives of urbanized capitalism. *Architecture and Behaviour, 2*(2), 107–122.
Krampen, M. (1979). *Meaning in the urban environment*. London: Pion.
Lane, M. (Ed.). (1970). *Structuralism: A reader*. London: Jonathon Cape.
Lavin, M. (1981). Boundaries in the built environment: Concepts and examples. *Man-Environment Systems, 11*(5/6), 195–206.
Lawrence, R. (1982). Domestic space and society: A cross-cultural study. *Comparative Studies in Society and History, 24*(1), 104–130.
Lawrence, R. (1986). *Le seuil franchi: Logement populaire et vie quotidienne en Suisse romande, 1860–1960*. Geneva: Georg Editeur.
Lawrence, R. (1987). *Housing, dwellings and homes: Design theory, research, and practice*. Chichester: Wiley.
Leach, E. (1970). *Lévi-Strauss*. London: Fontana.
Leach, E. (1976). *Culture and communication: The logic by which symbols are connected*. Cambridge: Cambridge University Press.
Leach, E. (1978). Does space syntax really "constitute the social"? In D. Green, C. Haselgrove, & M. Spriggs (Eds.), *Social organisation and settlement: Contributions from anthropology, archaeology and geography*. Oxford: BAR International Series (Supplementary No. 47).
Lebas, E. (1982). Urban and regional sociology in advanced industrial societies: A decade of Marxist and critical perspectives. *Current Sociology, 30* (Whole No. 1).
Lefebvre, H. (1972). *La pensée marxiste et la ville*. Paris: Casterman.
Lévi-Strauss, C. (1968). *Structural anthropology*. London: Allen Lane. (Translated from *Anthropologie structurale* by C. Jacobsen and B. Schoepf. Paris: Plon, 1958.)
Lüchinger, A. (1981). *Structuralism in architecture and urban planning*. Stuttgart: Karl Krämer Verlag. (Three-langauge edition.)
Lyons, J. (1973). Structuralism and linguistics. In D. Robey (Ed.), *Structuralism: An introduction*. Oxford: Clarendon Press.
Macksey, R., & Donato, E. (Eds.). (1972). *The structuralist controversy: The languages of criticism and the sciences of man*. Baltimore: John Hopkins University Press.
Malinowski, B. (1929). *The sexual life of savages in northwestern Melanesia: An ethnographic account of courtship, marriage, and family life among the natives of the Trobriand Islands, British New Guinea*. London: Routledge.
Markus, T. (1987). Buildings as classifying devices. *Environment and Planning B, 14*(4), 467–484.
Necipoğlu, G. (1981). Book review of G. Broadbent *et al.* Signs, symbols and architecture. (Chichester: Wiley, 1980). *Journal of the Society of Architectural Historians, 40*(3), 259–260.
Piaget, J. (1971). *Structuralism*. London: Routledge & Kegan Paul. (Translated from *Le Structuralisme* by C. Maschler. Paris: Presses universitaires de France, 1968.)
Preziosi, D. (1979). *The semiotics of the built environment: An introduction to architectonic analysis*. Bloomington: Indiana University Press.
Rapoport, A. (1977). *Human aspects of urban form: Towards a man-environment approach to urban form and design*. Oxford: Pergamon Press.
Rapoport, A. (1982). *The meaning of the built environment: A nonverbal communication approach*. Beverly Hills: Sage Publications.
Robey, D. (Ed.). (1973). *Structuralism: An introduction*. Oxford: Clarendon.
Sanders, D. (1986). *Behavioral conventions and archaeology: Methods for the analysis of ancient architecture*. Unpublished manuscript.

Saussure, F. de (1974). *Course in general linguistics.* Glasgow: William Collins. (Translated from *Cours de linguistique générale* by W. Baskin. Lausanne: Payot, 1916.)
Soja, E. (1980). The socio-spatial dialectic. *Annals of the Association of American Geographers', 70*(2), 207–225.
Soja, E. (1985). The spatiality of social life: Towards a transformative retheorisation. In D. Gregory & J. Urry (Eds.), *Social relations and spatial structures.* London: Macmillan.
Stiny, G., & Mitchell, W. (1978a). Counting Palladian plans. *Environment and Planning B, 5*(2), 189–198.
Stiny, G., & Mitchell, W. (1978b). The Palladian grammar. *Environment and Planning B, 5*(1), 5–18.
Sturrock, J. (Ed.). (1979). *Structuralism and since: From Lévi-Strauss to Derrida.* Oxford: Oxford University Press.
Tilley, C. (1982). Social formation, social structures and social change. In I. Hodder (Ed.), *Symbolic and structural archaeology.* Cambridge: Cambridge University Press.

3

Notes on Theory in Environment and Behavior Research

WILLIAM H. ITTELSON

The need for theory is a recurring theme in the environment–behavior–design literature. The reason for this need is clear. It is through theories that we make order out of a range of phenomena that otherwise would be unconnected and unrelated. Theories, in other words, make intelligible a domain of phenomena.

FOLK THEORIES

What is the domain of phenomena that we hope a theory or theories of environment and behavior will make intelligible? And how can we tell if a given theory is actually about that domain? These questions arise after reading two chapters entitled respectively "Conceptualizations" and "Theories" in environment–behavior–design research. One might reasonably expect these to present different theoretical or conceptual views of the same subject matter. Yet, the reader of the excellent chapters by Kaminski and Lawrence cannot avoid concluding that these are two separate approaches to two different domains of phenomena, in

William H. Ittelson • Department of Psychology, University of Arizona, Tucson, Arizona 85721.

spite of the similarities of the titles. Yet, at the same time, this is not completely true. Broadly speaking, both writers are interested in helping us to understand and explain the world of everyday experience as we go about living in, acting in, perceiving, constructing, and modifying the world within which we live. This is the world of shared experiences and actions, of people we interact with, and of natural places and constructed settings we inhabit, perceive, and modify. It is the world we see with our own eyes, touch with our own hands. It is also the world of vicarious experiences. It is the world of newspapers, books, and magazines, and of television and radio. It is the world of manipulating objects and reading instruments. It is the world of shared hopes and aspirations, fears, and expectations. It is the world within which we live and act. It is, in short, the world of everyday experience, and it is that world that we would like a theory or theories of environment–behavior–design research to be about.

At the same time, however, it is this world of everyday experience, of *observation,* that is also the final arbiter, the final court of appeals, that will ultimately pass judgment on our theories. We would like to think of that world as existing objectively, waiting impassively for us to observe it. But clearly, this is not the case. This world of shared, everyday experience with and in an environment, which we want our theories to explain, is itself the product of the way we go about imposing order on that world. In short, the world as we experience it is the product of, not the cause of, our perceptions, cognitions, and actions. Our world of everyday experience is not a necessary, "natural," inevitable, and universal product of being human. Rather, it is, in some general sense, a theoretical construct, a domain of phenomena defined by a set of theories—not scientific theories certainly, but theories nonetheless: informal, largely implicit, socially constructed and consensually validated theories. These theories can appropriately be labeled *folk theories,* although this term is currently being used in another context by other writers in slightly different ways (Churchland, 1986; Fodor, 1987; Stich, 1983). For these writers, folk theories are "commonsense" ways of explaining the phenomena of the natural world, including social and psychological phenomena. As such, they stand in opposition to, and hence must eventually be displaced by, scientific theories. In the position taken here, this is a misreading of the psychological role of folk theories. Folk theories do not explain the world of everyday experience; rather, they define it. Folk theories provide us with the world of experience and observation, the domain of phenomena, that we want our scientific theories to be about and to make intelligible.

SCIENTIFIC THEORIES

A theory is "a system of assumptions, accepted principles, and rules of procedure devised to analyze, predict, or otherwise explain a specified set of phenomena" (American Heritage Dictionary, 1974). Theories, as thus defined, have two distinct parts. First, we have the formal aspects of the theory: the assumptions, the principles, and the procedures that are used to analyze, predict, or explain. The second part is what is explained, that is, what the theory is about: a specified set of phenomena. (When reference is made to *theories* in this chapter, scientific theories will be meant. When folk theories are being referred to, it will be so stated.) Much of the writing on theories has been addressed to the first topic, that is, the formal aspects. The primary interest here, however, is to look at the second part. What does it mean for a theory to be *about* a specified set of phenomena? What does it mean, for example, for a theory to be about environment and behavior? Or more generally, what are theories about? Here, we find fairly general agreement that theories are about something that the theories themselves define. The phenomena of a theory are not known before the existence of the theory, because the theory is required to define the phenomena. This is equally true of scientific theories and of folk theories. Each defines the domain of phenomena that it is about.

Scientific theories are not about the phenomena of folk theories; that is, they are not about the phenomena of everyday experience. Two familiar and classical theories can illustrate this point. I still remember vividly my bewilderment as a young schoolboy being introduced to Euclidian geometry. I was fascinated by it, and still am, but I was mystified on being told that Euclidean plane geometry is not about the points and lines and triangles that I was drawing on a piece of paper. What Euclidean geometry is about, I was told, is dimensionless points, one-dimensional lines, and two-dimensional surfaces, which are totally abstract concepts that cannot be realized in the world of everyday experience and cannot be experienced by any perceptual process. Euclidean geometry, whatever it was about, was not about the lines that I was drawing on a piece of paper, far less was it about the corners of rooms and the location of trees in forests, or about outlines of plots of property, or about other items of everyday experience. I still find it difficult to understand that Euclidean geometry is about, and only about, dimensionless points, one-dimensional lines, and so forth, which are never found in everyday experience and have no physical representation.

Moving along a few years in the experience of this confused schoolboy, we come to Newtonian theory. What is Newtonian theory about? Not the moving trains, the levers, the beams, the projectiles that figured in the problems I was being asked to solve. It is about mass, it is about acceleration, about force and things of that sort. And again, I was bewildered on being told that mass, for example, is a property that we cannot feel, we cannot sense, we cannot observe. It is an abstract property defined by the theory that can never be palpably experienced through ordinary perceptual processes. There are, of course, many other examples. I have selected two that are of both personal and historical interest.

These thoughts inevitably come to mind on reading the preceding chapters. Kaminski talks about "natural units." We can acknowledge what is meant by this concept and at the same time recognize that behavior settings are not "natural units" if by that is meant that they can be encountered as part of everyday experience. The participants in a behavior setting do not and cannot experience the behavior setting. That is an abstract concept defined by the theory. To say this is in no way to denigrate the theory. Quite the contrary, it is simply another way of saying that it is, in fact, a theory.

When we say that a theory is about a specified set of phenomena, it is essential to bear in mind that the phenomena are defined by the theory, and that, as thus defined, they cannot be encountered outside the theory, in the world of everyday experiences or anywhere else. Theories have nothing to say about experience; they have things to say about themselves and about a domain of phenomena that they define, which could not be encountered before the development of the theory and have no meaning independent of the theory. Newtonian mass did not exist before Newtonian theory, although there were obvious historic antecedents. Newtonian mass is a property of Newtonian theory, and Newtonian theory is about those properties that it defines. The same is true of all theories, no less so of theories in the environment–behavior domain.

RELEVANCE OF SCIENTIFIC THEORIES

Where does the preceding discussion leave us in relating theories to the world of everyday experience? How does the domain of phenomena defined by scientific theory make contact with the domain of everyday experience defined by folk theories? That it does is evident. Euclidean theory may not be about the world of everyday experience, but it is extraordinarily useful if we are interested in lines on pieces of paper or the dimensions of rooms. Newtonian theory certainly has useful things to say if we are interested in where and what stones and missiles and bridges and rockets go and do. Ecological theory has greatly increased our understanding of how real people behave in and experience hospitals, schools, workplaces, and so on. Structural theory is where we turn if we want to understand settlement patterns as built and inhabited by real people. So how can one say that these theories are not about the objects of everyday experience and perception? The answer is clear. Theories are not *about* the everyday phenomena of ordinary experience, but they are *relevant* to these everyday phenomena. The proper question, therefore, is: What does it mean for a theory to be relevant to a specific domain of everyday phenomena, and more particularly, how do we determine what, if any, are the phenomena to which a theory is relevant?

Much of the writing about this issue is centered on the problems of verification and falsification of theories. However, the above argument suggests that the concepts of verification and falsification as applied to theories may be inappropriate. Theories may or may not be "well-formed." That is, they may, to varying degrees, conform to established formal criteria of theory. Some may be rather poor theories. They may, for example, not have well-stated principles, or the principles may be mutually contradictory and may therefore lead to mutually exclusive conclusions, or the procedures for drawing conclusions may be unclear, and so on. But whether well formed or not, a theory is about a domain of phenomena that the theory itself defines. A theory is a self-contained system whose reach cannot exceed its self-defined boundaries. A theory, in short, is about itself. Because verification and falsification imply a reference to some external criteria, a theory cannot be true or false. Or, more properly speaking, the concept of *truth* is irrelevant. Because a theory is talking about itself, it may be a poor theory, but it cannot be a false theory. To repeat, the concepts of *verification* and *falsification* of theories are inappropriate. They should be dropped from our lexicon and replaced by the concept of relevance.

Theories, as we have already seen, can be and often are relevant to everyday experience. What we can subject to a test is whether a particular theory is relevant to a particular domain of everyday experience. There are, then, two sets of phenomena: a domain of abstract phenomena defined by a scientific theory and a domain of everyday phenomena defined by folk theory. The question then arises, not how we verify or falsify the theory, but rather how we determine what phenomena of everyday experience are or are not relevant to the theory. I know of no set of formal rules that will accomplish this determination.

Historically, the attempt to provide just such a set of rules was the main thrust of the logical positivists, whose major effort was addressed to developing a set of formal procedures for relating the statements of theory to the phenomena of experience. Their formal rules have been abandoned, and nothing has arisen to take their place. At least, I know of no formal way that we can say, "Follow this set of procedures, rules, or whatever and you will arrive at the set of everyday phenomena to which the phenomena of the theory are relevant." We obviously do come to that conclusion, but it is, at best, an intuitive, circular, bootstrap type of process. There seem to be two general ways we go about making this decision. In one case, we start with a set of everyday phenomena in which we have become interested. Using unclear and unspecified inductive processes, we arrive at some generalizations about that set of phenomena. These generalizations are then abstracted, elaborated, and codified into a set of principles that define a new domain of theoretical phenomena. The theory is then believed to be relevant to the original domain of everyday phenomena that we observed and became interested in. Whether the theory is, in fact, relevant is an empirical question. This is, of course, the commonsense view of how theories are generated, and it is probably followed in many cases. It is, for example, almost certainly in broad outline, how ecological theory was developed.

In contrast to this process, we can start with an existing theory already developed and perhaps already shown to be relevant to a particular set of everyday phenomena. We can then seek other phenomena to which it is hoped the theory will be relevant. Or we can have a set of phenomena of interest and then search among existing theories for one hoped to be relevant. Again, relevance can be determined only empirically. This process is undoubtedly the genesis of the application of structural theory to the analysis of settlement patterns, for example.

In either event, there is no way of telling in advance what the relevant phenomena will turn out to be. In spite of appearances to the contrary, the theory can not tell us; only empirical evidence can do so. Again, probably the best known historical example is Newtonian physics. There is nothing in Newtonian physics that tells us that it is relevant to phenomena in a macrorange and not to phenomena in a microrange. Indeed, it was originally assumed that the distinction could be ignored. When it was discovered empirically that Newtonian theory does not satisfactorily predict or explain phenomena in a microrange, it was decided that Newtonian theory is not relevant to—that is, is not about—those phenomena. But no one could—and historically, no one did—arrive at that conclusion from the theory itself.

RELEVANCE: SUCCESS AND UTILITY

Clearly, we need to clarify what is meant when we say that the domain of phenomena to which a theory is relevant can be determined only empirically. Given a domain of everyday phenomena, there are two ways by which relevance can be empirically demonstrated: first, by formal observation, usually but not necessarily by means of experiment, and second, by application to technology. The distinction is not immediately obvious. All formal scientific observations require technology. It may be simple and primitive, perhaps a crude time-measuring device and a ruler, as in Galileo's famous inclined-plane experiments, or it may be extremely complex, perhaps a multibillion-dollar accelerator. But in either event, the technology is ancillary. The goal is the observation, and the theory must be both necessary and sufficient to account for the observation within acceptable limits of error of measurement. The observation, however, is always made in the domain of phenomena of everyday experience, not in the phenomena of the theory. If the experiment is a "success," the belief that the theory is relevant to the domain of everyday phenomena represented by the observation is strengthened. If the experiment is a "failure," belief in the relevance of the theory may be weakened, or it may remain unchanged through the addition of *ad hoc* or *ceteris paribus* clauses. Formal observation is of interest only to the community of scientists. Patterns of success and failure of observation ultimately serve to delimit a domain of everyday phenomena to which the theory is believed to be relevant. All this has been discussed in great detail elsewhere in the literature on verification and falsification (Lakatos & Musgrave, 1977; Popper, 1959; Schlich, 1979). The modification being suggested here is simply that theories cannot be shown to be true or false, only relevant or not to a specified domain of everyday phenomena. Failure to demonstrate relevance to one domain does not negate relevance to another.

The application of scientific theory to technology differs from formal observation in terms of the anticipated outcome. The outcome of formal observation is the observation. The outcome of technology is the achievement of a socially constructed and defined goal. For purposes of this discussion, *technology* is defined as "operating on the physical environment in order to change what we can do with the environment" or, to borrow a term from Gibson, 1979), "to change the affordance of the environment" (Ittelson, 1986). Scientific theory is necessary for at least a large range of technology, but no theory or set of theories is sufficient to prescribe in all its detail the design of any technology. A large literature on the history of technology attests to this fact and illustrates the variety of processes, methods, procedures, and other aspects of design that can be used to achieve the same technical outcome as well as the non-technical influences that enter into the design and acceptance of any technology (Bijker, Hughes, & Piach, 1987). In this large area broadly encompassed by the term *technology,* it is helpful to talk not about the success or failure of the theory, but about whether the theory is useful or not useful in helping to achieve the socially defined goal of the technology. There exists a large domain of everyday experience far removed from formal observation for which theory may be useful. In summary, then, a theory may demonstrate its relevance to a specific domain of everyday experience by its success in observation or by its usefulness in technology. These two kinds of relevance may be—and historically, have been on occasion—independent.

CONTACT BETWEEN SCIENTIFIC AND FOLK THEORIES

Scientific and folk theories are not about the same domains of phenomena. If they reside in separate and unconnected worlds, how is it possible for them to make any contact with each other, and how can one possibly be relevant to the other? The answer has to be that they cannot be totally unconnected; they must overlap at some point. To some extent, they must be embedded in the same domain of discourse. Where this domain lies can be found by reference to the definition of theory given earlier, which specifies that a theory is, among other things, "a system of assumptions." These assumptions, partly explicitly and partly implicitly, delineate a range of phenomena that constitute the scope of the "real and the possible" that is considered acceptable or allowable under that particular theoretical approach. Scientific and folk theories can communicate only to the extent that they share a common set of assumptions, that there is overlap between their separate understandings of the real and the possible. Feyerabend (1975) recounted tellingly some of the changes in assumptions that had to be made in the 17th century before the observations of science could make contact with or be incorporated into the folk-theoretical base of everyday experience. Over the intervening centuries, those assumptions have been full articulated and incorporated into a worldview that can be summarized as materialistic and mechanistic. In order to qualify as potentially real or possible, a phenomenon must, in other words, conform to the criteria of being both materialistic and mechanistic (Bohm, 1983). It is this shared worldview, this shared set of assumptions, that underlies both scientific theories and folk theories and that makes contact between them possible.

One question that researchers in environment and behavior need to ask themselves is whether the entire complex set of assumptions generically referred to by the terms *materialistic* and *mechanistic* (i.e., our current scientific and folk definitions of the real and the possible) constitute a suitable base for theories appropriate to dealing with the phenomena of interest. This question can be answered not in the abstract, but only through the long and difficult path of theory development and assessment.

If it should turn out that some of these assumptions need to change, we need not despair. The relationship between scientific and folk theories, through their common assumptions, is a dynamic, not a static, one. The nature of everyday experience can change, as can the nature of scientific phenomena. They act on each other by bringing about changes in their base of common assumptions. The challenge to a scientific theory of environment and behavior is the same as the challenge to all social and behavioral science: to develop scientific theories that are relevant to how we experience ourselves. There is no reason to believe in advance that this cannot be done. There may, however, be good reason to think that, in the process, the assumptions underlying both what we mean by *science* and how we experience our world may change.

THE ACCEPTANCE OF THEORIES

The acceptance of a scientific theory is a complex social process about which much has been written (Kuhn, 1962, 1977). It involves at least observational success and technological usefulness, as defined earlier, as well as many other issues. For example, the aesthetics of the theory, its simplicity, its symmetry, and its elegance, as well as its congruence with established modes of thought and the centrality of the issues addressed, may be of crucial importance. In addition, there are two separate "constituencies" for a theory: the scientific community and the larger society. To a certain extent each of these groups may accept or reject a theory independently of the other, although obviously, the scientific community is itself part of the larger society, which, in turn, is composed of many additional subgroups, each with its own and possibly diverse definitions of the real and the possible. No attempt will be made here to discuss all the issues implicated in the acceptance of theories. Instead, a few considerations of importance to environment–behavior–design theories suggested by the preceding two chapters will be touched upon.

The acceptance or rejection of a theory as being relevant to a specific domain of everyday phenomena is bound up in, though not entirely determined by, its success in formal observation and its usefulness in achieving societal goals. In general, the scientist is concerned with the outcome of formal observation. The larger society is interested in whether the theory is useful. However, although this distinction may appear clear in principle, in practice it is more often blurred or obscured. The opportunity even to pursue a particular line of scientific inquiry may depend both on society's belief that it is or will be useful and on science's belief that it is or may be congruent with currently accepted modes of thinking. These beliefs, in turn, may depend on the theory's effectiveness in molding the pattern of assumptions underlying the folk theories of the society. These considerations arise with respect to most efforts in the environment–behavior–design field. With very few exceptions, workers in this field have approached it as an "applied" field and have sought acceptance by demonstrating its technological usefulness in achieving social goals. In spite of some achievements reached largely through an *ad hoc* empiricism, in general these workers have not been notably successful in this endeavor. Certainly, there are not on the current scene any major theories that combine demonstrated observational success with technological usefulness.

Both ecological theory and structural theory are cases in point, although from slightly different perspectives. When we look at ecological theory, we see that it does define a limited and specific domain of phenomena that is relevant to certain aspects of environment and certain aspects of behavior. The *behavior setting,* as pointed out earlier, is an appropriate scientific concept. The behavior setting was not known before the theory, and the behavior setting is defined by the theory. Whether the behavior setting, and ecological theory in general, is scientifically relevant to aspects of everyday phenomena is a question of the success or failure of the theory, and one can say that it has been successful within at least a limited domain of phenomena. There is a domain of everyday phenomena to which ecological theory is relevant. Why, then, has it not met with widespread acceptance within either the scientific community or the society at large? One can only speculate that this is an example of the principle stated earlier that the acceptance of a theory depends on many factors in addition to its observational success. Ecological theory has not been widely accepted by the society at large, probably because it has not demonstrated its usefulness in achieving some socially defined set of goals. This failure, in turn, has led to a lack of interest on the part of a community of researchers who have largely based their hopes for acceptance on demonstrating usefulness and who, in addition, find ecological theory not to be congruent with their existing ways of thinking. This story is not unique to ecological theory. With some change in detail, it could be told for structural theory as applied to the environment–behavior–design field, as well as to most other existing theories or conceptual approaches to the field.

It is, of course, not true that ecological theory, or structural theory, or any of the other approaches, has received no acceptance. Quite the contrary. Each of these theories has been accepted by a group of competent and dedicated researchers who are intimately concerned—indeed, sometimes passionately concerned—with the specifics of that theory. Around such theories develops a small group actively engaged in discussing and arguing with each other, in training graduate students, in producing a seemingly endless array of studies pro and con over what appear from the outside to be trivial or peripheral issues, and in general, in providing all the elements of a microcosm of scientific activity. The aim of this activity is presumably to demonstrate a wider domain of relevant phenomena and a wider area of social usefulness. In the meantime, the rest of the scientific community and the public at large seem serenely indifferent.

ENVIRONMENT–BEHAVIOR–DESIGN THEORY: WHAT FOR?

If the above description is a fair representation of the current state of acceptance of environment and behavior theories, perhaps it suggests a need to examine the issues that these theories are intended to address. The value of a theory is directly dependent on the nature of the questions it asks and the way these questions are formulated. This is the most important and the most difficult aspect of any research endeavor. It is not unreasonable to suggest that the field of environment–behavior research has not been unified or clear about what questions it is trying to answer or how they should be formulated.

Several different, though not necessarily independent, orientations can be seen. Perhaps the most pervasive throughout the entire field, as already suggested, is the applied approach. Here, the effort is addressed at solving specific, immediate problems, usually through an *ad hoc* empiricism. Generalizations are few, and theory construction is not a likely outcome. A second approach is discipline-oriented. Here, the question is how environment–behavior issues can be incorporated into or added to the particular parent discipline. The basic questions and theories are usually those already existing in the discipline. In contrast, *interdisciplinary* has always been a major catchword for the field, and interdisciplinary research must be added as a third approach. However, this has proved to be easy to say and difficult to do, a problem not limited to this field. It is hard to find examples of true interdisciplinary work despite the earnest efforts of many researchers. Most of the successful work can probably be best described as disciplinary combinations. Disciplinary imperialism is not uncommon. The true interdisciplinary theory is certainly rare, and it is not completely clear how we would recognize one if we encountered it. Perhaps the most successful effort has been addressed to extracting and describing metatheoretical approaches that might cut across disciplines (Altman & Rogoff, 1987;Moore, Tuttle, & Howell, 1985; Stokols, 1987; Wapner, 1987).

This brief categorization is not in any way intended to be pejorative. Important, significant, and valuable work is being done in all of these areas, and one hopes it will continue. It would certainly be premature to suggest limiting the field in any way. Instead, expanding the scope of research by examining, enlarging, and clarifying the questions being addressed is suggested as a useful direction to pursue.

For all organisms, "environment and behavior" is shorthand for everything that we need to know in order to understand both organism and environment, or more properly the organism–environment complex, for that is indeed an indivisible whole. It is well recognized that this is a major undertaking that cannot and should not be approached from any arbitrarily limited number of directions. When we consider the human organism, the shorthand phrase has to be expanded into "environment, behavior, and design," reflecting the fact that humans represent the only organism that flexibly, self-consciously, and massively alters its environment. The people–environment complex always includes the question of design, broadly interpreted as the deliberate modification of the physical surroundings, or technological action, as previously defined. The relevance of design is true for the big questions of population, pollution, depletion of the biosphere, and catastrophic destruction, as well as for the little questions of dwellings, family life, personal goals, and even the functioning of the nervous system, together with all scales of questions in between. These must all be addressed in detail and from every possible perspective. We can only hope that in addressing these questions, individual researchers will always bear in mind that the phrase "environment, behavior, and design" describes the human condition. To develop theories that will make that everyday existence intelligible, that will have relevance to questions both great and small, and that will help us "analyze, predict, or otherwise explain" the conditions of human existence, and, in doing so, provide the means and the understanding necessary to direct the processes of change—that is the challenge underlying every little piece of environment–behavior–design research.

REFERENCES

Altman, I., & Rogoff, B. (1987). World views in psychology. In D. Stokols & I. Altman (Eds.), *Handbook of environmental psychology*. New York: Wiley.

American Heritage Dictionary of the English Language. (1974). New York: Dell.

Bijker, W. E., Hughes, T. P., & Piach, T. J. (Eds.). (1987). *The social construction of technological systems*. Cambridge, MA: MIT Press.

Bohm, D. (1983). *Wholeness and the implicate order*. London: ARK.

Churchland, P. S. (1986). *Neurophilosophy*. Bradford, MA: MIT Press.

Feyerabend, P. (1975). *Against method*. London: N.L.B.

Fodor, J. (1987). *Psychosemantics: The problem of meaning*. Cambridge, MA: MIT Press.

Gibson, J. (1979). *The ecological approach to visual perception*. New York: Houghton Mifflin.

Ittelson, W. H. (1986). *The psychology of technology*. Paper presented International Association of Applied Psychology, Jerusalem.

Kuhn, T. S. (1962). *The structure of scientific revolutions*. Chicago: University of Chicago Press.

Kuhn, T. S. (1977). *The essential tension*. Chicago: University of Chicago Press.

Lakatos, I., & Musgrave, A. (1977). *Criticism and the growth of knowledge*. Cambridge, MA: Cambridge University Press.

Moore, G. T., Tuttle, D. P., & Howell, S. C. (1985). *Environmental design research directions*. New York: Praeger.

Popper, K. R. (1959). *The logic of scientific discovery*. New York: Harper & Row.

Schlich, M. (1979). Meaning and verification. *Philosophical papers*. Boston: D. Reidel.

Stich, S. (1983). *From folk psychology to cognitive science*. Bradford, MA: MIT Press.

Stokols, I. (1987). Conceptual strategies of environmental psychology. In D. Stokols & I. Altman (Eds.), *Handbook of environmental psychology*. New York: Wiley.

Wapner, S. (1987). A holistic, developmental systems-oriented environmental psychology. In D. Stokols & I. Altman (Eds.), *Handbook of environmental psychology*. New York: Wiley.

II

ADVANCES IN PLACE RESEARCH

4

Environment, Behavior, and Design Research on Urban Forests

HERBERT W. SCHROEDER

To many people the term *urban forest* seems incongruous and contradictory. But a look around any major city quickly reveals that trees and other vegetation are an important feature in many urban settings. For example, an aerial photo survey of Dayton, Ohio, showed that 22% of the city's land area is covered with trees, and that 35% is covered with other kinds of vegetation (Sanders & Stevens, 1984). According to one overall estimate, 30% of the average city in the United States is covered with trees, a proportion larger than the average tree cover for countryside (Dwyer, Deneke, Grey, & Moeller, 1983).

Some professionals prefer to identify as urban forests only those areas of vegetation that approximate the traditional image of a forest, that is, essentially natural ecosystems located within urban surroundings (Figure 1). Even with this restricted definition, urban forests are a major component of some metropolitan areas. For example, the forest preserve system surrounding Chicago occupies over 10% of the land area in a county with almost 6 million people (Wendling, Gabriel, Dwyer & Buck, 1981).

In this chapter, I use a broad definition of the urban forest, to

Herbert W. Schroeder • North Central Forest Experiment Station, U.S. Department of Agriculture Forest Service, Chicago, Illinois 60646.

FIGURE 1. Some urban forests, such as this forest preserve in the city of Chicago, closely resemble the traditional image of a natural forest.

include virtually any occurrence of woody vegetation in populated places (Grey & Deneke, 1986). Besides city parks (Figure 2), this may include trees:

> . . . along city streets and sidewalks, riverbanks and transportation corridors, powerlines and pipelines, in forest preserves, cemeteries, golf courses, nature centers, vacant lots and fields, arboreta, conservatories, yards and estates, and municipal watersheds. (Dwyer, 1980, p. 11)

Management of urban forest environments is complicated by the diverse character of the settings; the multiplicity of public and private landowners, planners, and agencies; and the intricacies of local politics and public involvement (Dwyer, 1980; Dwyer & Schroeder, 1982). In recent years, urban forest managers have been caught between the increasing recreational use of urban forest resources and the decreasing budgets for managing those resources. This dilemma has created a need for more efficient ways to manage urban forests for the benefit of urban residents. Researchers in the social science and design disciplines have responded to this need with studies of the human perceptual and behavioral aspects of the urban forest.

FIGURE 2. Although city parks such as this one are not natural ecosystems, they may still be thought of as urban forests.

In this chapter, I summarize recent (since 1980) environment, behavior, and design research on urban forests, and I suggest some future directions that this research might take. I have included work that focuses specifically on urban forest and open-space settings and that deals with human perception, preference, and behavior with respect to these settings. I have not included references to nonurban natural settings or nonvegetated urban settings, or to the physical and biological aspects of urban forest management, except where these are particularly relevant to human–environment issues in urban forests.

MAJOR RESEARCH TOPICS

The literature pertaining to the human aspects of urban forestry is diverse and multidisciplinary. To impose order on this diversity, I have identified some major topics that have characterized the last several years of work in this field. These topics are not independent, and references cited under one topic are often relevant to other topics as well.

Benefits and Values Associated with Urban Forests

To manage forests effectively for public benefits, urban foresters need to know what those benefits are. Failure to identify all the benefits provided by urban forests may result in some benefits' being lost through misguided management. Also, to justify budgets and expenditures for urban forest management, managers need to show that the benefits provided by the forest are important to the public. Urban forestry budgets must compete for scarce funds with city services, such as street maintenance, police, and fire protection. Without specific information on the importance of urban forest benefits, local governments may simply assume that urban forests are not very important.

Identifying Kinds of Benefits. Many of the benefits of urban forests can be measured in physical or biological terms. Moderation of temperature extremes, improved air quality, noise abatement, and erosion control are frequently cited benefits of urban vegetation (Grey & Deneke, 1986). But the most important benefits of urban trees may, in fact, be found in the effects they have on people's subjective experiences, moods, and feelings.

The literature on landscape assessment consistently shows that environments with natural elements such as trees and water are highly preferred to urban scenes lacking these elements (Ulrich, 1986). The presence of natural features in urban environments enhances residents' overall satisfaction with their living conditions (Kaplan, 1982; Marans & Fly, 1981). Recent studies have attempted to learn more precisely how vegetation affects people's moods and emotional states. For example, when viewers' responses were measured by mood and feeling rating-scales in one recent study, nature scenes resulted in higher levels of attentiveness and interest, whereas urban scenes tended to increase sadness and fear (Ulrich, 1981). Studies of physiological responses to the visual environment (Ulrich, 1981; Ulrich & Simons, 1986) have shown that scenes with vegetation promote more relaxed and less stressful states than do urban scenes lacking vegetation (Figure 3).

Perceived park benefits seem to fall into two distinct categories: *passive* natural benefits obtained from viewing aesthetic elements such as trees; and *active* recreational and social benefits obtained from using environments for sports and play (Hayward & Weitzer, 1984; Ulrich & Addoms, 1981). Park programs frequently emphasize active sports, but passive aesthetic enjoyment may actually be the greatest benefit of urban forests (Figure 4). This enjoyment can provide a sense of relief and escape from urban life for city residents (Hayward & Weitzer, 1984; Ulrich, 1986). Infrequent users of parks and nonusers also obtain sub-

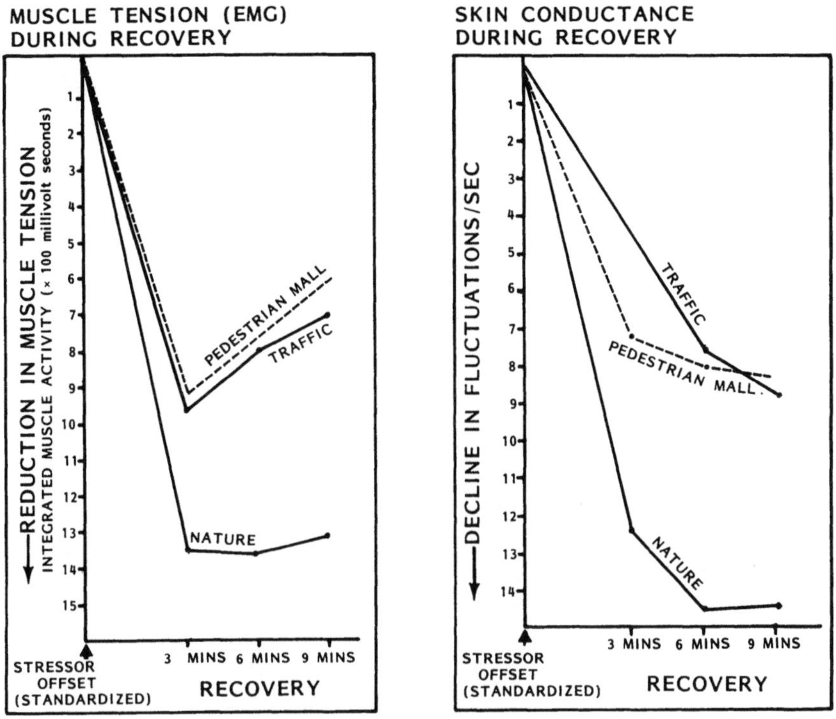

FIGURE 3. Physiological measurements show that nature scenes promote recovery from stress much better than scenes with mainly artificial features (Ulrich & Simons, 1986).

stantial benefits by observing parks from the street or through windows, and by just knowing the parks are there to use if needed (Kaplan, 1980, 1981; Ulrich & Addoms, 1981). The importance of window views of urban nature is suggested by a study in which hospital patients with a view of a wooded scene recovered from surgery more quickly and with fewer complications than patients whose windows looked out onto a brick wall (Ulrich, 1984).

A survey of gardeners (Kaplan, 1983) showed that the most important perceived benefit of gardening was the feeling of peacefulness and tranquility it produced. At the same time, gardening provided people with the fascination of being involved in a natural process. Observing and feeding wildlife, which is often associated with urban forests, probably provides similar opportunities for fascination and involvement. In one survey, almost half the respondents reported engaging in wildlife-related activities (mainly bird feeding) close to home (Shaw, Mangun, & Lyons, 1985).

FIGURE 4. Many park programs are structured around athletic activities and facilities (A). Relaxation and refuge from city surroundings, however, may be the most important value of urban forests (B).

In a more theoretical vein, Kaplan (1984c) suggested that two characteristics of natural settings that may foster restorative effects on human beings are "fascination" (that is, the ability to promote effortless attention) and "coherence" (or the property of making sense). To the extent that urban forests possess these characteristics, they may be able to provide benefits usually associated with more remote natural sites.

Aside from improving the quality of neighborhoods, active involvement in tree-planting programs seems to enhance a community's sense of social identity, self-esteem, and territoriality, and it teaches residents that they can work together to choose and control the condition of their environment (Ames, 1980).

Economic Measures of Benefits. Urban managers faced with the hard realities of municipal budgets often find it useful to place dollar values on the benefits produced by urban forests. For insurance and tax purposes, formulas exist for appraising the value of a tree based on the size, species, and condition of the tree (Grey & Deneke, 1986; Morales, Micha, & Weber, 1983). These formulas, however, have no theoretical or empirical basis and do not provide valid economic measures of the benefits produced by urban forests.

Three economic methods have been used for measuring the value of urban forest benefits. The hedonic pricing method infers the value of urban forest resources by observing the changes in residential property values associated with proximity to a park (Hagerty, Stevens, Allen, & More, 1982) or with trees on the property (Anderson & Cordell, 1985; Morales *et al.*, 1983). The travel cost method measures the costs that users incur to reach a site, and it infers the value of the site from those costs (Darragh, Peterson, & Dwyer, 1983). The contingent valuation or direct-questioning method has been used to ask people to state the amount they would be willing to pay to preserve vegetation in a land-use-planning context (Coughlin & Strong, 1983).

Each of these three methods has its shortcomings (Allen, Stevens, & More, 1985). For example, the travel cost method does not count aesthetic benefits to nearby residents who do not actually enter the park; the hedonic pricing method does not measure benefits to users living far from the park; and the contingent valuation method is subject to difficult psychometric measurement and validation problems, because of the hypothetical nature of the questions posed to the respondents.

Perception of Urban Forest Environments

Environmental perception studies have sought to identify the characteristics and features that enhance the perceived quality of urban forest sites. As noted above, landscapes with natural elements such as trees and water are generally preferred over artificial landscapes (Ulrich, 1981, 1986). Viewers' preference ratings of photographs of urban scenes show that preference decreases with increasing development intensity. Increasing the amounts of vegetation and improving the quality of landscape maintenance increase preference (Anderson & Schroeder, 1983). Views of parks with natural features are consistently preferred to residential and commercial landscapes (Herzog, Kaplan, & Kaplan, 1982; Ulrich & Addoms, 1981).

Several studies have found that the most preferred features of urban parks and forest preserves include trees and forested areas, water, good maintenance, and peace and quiet (Schroeder, 1982; Schroeder & Anderson, 1984; Young & Flowers, 1982). Features that detract from the attractiveness of a park include manufactured objects (e.g., fences and parking lots), poor condition of vegetation, urban surroundings adjoining the park, litter, graffiti, crowding, and large, monotonous fields (Schroeder, 1982; Schroeder & Anderson, 1984).

Vegetation may be a universally preferred environmental feature, but not all kinds of vegetation are equally preferred. Perhaps the most universally liked setting is an open stand of large trees, with evenly mowed grass beneath. This configuration is typical of many American community parks, and it seems to be aesthetically appealing to a wider range of people than are either very open scenes or dense impenetrable forests (Kaplan, 1984a). Consistent with this finding, a study of perceived tree density in municipal parks (Schroeder & Green, 1985) showed that intermediate tree densities of 40–65 trees per acre were more attractive than higher or lower densities (Figure 5). Vegetation in a park may also influence the perception of the size or spaciousness of the site. Seemingly spacious environments can be created in relatively small parks by using vegetation to mask surrounding urban features and by providing paths to allow movement into the area (Talbot & Kaplan, 1986).

Besides vegetation, urban forests usually contain artificial structures and facilities. Although these often detract from the aesthetic quality of a landscape, not all types of artificial features are equally unattractive. A high degree of contrast or incongruity between a structure's color, texture, size, or shape and its natural setting were found to decrease the preference for park scenes in the Northeast (Wohlwill & Harris, 1980). Similarly, incongruity and crowding of structures decreased the attractiveness of forested subdivisions in Colorado (Vining, Daniel, & Schroeder, 1984).

Vegetation can significantly enhance the quality of more developed

FIGURE 5. Most people prefer park settings with intermediate tree densities (B) over settings with very few trees (A) or too many trees (C).

FIGURE 5. (*Continued*)

urban settings, such as parking lots (Shaffer & Anderson, 1985) and residential streets (Buhyoff, Gauthier, & Wellman, 1984; Lien & Buhyoff, 1986; Schroeder & Cannon, 1983). Larger trees on streets are preferred over smaller trees, perhaps because of the overarching canopy and shade they provide (Figure 6). Open, parklike stands of large trees were also preferred in surveys of nature preference at multiple-family housing sites (Kaplan, 1982, 1985). Residents of these sites also expressed a liking for natural, unmanaged settings close to their homes.

Most studies of urban forest perception have focused on the visual landscape, but sounds and noises are also an important feature of most urban environments. The physics of noise abatement due to vegetation has been well studied, but there is little information on the perceived relationships between noise and vegetation. A recent study suggested that fittingness or congruity plays a role, too (Anderson, Mulligan, Goodman, & Regen, 1983). Natural sounds, such as birdsong and wind, were perceived to enhance the quality of forested sites, whereas unnatural sounds, such as traffic, detracted. For unvegetated, downtown sites, however, traffic was perceived as the most enhancing sound. Heavily vegetated sites appear to be much more sensitive to both the enhancing and the detracting effects of sounds, perhaps because they arouse stronger expectations about what kinds of sounds are appropriate.

FIGURE 6. Streets with large trees and an over-arching canopy (A) are perceived as much more attractive than streets with small trees or no trees (B).

On-Site Behavior in Urban Forests

Studies of overt behavior in urban parks and forests require very different research methods from studies of perception and preference. One approach is to look for indirect indicators of use, such as amount of wear and width of trails (More, 1980), which give a rough idea of how many people use an area. Questions on surveys may also provide some insight into how people use forest sites. For example, a survey of users in Chicago-area forest preserves (Young & Flowers, 1982) revealed surprisingly short visits: 53% of the respondents said that they planned to stay an hour or less. Most visitors were frequent users of the forest preserves, and the most common activity reported was simply "relaxing."

The best way to obtain accurate information about numbers of visitors and kinds of behaviors, however, is through direct observation. Observation studies require careful planning and long periods of time in the field, but the information gained may be well worth the effort (More, 1984). Observations can provide important insights into the group structure and context of recreation activities in urban parks that may be missed in surveys about recreation participation (Hutchison & Fidel, 1984).

A detailed study of two northeastern parks (More, 1985a) revealed not only that the parks were heavily used, but that they were used at all times of the day and night and in all kinds of weather. The users included a very broad cross section of the urban populace. Although a wide variety of activities was observed, the most common activities were ordinary ones such as walking, talking, and observing.

Observation of vehicles entering Chicago-area forest preserves (Dwyer, Schroeder, & Buck, 1985) revealed very stable patterns of use over time (Figures 7 and 8). About 75% of the variance in total daily use was explained by the month of the year, the day of the week, and the weather. Stable patterns of use over different times of day were also observed, although the patterns were different for different locations and for different days of the week.

It is difficult to generalize from the few observational studies of urban forest behavior that have so far been completed, especially given the tremendous diversity of activities and users that characterizes urban parks. One promising suggestion is to define categories of parks that display similar use and behavior patterns. More (1985b) suggested that such categories could be based on the parks' size, activity resources, amenity resources, and aesthetic quality.

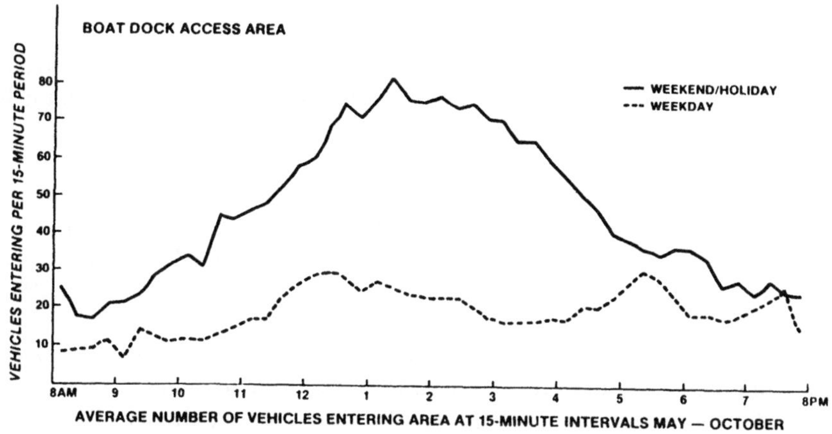

FIGURE 7. The number of vehicles entering a Chicago-area forest preserve shows a stable pattern over the day (Dwyer, Schroeder, & Buck, 1985).

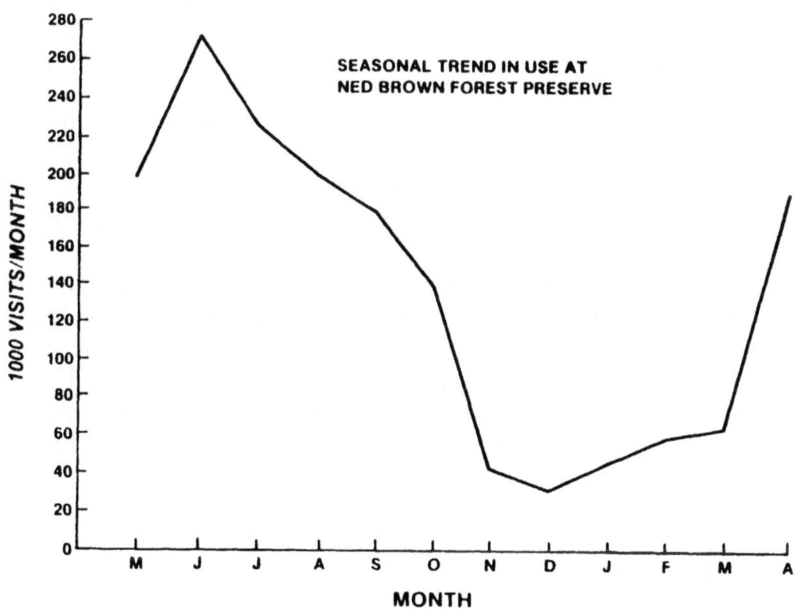

FIGURE 8. Numbers of visits to a Chicago-area forest preserve increase rapidly during the spring and peak in early June (Dwyer, Schroeder, & Buck, 1985).

Destination Choice Behavior

Several studies have examined how people decide which urban recreation sites they will visit. Travel time or distance to the site seems to be a primary consideration. For example, 37% of Chicago-area forest-preserve visitors travel 2 miles or less from home or work to the forest, and about half reported a travel time of 10 minutes or less (Young & Flowers, 1982). Distance has consistently appeared as a dominant factor in models predicting choice of Chicago-area urban-forest recreation sites (Darragh, Peterson, & Dwyer, 1983; Peterson, Dwyer, & Darragh, 1983) and urban forest trails (Lieber & Allton, 1983).

The perceived attributes of urban forest sites also influence choices, but the process is difficult to model. In one study, some attributes had a consistent effect on the probability of a park's being chosen, but other attributes had effects that varied according to the type of site (Peterson *et al.*, 1983). This finding suggests that there are different categories of urban forest sites, and that the choice process is "nested" (Peterson, Stynes, Rosenthal, & Dwyer, 1985). That is, a visitor first decides which type of site to visit (e.g., forest preserve, botanical garden, or indoor conservatory) and then chooses an individual site within the category.

Choices among urban-forest recreation sites are influenced by the amount of information that people have about the sites available to them. Many people are unaware that certain parks exist and are unfamiliar with the locations, features, and facilities of the parks they are aware of (Spotts & Stynes, 1985). Surveys have shown an especially low level of awareness for distant parks, new parks, and less developed parks. Many respondents in these surveys were unfamiliar with the locations and facilities of even some major city parks (Spotts & Stynes, 1984).

Widespread lack of information about urban parks complicates the task of modeling the choice process. When knowledge variables are incorporated into park choice models, it appears that distance from the park is the best predictor of both awareness and visitation rates. Thus, it is possible that the effect of distance on urban park trips is due as much to differences in awareness as to differences in travel cost (Stynes, Spotts, & Strunk, 1985).

Safety, Crime, and Conflict

Although crime and social conflict are problems in all kinds of recreation areas, they are often a particularly acute concern in urban parks and forests. One survey, for example, identified drinking and drug use, fear of crime, teenagers, and clashes with rangers as significant concerns of Chicago forest preserve users (Young & Flowers, 1982). Social conflict includes a wide range of behaviors, from violent crimes such as murder and armed robbery, to "nonviolent" offenses such as vandalism and drug use, to behaviors that, although not illegal, may be threatening or offensive to other users (Chubb & Westover, 1980).

Design and behavior research has focused on people's subjective perceptions of crime and safety. Interviews of users in three regional parks in Michigan and Minnesota showed that most users were satisfied with park rules and with the level of compliance and enforcement (Westover, 1985b). Minorities, low-income visitors, frequent visitors, and young males were more likely to perceive rules as unfair, and to report observing rule violations. Exposure to rowdy or inappropriate behavior decreased the perceived effectiveness of law enforcement. Although most users felt the parks were generally safe, over 40% reported experiencing fear or engaging in avoidance behavior at some time (Westover, 1985a). It appeared that avoidance behaviors were successful in reducing the visitors' level of fear.

Viewers' ratings of park photographs showed that features of the physical park environment may influence levels of perceived risk (Schroeder & Anderson, 1984). Long-view distances, open grassy areas, and water were associated with high safety, and dense vegetation, graffiti, and litter decreased perceived safety (Figure 9). Undeveloped, heavily forested environments were perceived as the most scenically attractive but the least safe sites; and open athletic fields were perceived as least scenic but most safe. Parks and picnic areas that had substantial tree canopy overhead, but little foliage at eye level to interfere with visibility, were perceived as both safe and scenic (Figure 10).

A similar study of urban parking lots showed that lots with vegetation were perceived as safe only when the vegetation was well maintained and was part of a landscape design (Shaffer & Anderson, 1985). Perceived safety was also enhanced when vegetation did not block the visibility of entrances to buildings.

On urban trails, perceptions and concerns about safety were related to the level of interaction that users expected to experience with other people (Lieber & Fesenmaier, 1985). Where there was a possibility of threatening encounters with other people, trail users indicated that the most effective means of allaying fear would be to increase the presence of law-enforcement personnel and to move trail heads away from parking areas. Cutting vegetation to increase visibility was seen as a very undesirable policy.

Recent field experiments have tested whether depreciative behav-

FIGURE 9. Graffiti and poor maintenance create the perception that a park is dangerous.

FIGURE 10. A park with a substantial tree canopy but little vegetation blocking the view at eye level may be perceived as both safe and aesthetically attractive.

iors can be controlled by management actions and environmental conditions. One study showed that appeals to campers to help reduce littering led to an increase in antilittering behavior (Christensen & Clark 1983). Another study showed that carving on picnic tables was more frequent at sites where there was previous carving and where authority was less salient (Samdahl & Christensen, 1985). These experiments were conducted in nonurban campgrounds, but they suggest approaches that could also be used to address these problems at urban forest sites.

Sources of variation in Preference and Behavior

It is often said that "beauty is in the eye of the beholder," meaning that different kinds of people have different standards of preference. But environmental preference studies often show a very high level of agreement between very diverse groups of people. For example, in one study, a group of white male members of a rural Georgia civic club and a group of predominantly young black female temporary employees of an urban Atlanta nature center gave very similar preference ratings to a set of urban scenes (Anderson & Schroeder, 1983).

More detailed examination of environmental preferences, however, has revealed that some individuals do vary from the consensus (Kaplan, 1984a; Schroeder, 1983). For example, in a study of the perceived safety and the scenic quality of urban parks (Schroeder & Anderson, 1984), the majority of raters thought that natural-appearing scenes with dense vegetation were the most attractive, but some raters gave their highest ratings to more developed and manicured parks. Also, the majority thought that heavily wooded parks were the most dangerous, but a minority felt that densely wooded parks were among the most safe.

It is natural to wonder whether variations in preference and behavior are related to identifiable characteristics of urban residents. If different preference and behavior patterns can be identified for different segments of the urban population, then managers could more effectively manage urban forest resources for the benefit of their users. This section reveiws what has been learned recently about the sources of variation in urban forest perception and behavior.

Ethnicity. Blacks are the largest ethnic minority in American cities. There have been several studies about their recreation activity patterns (O'Leary & Benjamin, 1982; Stamps & Stamps, 1985; Wendling, 1980), but most of this research has not focused on the environments in which the activities take place.

In general, black Americans appear to be less interested than white

Americans in nature and the outdoors, and to have less knowledge about environmental concerns and wildlife (Kellert, 1984). Several studies seem to suggest that blacks are more oriented to urban environments than to natural ones. An Indiana survey showed that blacks used city and town rather than forest, farm, and water environments for most of their recreation (O'Leary & Benjamin, 1982). Blacks in Lansing, Michigan, were found to be less aware than whites of the parks available for outdoor recreation in that city (Spotts & Stynes, 1984). In a survey of fifth-grade children attending an urban nature center, black children emphasized the fun and play aspects of the experience, and white children emphasized natural qualities of the environment such as plants and animals (Metro, Dwyer, & Dreschler, 1981). Black visitors at an interpretive center were more interested than whites in attending a program that placed nature within the context of the urban environment (Fritschen & Stynes, 1980).

Interviews with low- and moderate-income blacks (Talbot & Kaplan, 1984) suggested that blacks differ from whites not so much in a lack of interest in enjoying the outdoors, as in the kind of outdoor environments they enjoy. Blacks preferred orderly, well-maintained areas with built features over more natural and densely wooded areas, a preference that may have been associated with fear of physical danger.

Hispanic Americans are another significant minority in many cities, but there has been very little research on their use of urban parks. Onsite observations of parks in Chicago neighborhoods (Hutchison & Fidel, 1984) revealed that Mexican-Americans did less jogging, walking, and biking, and more stationary and sports activities, than Anglo-Americans. The Mexican-Americans' use of parks was characterized by large multifamily groups with a mixture of ages. This pattern may be related to the family- and group-oriented culture of the Mexicans, and to the traditional use of the plaza in Mexican cities.

Residential Background. Differences in recreational use and preference for outside environments may be related to differences in the residential backgrounds of different urban groups. A national survey showed that residents of large cities had a lower level of understanding of and personal interest in wildlife and the outdoors than suburbanites (Kellert, 1984). An analysis of college students' ratings of photographs of urban parks and forests (Schroeder, 1983) revealed that the raters with urban backgrounds were more likely to prefer developed city parks, and that the raters with suburban and rural backgrounds were more likely to prefer natural-appearing forests. In open-ended responses to photographs of urban parks and forests (Schroeder, 1982), urban respondents mentioned vegetation, water, and fields less frequently as preferred features than did nonurban respondents, and they were less likely to mention artificial features as undesirable. Urbanites also expressed more interest in the ability to engage in activities at the sites. This response is consistent with that in a survey (Ulrich & Addoms, 1981) showing that active recreation and socializing were more important as park benefits for people with urban backgrounds than for nonurban people.

Gender. Male and female users differ in their use of urban parks (Westover, 1986). Observational studies have shown that males are more frequent users of central-city parks than are females, and that the proportion of female users changes during the day, being highest during midday and lowest in the morning and evening (More, 1985a). Males are more likely to engage in active sports, and females are more likely to engage in passive activities, such as sitting and reading (Ulrich & Addoms, 1981). Differences in use patterns between males and females may be influenced by fear of crime. Females are more likely than males to report that city parks are unsafe and to express more fear of crime, a finding suggesting that females require more visible, formal law enforcement in order to feel secure in parks (Westover, 1985b).

Age. Some research results suggest that nature and urban forests are less important to older than to younger urbanites. In a national survey, people over 35 years old expressed a lower interest in and affection toward the outdoors than did people aged 18–35 (Kellert, 1984). In a study of the visual quality of residential streets, the level of upkeep of the street was the most important predictor of elderly people's preference, and natural elements were not a significant factor (Nasar, 1981).

Other research, however, suggests that urban forests may, in fact, be important to the elderly. Interviews of older park users (Godbey & Blazey, 1983) showed that parks may help create a positive state of mind among old people. Their park use is often part of a regular routine rather than a break from daily routine. Elderly users tend to accommodate their use of the park to conditions such as safety and special events, rather than fitting their use pattern to their own personal schedules.

At the other end of the age continuum, forest experiences and preferences of schoolchildren visiting an urban nature center were studied by Metro *et al.* (1981). Most of the children had been in a forest before with family or another group and had enjoyed the experience. Exploring was the most frequently reported experience. Most children said they felt safe in the forest. The most frequently perceived danger was from wild animals.

A comparison of preferences of senior citizens and sixth-grade children in a Minnesota town (Stiegler, 1985), suggested that urban forests are perceived quite differently by elderly people and children. The seniors saw urban forests as more similar to natural forests than did the children. The majority of the children preferred natural forests to urban forests, but the majority of the seniors preferred urban forests.

Education. College education seems to be associated with greater concern about, appreciation of, interest in, and understanding of animals and nature (Kellert, 1984). Professional training and experience in resource or park management may also influence perceptions and preferences. In a study of preferred tree density in parks (Schroeder & Green, 1985), staff members at an arboretum preferred higher tree densities than did other groups of raters, and employees of suburban park districts preferred the lowest density. Landscape architecture students' ratings of urban scenes have also been observed to diverge somewhat from other groups' preferences (Anderson & Schroeder, 1983).

Activities. It seems obvious that people's preferences for urban park environments should depend on what they intend to do there, but there has not been much research relating environmental preferences to specific recreation activities. In a survey of urban-forest trail-users, clear differences were found between people using the trails for different activities. Bicyclists strongly preferred paved trails, and hikers, joggers, and cross-country skiers preferred wood chips. Skiers liked hilly trails, and joggers disliked them. Skiers and hikers liked frequent changes in trailside scenery, and joggers had a strong preference for trails close to home (Allton & Lieber, 1983; Lieber & Allton, 1983). More information of this sort concerning other kinds of urban forest sites would be useful for designing environments to serve specific activities.

Disabled Users. Urban forests may be important in serving the outdoor recreation needs of disabled people because they are potentially more accessible than backcountry forest sites. There has been very little research on the general environmental preferences of disabled people, but some recent studies provide useful insights into the question of the accessibility of recreation resources to the disabled. Attempts to make urban parks more accessible have focused on physical barriers to park use, but social barriers may be just as serious a problem (West, 1981). In a survey of disabled park users, about half of the respondents who perceived negative reactions to their disability subsequently reduced their level of recreation participation. These individuals might be helped by programs to develop their coping strategies and maintain their positive self-concepts despite social stigma experienced in public parks (West, 1984).

APPLICATION OF RESEARCH TO MANAGEMENT SYSTEMS

Research on urban forests is useful only to the extent that it can contribute to the planning and management of vegetation in cities. At its most basic, urban forest management involves the planting, maintenance, protection, and removal of trees in cities (Grey & Deneke, 1986). These activities occur not in isolation, however, but in a broader context of land use and recreation planning, which has the goal of improving the quality of life by relating human needs to environmental and spatial resources (Gold, 1983a). In this section, I review examples of how design and behavior research has been or could be used in the process of planning and managing urban forests.

Public surveys can provide information on how the public perceives the importance of management objectives and the performance of the agency in meeting those objectives (O'Leary, 1982). Surveys have been used to document the importance of urban trees to residents; to identify opinions about where trees should be located, what kinds of trees are preferred, and what services should be provided; and to reveal sources of dissatisfaction with tree management programs (Getz, Karow, & Kielbaso, 1982; Schroeder & Appelt, 1985).

Surveys not only can provide guidance for designing new parks and facilities but can also assess the success of the design after the park is completed (Kaplan, 1980, 1981). Surveys using photographs and other visual simulations have proved useful in obtaining public input into the design process for community redevelopment and in evaluating the public response to alternative design proposals (Smardon, 1985). Knowledge about public motives and perceptions has provided a basis for publicity efforts to correct lack of information and misinformation in the public's image of a park (Hayward & Weitzer, 1983), and to broaden the appeal of interpretive programs (More, 1983).

Research can identify weaknesses in established planning mechanisms. For example, a survey of residents of urban neighborhoods (Talbot, 1982) showed that certain physical characteristics of neighborhoods, including tree density, affected people's satisfaction with where they lived. Most of these characteristics, however, were not among the standard variables used for zoning urban land uses. Coughlin, Mendes, and Strong (1984) observed that many communities have programs for protecting trees, but that these are concerned only with maintaining adequate numbers of trees and not with placing trees in locations where they will create the greatest community benefit. These planning mechanisms could be improved by taking advantage of what has been learned in research on public percpetions and the benefits of urban forests.

A case study of land-use planning in a Pennsylvania township (Coughlin & Strong, 1983) demonstrates how the measurement of vegetation benefits can be an integral part of the planning process. Scenic areas of the township were identified by the residents, and the value of vegetation was measured with a contingent valuation survey. This study was concerned with the urbanization of forest and farmland and assumed that the benefits of vegetation on a parcel of land would be completely lost if the parcel was developed. This assumption could be revised based on recent research, to include the benefits of preserving and managing vegetation on developed parcels of land.

Recently, several urban-forestry-management systems have been proposed that would integrate vegetation management into the routine planning process. These systems would set standards for environmental characteristics in specific areas of a city (Sanders & Rowntree, 1984), would monitor and map tree diversity levels to prevent vulnerable single-species populations of trees (Sanders, 1981), and would estimate and map popular satisfaction rates for vegetation across city neighborhoods (Sanders, 1984). All these proposals could take advantage of the currently available procedures for measuring public perceptions of the urban forest. Palmer (1984) proposed that social and perceptual assessments of neighborhoods be combined with physical and biological data to define planning units in the urban forest. These units would play the same role in urban planning that timber stands play in forest planning.

Microcomputer data bases, developed for inventorying urban forest resources, are becoming increasingly important as management tools in urban forestry. Inventories that record the location, size, and species of street trees can provide data for predicting the aesthetic quality of urban streets (Lien & Buhyoff, 1986). A "scenic beauty equation" could easily be built into a street-tree data base to allow managers to locate streets in need of aesthetic improvement and to predict the visual impact of changes in street tree populations. Inventory methods that will assist in the management of trees in parks are also being developed, and these can be linked to perceptual models based on research about public preferences for trees in landscapes (Schroeder & Green, 1985).

IMPLICATIONS FOR FUTURE RESEARCH AND APPLICATION

Trees and other vegetation provide an effective means of improving the perceived quality and the livability of urban environments. Continuing research on the environment, behavior, and design aspects of urban forestry should emphasize those topics and issues that will most enhance the ability of urban forest managers to use vegetation to improve urban living.

Research on benefits suggests that the most significant benefits of urban forests may be "passive," that is, benefits of aesthetic enjoyment and enhanced environmental quality obtained from simply observing the forest or just knowing it is there. More research is needed if we are to understand the importance and significance of experiencing urban vegetation in people's daily lives. This research should be ecological in character; that is, it should study people's behavior and experience in their real-world daily activities. Even brief and casual encounters with vegetation may have significant impacts on the quality of everyday life, and innovative research approaches will be needed to study these effects.

Qualitative studies of the significance of urban trees for individuals would help to fill out our understanding of urban forest benefits. The quality of life is partly a function of the emotional attachments that people form to specific places and to special features of their surroundings. Individual trees may be associated with memories of significant places and events and may thus become the objects of strong feelings (Tyznik, 1981). In-depth interviews with individuals about the meaning and significance of trees to them could be both fascinating and useful.

At the same time, research that identifies and measures tangible indicators of urban forest benefits would greatly strengthen the arguments for spending public funds to manage urban trees. More research is needed on the physiological and health effects of vegetation, and on the relation of vegetation to indicators of social welfare, such as crime, vandalism, and mental illness.

Further research is needed on variations in preferences for outdoor environments across different segments of the urban population. One of the most important dimensions of variation in urbanites' preferences for parks appears to involve the developed versus the natural character of a site. Different people prefer different levels of development, and parks with different levels of development may provide distinctly different kinds of benefits (Figure 11). The Recreation Opportunity Spectrum (ROS), currently used for forest-recreation planning in the United States, classifies recreation opportunities on a continuum from primitive to urban (Clark & Stankey, 1979). The ROS, however, was developed in the context of nonurban forest planning, and it does not capture the variation in the character, uses, and values of parks and forests within the urban context (Kaplan, 1984b). Further research on the perceived naturalness of urban forest sites, the preferred levels of development, and the benefits of natural versus developed sites could lead to a system similar to the ROS, but more suited to the urban context.

FIGURE 11. Some people prefer to picnic in a rustic and secluded forest location (A). Others may prefer the same activity in a more highly developed and heavily used city park (B).

The role of researchers in urban forestry should not be only to provide information to managers, but also to provide managers with methods for obtaining their own information. Urban resources and urban populations are so diverse and so rapidly changing that the research results from one situation may often not generalize to other situations. Researchers should try to get managers directly involved in doing research on their own sites and users, so that they can answer specific questions and evaluate the consequences of alternative management actions. Field experiments that test the effectiveness of management actions—for example, to control litter and vandalism—would be a good context for cooperative work between researchers and managers.

The growing use of microcomputers in management and research offers many opportunities for making research results usable by and accessible to managers. Interactive programs that predict levels of use at urban forest preserves (Dwyer *et al.*, 1985) and the distribution of choices over sets of parks (Schroeder & Louviere, 1986) have proved effective in conveying research results to managers who are unfamiliar with the technical aspects of regression and logit models. Researchers should be alert for opportunities to tie their research into existing or proposed computerized-management-and-planning systems for urban forests.

Research on urban forest managers themselves may be useful in uncovering the institutional structures within which urban forest management takes place. Urban forestry is carried out in a multitude of municipal governments, and we do not have a clear overall picture of who the managers are, what their background and training are, what problems and conflicts they encounter in doing their work, and how they resolve those conflicts. A better knowledge of the urban forest managers would help us to see how research can fit into the overall picture of urban forest management.

Researchers and managers may also be able to learn by observing management practices in other countries. Many European countries with large urban populations and limited forest resources have had centuries of experience in managing forest environments. Studies of forestry practices in Germany (Schabel & Dwyer, 1985) and the Netherlands (Dwyer & Strong, 1986) suggest that some of the intensive management practices used in those countries could also be applied in American cities (Figure 12).

Not all interactions between trees and people are pleasant. Falling branches and trees can cause property damage, severe injury, and death. The problem of liability at public recreation sites is becoming a serious concern for urban forest managers. Public agencies, no longer

FIGURE 12. Strict land-use controls and intensive planning in the Netherlands have led to innovative uses of trees in heavily populated landscapes.

protected by the doctrine of sovereign immunity, must undertake risk management programs to reduce the danger of severe injury at their sites (Gold, 1983b). Damage and injury from falling trees and branches may lead to staggering court settlements against public agencies. Urban foresters need to understand their legal responsibilities for inspecting and removing potentially hazardous trees from public parks and rights-of-way (Anderson & Eaton, 1986). Research on the legal aspects of urban forest management would be very timely now.

In this chapter, I have reviewed recent progress in several areas of research on design, perception, and behavior in urban forest environments, and I have tried to suggest some useful directions for further work. The ultimate test of the value of this research will be whether it leads to better environments for people. It is still not clear what impact environment, behavior, and design research has had on the quality of urban forest environments, but at the very least, research has helped to increase managers' awareness of the importance of the human and social aspects of managing these resources. The best future prospects for truly useful work in this area will be found in a close partnership between researchers and managers that identifies, studies, and ultimately solves problems of urban forest management.

REFERENCES

Allen, P. G., Stevens, T. H., & More, T. A. (1985). Measuring the economic value of urban parks: A caution. *Leisure Sciences, 7*(4), 467–477.

Allton, D. J., & Lieber, S. R. (1983). Attributes of Chicago trail areas. *Leisure Sciences, 5*(3), 197–220.

Ames, R. G. (1980). Urban tree planting programs: A sociological perspective. *HortScience, 15*(2), 135–137.

Anderson, L. M., & Cordell, H. K. (1985). Residential property values improved by landscaping with trees. *Southern Journal of Applied Forestry, 9*(3), 162–166.

Anderson, L. M., & Eaton, T. A. (1986). Liability for damage caused by hazardous trees. *Journal of Arboriculture, 12*(8), 189–195.

Anderson, L. M., & Schroeder, H. W. (1983). Application of wildland scenic assessment methods to the urban landscape. *Landscape Planning, 10*, 219–237.

Anderson, L. M., Mulligan, B. E., Goodman, L. S., & Regen, H. Z. (1983). Effects of sounds on preferences for outdoor settings. *Environment and Behavior, 15*(5), 539–566.

Buhyoff, G. J., Gauthier, L. J., & Wellman, J. D. (1984). Predicting scenic quality for urban forests using vegetation measurements. *Forest Science, 30*(1), 71–82.

Christensen, H. H., & Clark, R. N. (1983). Increasing public involvement to reduce depreciative behavior in recreation settings. *Leisure Sciences, 5*(4), 359–379.

Chubb, M., & Westover, T. N. (1980). Antisocial behavior: Typology, messages, and implications for recreation resource managers. In *Proceedings of Land-Use Allocation: Processes, People, Politics, Professionals*. Spokane, WA.

Clark, R. N., & Stankey, G. H. (1979). *The recreation opportunity spectrum: A framework for planning, management, and research* (USDA Forest Service General Technical Report PNW-98). Portland, OR: Pacific Northwest Forest and Range Experiment Station.

Coughlin, R. E., & Strong, A. L. (1983). *Forests, fields, and urban development: Planning as though vegetation really mattered* (Research Report Series no. 2). Philadelphia: University of Pennsylvania, Department of City and Regional Planning.

Coughlin, R. E., Mendes, D. C., & Strong, A. L. (1984). *Private trees and public interest: Programs for protecting and planting trees in metropolitan areas* (Research Report Series no. 10). Philadelphia: University of Pennsylvania, Department of City and Regional Planning.

Darragh, A. J., Peterson, G. L., & Dwyer, J. F. (1983). Travel cost models at the urban scale. *Journal of Leisure Research, 15*(2), 89–94.

Dwyer, J. F. (1980). Managing urban forests for recreation. *Trends, 17*(4), 11–14.

Dwyer, J. F., & Schroeder, H. W. (1982). Urban river recreation: New challenges and opportunities. *Naturalist, 33*(Summer), 6–10.

Dwyer, J. F., & Strong, A. L. (1986). Urban-rural forestry in the Netherlands. *Journal of Forestry, 84*(2), 57–59.

Dwyer, J. F., Deneke, F. J., Grey, G. W., & Moeller, G. H. (1983). Urban forests: Where trees and people go together. In *1983 USDA yearbook of agriculture*. Washington, DC: U.S. Department of Agriculture.

Dwyer, J. F., Schroeder, H. W., & Buck, R. L. (1985). Patterns of use in an urban forest recreation area. In *Proceedings of the 1985 National Outdoor Recreation Trends Symposium*, Vol. 2. Myrtle Beach, SC.

Fritschen, J. M., & Stynes, D. J. (1980). Interpretation for urban audiences. In *Proceedings of the Association of Interpretative Naturalists Workshop*. Cape Cod, MA.

Getz, D. A., Karow, A., & Kielbaso, J. J. (1982). Inner city preferences for trees and urban forestry programs. *Journal of Arboriculture, 8*(10), 258–263.

Godbey, G., & Blazey, M. (1983). Old people in urban parks: An exploratory investigation. *Journal of Leisure Research, 15*(3), 229–244.

Gold, S. M. (1983a). A human service approach to recreation planning. *Journal of Park and Recreation Administration, 1*(1), 27–37.

Gold, S. M. (1983b). Risk management in public playgrounds. *Journal of Park and Recreation Administration, 1*(3), 1–10.

Grey, G. W., & Deneke, F. J. (1986). *Urban forestry* (2nd ed.). New York: Wiley.

Hagerty, J. K., Stevens, T. H., Allen, P. G., & More, T. (1982). Benefits from urban open space and recreational parks: A case study. *Journal of the Northeastern Agricultural Economics Council, 11*(1), 13–20.

Hayward, D. G., & Weitzer, W. H. (1983). Understanding urban park users: A key to effective planning and management. *Parks and Recreation Resources, 2*(2), 24–27.

Hayward, D. G., & Weitzer, W. H. (1984). The public's image of urban parks: Past amenity, present ambivalence, uncertain future. *Urban Ecology, 8*, 243–268.

Herzog, T. R., Kaplan, S., & Kaplan, R. (1982). The prediction of preference for unfamiliar urban places. *Population and Environment, 5*(1), 43–59.

Hutchison, R., & Fidel, K. (1984). Mexican-American recreation activities: A reply to McMillen. *Journal of Leisure Research, 16*(4), 344–349.

Kaplan, R. (1980). Citizen participation in the design and evaluation of a park. *Environment and Behavior, 12*(4), 494–507.

Kaplan, R. (1981). *Evaluation of an urban vest-pocket park* (USDA Forest Service Research Paper NC-195). St. Paul: North Central Forest Experiment Station.

Kaplan, R. (1982). Managing greenspace in multiple-family neighborhoods. In *Proceedings of the 1982 Convention of the Society of American Foresters*. Cincinnati, OH.

Kaplan, R. (1983). The role of nature in the urban context. In I. Altman & J. F. Wohlwill (Eds.), *Behavior and the natural environment*. New York: Plenum.

Kaplan, R. (1984a). Dominant and variant values in environmental preference. In A. S. Devlin & S. L. Taylor (Eds.), *Environmental preference and landscape preference*. New London: Connecticut College.

Kaplan, R. (1984b). Human needs for renewable resources and supportive environments. In G. A. Bradley (Ed.), *Land use and forest resources in a changing environment: The urban/forest interface*. Seattle: University of Washington Press.

Kaplan, R. (1984c). Impact of urban nature: A theoretical analysis. *Urban Ecology, 8*, 189–197.

Kaplan, R. (1985). Nature at the doorstep: Residential satisfaction and the nearby environment. *Journal of Architectural and Planning Research, 2*, 115–127.

Kellert, S. R. (1984). Urban American perceptions of animals and the natural environment. *Urban Ecology, 8*, 209–228.

Lieber, S. R., & Allton, D. J. (1983). Modeling trail area evaluations in metropolitan Chicago. *Journal of Leisure Research, 15*(3), 184–202.

Lieber, S. R., & Fesenmaier, D. R. (1985). Physical and social conditions affecting recreation site preferences. *Environment and Planning A, 17*, 1613–1625.

Lien, J. N., & Buhyoff, G. J. (1986). Extension of visual quality models for urban forests. *Journal of Environmental Management, 22*(3), 245–254.

Marans, R. W., & Fly, J. M. (1981). *Recreation and the quality of urban life* (Research Report Series). Ann Arbor: The University of Michigan, Institute for Social Research.

Metro, L. J., Dwyer, J. F., & Dreschler, E. S. (1981). *Forest experiences of fifth-grade Chicago public school students* (USDA Forest Service Research Paper NC-216). St. Paul: North Central Forest Experiment Station.

Morales, D. J., Micha, F. R., & Weber, R. L. (1983). Two methods of evaluating trees on residential sites. *Journal of Arboriculture, 9*(1), 21–24.

More, T. A. (1980). *Trail deterioration as an indicator of trail use in an urban forest recreation area* (USDA Forest Service Research Note NE-292). Broomall, PA: Northeastern Forest Experiment Station.

More, T. A. (1983). The nonusers of an urban forest interpretive center. *Journal of Interpretation, 8*(1), 1–9.

More, T. A. (1984). A practical guide to the use of observation in the study of urban parks. In J. D. Peine (Ed.), *Proceedings of a workshop on unobtrusive techniques to study social behavior in parks.* Atlanta: U.S. Department of the Interior, National Park Service.

More, T. A. (1985a). *Central city parks: A behavioral perspective.* Burlington: University of Vermont, School of Natural Resources.

More, T. A. (1985b). Evaluating and interpreting use data in urban park settings. In *Proceedings of the 1985 National Outdoor Recreation Trends Symposium,* Vol. 2. Myrtle Beach, SC.

Nasar, J. L. (1981). Visual preferences of elderly public housing residents: Residential street scenes. *Journal of Environmental Psychology, 1,* 303–313.

O'Leary, J. T. (1982). Managing urban river corridors: Implications for urban forestry. In *Proceedings of the 1982 Convention of the Society of American Foresters.* Cincinnati, OH.

O'Leary, J. T., & Benjamin, P. J. (1982). *Ethnic variation in leisure behavior: The Indiana case* (Station Bulletin no. 349). West Lafayette, IN: Purdue University, Department of Forestry and Natural resources, Agricultural Experiment Station.

Palmer, J. F. (1984). Neighborhoods as stands in the urban forest. *Urban Ecology, 8,* 229–241.

Peterson, G. L., Dwyer, J. F., & Darragh, A. J. (1983). A behavioral urban recreation site choice model. *Leisure Sciences, 6*(1), 61–81.

Peterson, G. L., Stynes, D. J., Rosenthal, D. H., & Dwyer, J. F. (1985). Substitution in recreation choice behavior. In *Proceedings of symposium on recreation choice behavior* (USDA Forest Service General Technical Report INT-184, pp. 19–30). Ogden, UT: Intermountain Research Station.

Samdahl, D. M., & Christensen, H. H. (1985). Environmental cues and vandalism: An exploratory study of picnic table carving. *Environment and Behavior, 17*(4), 445–458.

Sanders, R. A. (1981). Diversity in the street trees of Syracuse, New York. *Urban Ecology, 5,* 33–43.

Sanders, R. A. (1984). Estimating satisfaction levels for a city's vegetation. *Urban Ecology, 8,* 269–283.

Sanders, R. A., & Rowntree, R. A. (1984). Environmental management through urban forestry on the hillsides of Cincinnati, Ohio. *Journal of Environmental Management, 19,* 161–174.

Sanders, R. A., & Stevens, J. C. (1984). Urban forest of Dayton, Ohio: A preliminary assessment. *Urban Ecology, 8,* 91–98.

Schabel, H. G., & Dwyer, J. F. (1985). Institutional aspects of forest recreation resource management in West Germany. *Landscape Journal,* 4(1), 1–6.

Schroeder, H. W. (1982). Preferred features of urban parks and forests. *Journal of Arboriculture, 8*(12), 317–322.

Schroeder, H. W. (1983). Variations in the perception of urban forest recreation sites. *Leisure Sciences, 5*(3), 221–230.
Schroeder, H. W., & Anderson, L. M. (1984). Perception of personal safety in urban recreation sites. *Journal of Leisure Research, 16*(2), 178–194.
Schroeder, H. W., & Appelt, P. (1985). Public attitudes toward a municipal forestry program. *Journal of Arboriculture, 11*(1), 18–21.
Schroeder, H. W., & Cannon, W. N. Jr. (1983). The esthetic contribution of trees to residential streets in Ohio towns. *Journal of Arboriculture, 9*(9), 237–243.
Schroeder, H. W., & Green, T. L. (1985). Public preference for tree density in municipal parks. *Journal of Arboriculture, 11*(9), 272–277.
Schroeder, H. W., & Louviere, J. J. (1986). A model for predicting distribution of recreational use over a system of parks. In *Proceedings of the Forestry Microcomputer Software Symposium*. Morgantown: West Virginia University, Division of Forestry.
Shaffer, G. S., & Anderson, L. M. (1985). Perceptions of the security and attractiveness of urban parking lots. *Journal of Environmental Psychology, 5*, 311–323.
Shaw, W. W., Mangun, W. R., & Lyons, J. R. (1985). Residential enjoyment of wildlife resources by Americans. *Leisure Sciences, 7*(3), 361–375.
Smardon, R. C. (1985). A visual approach to redesigning the commercial strip highway. *Transportation Research Record*, No. 1016, 1–6.
Spotts, D. M., & Stynes, D. J. (1984). Public awareness and knowledge of urban parks: A case study. *Journal of Park and Recreation Administration, 2*(4), 1–12.
Spotts, D. M., & Stynes, D. J. (1985). Measuring the public's familiarity with recreation areas. *Journal of Leisure Research, 17*(4), 253–265.
Stamps, S. M. Jr., & Stamps, M. B. (1985). Race, class, and leisure activities of urban residents. *Journal of Leisure Research, 17*(1), 40–56.
Stiegler, J. E. (1985). *Public perception of the urban forest.* Unpublished master's thesis, University of Minnesota, Graduate School, St. Paul.
Stynes, D. J., Spotts, D. M., & Strunk, J. R. (1985). Relaxing assumptions of perfect information in park visitation models. *Professional Geographer, 37*(1), 21–28.
Talbot, J. F. (1982). Zoning reconsidered: The impacts of environmental aesthetics in urban neighborhoods. In P. Bart, A. Chen, & G. Francesato (Eds.), *Proceedings of EDRA 13: Knowledge for Design*. Washington, DC: Environmental Design and Research Association.
Talbot, J. F., & Kaplan, R. (1984). Needs and fears: The response to trees and nature in the inner city. *Journal of Arboriculture, 10*(8), 222–228.
Talbot, J. F., & Kaplan, R. (1986). Judging the sizes of urban open areas: Is bigger always better? *Landscape Journal, 5*(2), 83–92.
Tyznik, A. (1981). Trees as design elements in the landscape. *Journal of Arboriculture, 7*(2), 53–55.
Ulrich, R. S. (1981). Natural versus urban scenes: Some psychophysiological effects. *Environment and Behavior, 13*(5), 523–556.
Ulrich, R. S. (1984). View through a window may influence recovery from surgery. *Science, 224*, 420–421.
Ulrich, R. S. (1986). Human responses to vegetation and landscapes. *Landscape and Urban Planning, 13*, 29–44.
Ulrich, R. S., & Addoms, D. L. (1981). Psychological and recreational benefits of a residential park. *Journal of Leisure Research, 13*(1), 43–65.
Ulrich, R. S., & Simons, R. F. (1986). Recovery from stress during exposure to everyday outdoor environments. In *Proceedings of EDRA 17*. Washington, DC: Environmental Design and Research Association.

Vining, J., Daniel, T. C., & Schroeder, H. W. (1984). Predicting scenic values in forested residential landscapes. *Journal of Leisure Research, 16*(2), 124–135.
Wendling, R. C. (1980). Black/white differences in outdoor recreation behavior: State-of-the-art and recommendations for management and research. In *Proceedings of Social Research in National Parks and Wildland Areas.* Gatlinburg, TN.
Wendling, R. C., Gabriel, S. J., Dwyer, J. F., & Buck, R. L. (1981). Forest Preserve District of Cook County, Illinois. *Journal of Forestry, 79*(9), 602–605.
West, P. C. (Ed.). (1981). *Vestiges of a cage,* Vol. 1. Ann Arbor: University of Michigan, Natural Resource Sociology Research Lab.
West, P. C. (1984). Social stigma and community recreation participation by the mentally and physically handicapped. *Therapeutic Recreation Journal, 18*(1), 40–49.
Westover, T. N. (1985a). Perceptions of crime and safety in three midwestern parks. *Professional Geographer, 37*(4), 410–420.
Westover, T. N. (1985b). Perceptions of rule compliance and law enforcement in urban and suburban parks. *Recreation Research Review, 12*(2), 22–29.
Westover, T. N. (1986). Park use and perception: Gender differences. *Journal of Park and Recreation Administration,* 4(2), 1–8.
Wohlwill, J. F., & Harris, G. (1980). Response to congruity or contrast for man-made features in natural-recreation settings. *Leisure Sciences, 3*(4), 349–365.
Young, R. A., & Flowers, M. L. (1982). *Users of an urban natural area: Their characteristics, use patterns, satisfactions, and recommendations* (Forestry research report 82-4). Urbana-Champaign: University of Illinois, Department of Forestry, Agricultural Experiment Station.

5

Environment and Behavior Research for Developing Countries

GRAEME JOHN HARDIE

Why, one might ask, is there a chapter in this series devoted specifically to research in and for developing countries? Is it not inappropriate to think that environment and behavior research should be different in developing countries from in the countries where the field is more established? Surely, environment and behavior research issues are the same anywhere in the world? Hopefully, as this chapter unfolds, it will make clear the differences and the need for the chapter. Briefly, one could argue that the issues are the same; yet, they have a particular bias and orientation that makes the work different from work undertaken in First World situations. In the main, this difference is due to different political, social, and economic conditions and the nature of the situation existing in this context, which affect the research undertaken.

This chapter begins by describing the context of environment and behavior research in developing countries focusing on such questions as who does it, who pays for it, and who owns it. These are issues that affect the research orientation and direction. The main concentration of research in this context is housing. The second section discusses the dominant themes related to this issue, such as research related to house design, infrastructure planning, participation approaches, and issues of

Graeme John Hardie • Center for Design Research and Service, School of Design, North Carolina State University, Raleigh, North Carolina 27695.

policy. Fieldwork techniques that have been used in investigating the connection between environment and behavior in developing contexts are then presented. Finally, the chapter considers direction for future research. It is the author's intention that this chapter may serve as a way of alerting researchers who want to involve themselves in developing countries to the issues they need to consider. Perhaps of even greater importance, this chapter highlights areas that deserve consideration and research in the future.

At the outset, it must be stated that the designation *developing countries* is taken to include broadly all countries other than Australia, Israel, Japan, New Zealand, and those in Europe and North America. The term *Third World* could be and is used here, but this term has become complicated by the new designations *Fourth World* and *Fifth World.* Perhaps a better definition of the countries under consideration is that they are those where the field of environment and behavior research is itself developing. Some researchers may feel affronted that their country has been regarded in this chapter as "developing." In advance—apologies. Further, some researchers may feel more deeply about the fact that their research has not been noted. Here, too, an apology is called for, but this omission is due, in part, to the difficulty of obtaining the published material of work undertaken in developing countries, as so much of this material appears only in reports prepared for government or private agencies and, therefore, does not readily enter the public domain. If published, this material may also not be available in English.

THE CONTEXT OF DEVELOPING COUNTRIES

In a review of the literature reporting on research undertaken in developing countries, a number of features stand out, for instance, who carries out research in developing countries, who funds it, and who owns it. These questions influence the kind of research undertaken in developing countries, which is most often applied in nature. For the most part, the researchers are architects who have an interest in understanding the human implications of their work and thus undertake research in order to be more effective in what they create. For those working in a foreign country or among people with different cultural values, this is particularly pertinent because they may realize that it is difficult to make assumptions about what is of value to the people being served.

Not surprisingly, of the social sciences, the field most represented by researchers is anthropology, perhaps because much of the work is focused on communicating cultural values and perceptions and their physical expression to planners and designers who are not of the same cultural group as those for whom they are planning.

Where psychologists have become involved, they have mostly been locally based researchers—for instance, in Venezuela and Mexico, where research is being done by some researchers from those countries who have been trained in Europe and the United States. Sanchez, Wiesenfeld, and Cronick (1987a) pointed to the difficulty of their academic background because "there is a tendency to focus attention on problems that have been identified in environments outside Latin America instead of using the information obtained outside to come to terms with the problems that are affecting the area" (p. 1341). The recent establishment of collaboration between the environmental psychology program of the University of Arizona and the University of Lima, Peru, will obviously influence the work and orientation in that country, and it will be interesting to see how these associations deal with the contextual and orientation problems raised by Sanchez *et al.* Similarly, the School of Architecture and Urban Planning at the University of Wisconsin–Milwaukee has established a collaborative program with the Gadjah Mada University in Yogyakarta, Indonesia.

In some instances, human geographers have become involved in these issues in both basic and applied research (Duncan, 1983; Hart, 1984). In summary, though, the disciplinary orientation of environmental researchers in developing countries is, for the most part, due to the fact that agencies funding research on a contract basis have directed their support to architects and anthropologists.

The problems facing professional planners in developing countries are well summarized by Bor (1982) and show clearly the need for research that will give a full understanding of the context to the practitioner. Bor maintained that practitioners from developed countries working in the Third World have to, first, gain a thorough understanding of the cultural, social, economic, and political background of the developing country in question, and of the aspirations of its people. They must also, he suggested, rethink and adapt their own expertise to suit local circumstances.

This is a tall order, but it makes clear the need for research in the environment and behavior field. Without gaining insight other than a physical appreciation of form, the planner or architect will remain a stranger in a very strange land, and what is produced will be inappropriate. The planner in this situation can take nothing for granted because the environmental form is an expression of values alien to the planner. It then behooves the planner to understand why people take actions, organize their environment, and carry out behaviors in particular spatial forms that seem strange and sometimes meaningless. For the locals, the reasons for the particular spatial expression are most often totally rational, but they are based on different operating principles from those with which the planner is familiar.

How to understand these principles and to facilitate the communication of them then becomes an important component of the planning and design process. It requires not only the encouragement of the participation of the users in the process but also the need to develop ways by which understanding is reached and communication is facilitated. This approach can assist the users in making explicit what, for them, is implicit in their conceptual environmental structuring and what underlies their preferences and suggestions. A number of researchers (Bentz, 1981; Hardie, 1980, 1988; Hart & Hardie, 1985; Sanoff, 1979, 1988; Stea, 1981) have paid special attention to these issues and have developed ways and means of investigating them. To meet this need in terms of training, Stea established a training program in Santa Fe, New Mexico, for built-environment professionals who will operate in intercultural situations. This perspective has also influenced the type of researchers involved in this field, who often have an anthropological perspective and orientation, for it is necessary to become aware of the social, cultural, and ecological context in order to make action appropriate, whether in designing actual facilities or in establishing policies and guidelines.

If it is possible to separate applied from basic research, then in developing countries research can have an immediacy and potential for application not often encountered among First World researchers. Government agencies and planners often require insight into particular issues in order to guide their planning. A new town may be in the process of being planned or a housing scheme of being developed, and information is required about what people prefer and how they will operate in these environments. Issues could also relate to various aspects of the country's housing policy; to consideration of appropriate approaches to housing finance; to infrastructure, such as the supply of water or the disposal of sewerage; or even to the issue of acceptable housing densities and the community's attitudes toward privacy. In this situation, the researcher's role gains in impact, and the results of the investigation can have a direct bearing on what is being designed and planned.

Certainly, not all research undertaken in developing countries is applied. This will depend on whether the researchers, particularly those who are foreigners, are given funding for independent research or are hired as consultants. Neither should the applied nature of research exclude the potential of the results' having a bearing on basic research in the environment and behavior field. Although the first intent is to solve a specific problem, well-conceived research (such as Guinness's on an Indonesian spontaneous settlement, 1986) is more than likely to show that the findings can influence the field as a whole.

It is the author's contention that, when basic research is undertaken in developing countries, the research is usually owned and controlled by the researcher; as it is without contractual implication, it will more than likely have little impact on the local community. Most often, the issues investigated in this way are of the researcher's creation and do not emerge from a need within the local context. Usually, the research issues considered in this way relate to investigations that the researcher has undertaken in other contexts, or the researcher may have developed a particular hypothesis that will have no impact on or little connection with the needs of the local situation or community (for a notable exception, see Setiawan, Subanu, & Hermanislamet, 1987).

This aspect raises the issue of the accountability of the researcher to the people involved in the research endeavor. Sanchez, Cronick, and Wiesenfeld (1987c) argued that the researcher cannot be separated from the object of the research. The act of enquiry, they suggested, is itself a contamination. This particular group of researchers in Venezuela share their findings with the community they are researching. Sadly, this is not the approach adopted by most researchers. In much contract work, the client is a government agency, and research accountability is not with the people being researched, nor is it viewed as being a resource for their use.

Should a researcher be afraid of undertaking contractual research in a developing country? The issue is perhaps no different from that related to contract work undertaken anywhere in the world. From personal experience, having for many years undertaken contractual research in developing countries, I can fortunately say that the contractual research for me has never been repetitive; rather it has forced me as a researcher in my particular speciality of housing to consider a much wider range of issues than those that my own research agenda would have included. This circumstance has considerably broadened my perspective from the narrow focus that I had established for myself, and this broadening has been to my long-term advantage.

HOUSING: THE PRIMARY FOCUS OF ENVIRONMENT–BEHAVIOR RESEARCH IN DEVELOPING COUNTRIES

On consideration of the environment and behavior research undertaken in a number of countries, such as in the Middle East, Africa, parts of Southeast Asia, and South America, it is clear that the major focus is on issues related to housing, settlement planning, and infrastructure. For instance, the dissertations from the Centre for Architectural Research and Development Overseas (CARDO) at the University of Newcastle-upon-Tyne, which has had broad experience in developing countries, have, for the most part, been housing-related. The topics have included socialist villages in Algeria, Arab domestic architecture, ventilation problems in Bangladesh, water supply in Bolivia, indigenous housing in Burma, social housing in Chile, housing design in hot-humid climates (India), workers' housing in Indonesia, earthquake-resistant housing in Iran, urban spaces in Iraq, cultural resources for housing in Jordan, housing finance in Mexico, contractors required for housing in Nigeria, building materials in Sri Lanka, squatter housing in Sudan, rural housing in Tanzania, education of slum dwellers in Thailand, rural development in Venezuela, and housing policy in Zimbabwe (Centre for Architectural Research and Development Overseas, 1987).

In a recent review of the projects being undertaken by the Psychology Department of the Central University of Venezuela, 7 of 11 projects were focused on housing (Sanchez, Cronick, & Wiesenfeld, 1987b). These consisted of evaluations of single- and multifamily unit housing, the housing satisfaction of different types of housing, issues of residential density, locus of control and crowding perceptions, and investigations of privacy in different housing projects. A review of master and doctoral theses in the environment–behavior field in South Africa (van Staden, 1986) has shown, by contrast, that housing accounted for only 20% of the research undertaken between 1977 and 1986. On the other hand, environmental cognition accounted for 27% of the theses, the general investigation of behavior and environment accounted for 25%, and investigations into environmental stress, environmental education, environmental attitudes, and personal space accounted for the remainder of research themes (28%). It is of note that only 15% of the investigations undertaken were in what might be considered the Third World environment of South Africa, as most of the studies were focused on investigating the attitudes and perceptions of only 10% of the population: the white minority.

It is of note that there are rarely reports of investigations in developing countries into specialized types of housing, such as housing for the elderly. In Africa, for example, the elderly, by custom, are accommodated at home with their children or even on their own with the aid of a grandchild; no specialized facilities exist. However, these kinds of needs will have to be considered in the future as urbanization increases and perhaps brings changes to the extended-family structure.

FIGURE 1. A new town emerges to develop a natural resource with housing built by the government in Selibe Pikwe, Botswana.

FIGURE 2. Typical informal housing built on the fringe of an urban center in Malakasi, Durban, South Africa.

FIGURE 3. A *favella* in the Bay of Rio de Janeiro, Brazil.

Why is there an emphasis on housing in developing countries? Obviously, the burgeoning population of the developing world is a major reason, in tandem with the ever-increasing size of the cities as people who have traditionally subsisted on their agricultural holding move to seek a position in the commercial and industrial life of their country. Often, job opportunities are located in new towns, which are created in order to develop a natural resource. Typically, the government builds the housing. According to Blitzer, Hardoy, and Satterthwaite (1981), if one puts together all recent reports on national urban housing conditions, it is not unlikely that at least half the population in most African, Asian, and Latin American urban centers is housed in substandard living environments—in overcrowded conditions in inner city slums or deteriorating units, in squatter settlements or illegal subdivisions with no or (at best) grossly inadequate basic services. These pressing human conditions, unknown to researchers in the developed world, thus eclipse, or should eclipse, all other issues that might interest the researcher in this context.

HOUSE DESIGN

Many of the social catastrophes related to housing experienced in the developed world, such as at Pruitt Igoe in St. Louis, Missouri, are being repeated in developing countries. Sadly, some of these have been perpetrated by sincere and earnest individuals who seek technological solutions to the housing crisis but are not always sensitive to the people's attitudes, values, and behavior. Fathy (1973), the Egyptian architect, has, for example, been extolled for the town he built in Gourna. The buildings were of low cost, were built out of locally available material, and used the traditional Mesopotamian system of building arches. However, an opportunity was missed because the project creators did not adequately investigate what was valued by the people. Visiting the site today, one is impressed by both the architecture and even the way in which many facets of social traditions were expressed in the buildings, but one is also depressed by the state of disrepair and the low value placed on the buildings by the local people living in the settlement, who wanted buildings that looked modern. The project was technically appropriate but not in harmony with the human environment. One reviewer, commenting on the project, stated that the planning intentions far exceeded the rural population's frame of social reference, while overestimating the community's ability to upgrade itself (Rastorfer, 1985).

Alexander's approach in Mexicali, Mexico (1985), showed one way of addressing this issue, but the resulting physical form, although functionally appropriate, was not stylistically appropriate in terms of the owners' conception of how a house should look.

The following discussion presents research that has attempted to bridge the gap between providing functional or technical shelter and a "home" that meets and supports the household's values and symbolic requirements. A crucial question in this regard, and most relevant to developing countries, is the ability to understand the impact of changing and "modernizing" values on traditional values and the resulting impact on the organization of the physical environment.

Rapoport (1983) proposed that designing environments in developing countries is no different in principle from designing in any other kind of environment. For two reasons, though, designing environments in developing countries can serve as an excellent example of the process for providing a paradigm for design generally. Rapoport believes, first, that the process has intrinsic value because there is a need for better solutions to improving environments and, second, that it has extrinsic importance in order to be appropriate to the people whereby the sociocultural variables become critical to the process. In addition, there is the characteristic of cultural change and the stress that results from this process and thus the need for the design of supportive environments. Rapoport proposed that, for the environment to be supportive in such a dynamic situation, it is necessary to consider a form of cultural synthesis. He stated that it thus becomes necessary to identify the cultural specifics involved, over and above any more general aspects of supportive environments related to biosocial and psychological variables. "The task," stated Rapoport (1983), "then becomes to establish some crucial relations between aspects of culture and of supportive environments" (p. 256). Further, Rapoport proposed that, when a traditional culture is confronted with alternative (generally Western-based) cultural values, certain core characteristics of the traditional culture may remain in place.

In order to test this hypothesis in the field, he and I undertook fieldwork in one town in Southern Africa where different levels of change are evident in settlements of different ages and types. Our preliminary findings indicated that certain traditional and culturally based uses of space are maintained and are physically evident in all settlements except those developed for the elite. However, even among the elite, where the house form expresses Western values and gives little attention to traditional concepts, the occupants still managed to manipulate the environment to serve traditional purposes. For instance, the house front presents a decorative yet nonparticipatory view to the street, disregarding traditional approaches, where the street serves as a communal social space. The people's solution to this problem is to turn the garage into a place where the household can sit in the shade of the door and watch the action in the street.

Alnowaiser (1985) compared traditional and modern settlements in Saudi Arabia and presented a different view of investigating the differences between them. Rather than seeking to understand connections and continuities, he focused on the differences, trying to understand the substantive and instrumental qualities of spaces in each setting. In this way, he sought to see what qualities differentiate the two settings. He proposed that the substantive place qualities are rooted in the cultural (social, religious, and traditional) norms and the environmental (climatic and topographical) systems. These substantive qualities include symbolic, authentic, social, sacred, communal, organic, and unself-conscious places that exist mostly in traditional physical environments. On the other hand, instrumental place qualities include spatial features resulting from individualism, materialism, acquired consumption habits, and the economic concept of efficiency. He suggested that characteristics of each may exist in both traditional and modern environments, but in different degrees. For example, instrumental qualities are more dominant in modern urban forms than in traditional ones, as the former emphasizes efficiency of movement and use whereas the latter emphasizes quality. This perspective, however, is instructive only in suggesting the degree to which one settlement is either modern or traditional. It does not seek to understand what the process is itself and how to design for it appropriately.

Many other researchers have investigated cultural influences on house form. In India, for example, Chandhoke (1980) focused on the various factors that shape and sustain settlement and housing patterns. He considered social, economic, and natural environmental impacts and then questioned the reason that settlement patterns can vary within the same region when the variables are alike and yet different settlement types exist. The two extremes that he found in India are the clustered and the dispersed patterns, and there is a wide variation in between. He then considered the reasons for the difference in patterns and house types and concluded that the habitation "reflects" the lifestyle and culture of its inhabitants. . . . A house is an institution, not just a structure. It reflects the thinking, feelings, beliefs and aspirations of its occupants" (p. 299).

Another researcher (Ahuja, 1980) investigated vernacular settlements and their long process of evolutionary growth as compared to those settlements designed on the drawing board. Designed settlements, Ahuja suggested, are developed through hours of toil, whereas vernacular settlements generate their design and form through social organization and social control, where the social order is "interposed between human participants, providing potential forces that produce a coherent system" (p. 295). Another researcher in India (Dandekar, 1981) focused on the social and spatial constraints operating on women in the house and settlements in India and painted a grim picture of the life of women in the village of Sugao. The social structure and the physical environment impose great strictures on the woman and her life, often with little spatial privacy.

In Tunisia, Boukraa (1981) focused on aspects of the house and concluded that the Islamic house "is an entity that links its architectural structure with the economic system and social order, and reflects the basic characteristics of the Islamic civilization" (p. 427). The author described various facets of the house and how these reflect Islamic beliefs and social values. These values and beliefs influence such issues as privacy, the screening of the public from the family quarters in the house, the authority of the male household head, and the way in which these values are expressed spatially. The author then compared the physical features of the traditional house with those in the modern villa and how the latter conflicts with traditional approaches. In the light of this study, the research of Enders (1981) stands in stark contrast. His work shows how the Tunisian government's attempts to improve the rural quality of life, which resulted in resettlement villages, was in conflict with what the people themselves considered an improvement. Economic efficiency on the part of the government resulted in a settlement pattern and house design that were unsympathetic to the people's values.

From the African perspective, Prussin (1969), in an early and significant book, considered six villages in northern Ghana and how they differ in settlement layout. To some extent, the differences reflect topographical and climatic variation, but more fundamentally, they reflect different kinship organizations, cosmological perspectives, and traditions. Prussin pointed out that it would be eclectic and romantic simply to imitate the forms she found in rural northern Ghana in other countries. She concluded her book by stating that cultural change is inevitable, and that changes produce new architectural forms, forms that result from a synthesis between existing cultures and cultures from elsewhere with which they are rapidly establishing a liaison: "The study of how architectural change has evolved in time can, however, provide the understanding that is requisite for contending with the increasingly rapid changes the future holds" (p. 118). Her words were somewhat prophetic, because it is these issues that continue to hold the attention of researchers in developing countries and particularly in Africa.

More recently, in Ghana, Pellow (1986) considered continuity in change in architectural form by focusing on the evolution of housing among a migrant community in urban Ghana. She assessed the influences on the evolving forms from three major cultural traditions—Hausa, southern Ghana traditions, and Western influences; and she looked specifically at the relationship between spatial structures and the ethos of sexual segregation.

Continuity and change were also the theme explored by Hardie (1980, 1986), focusing on the houses and settlements of the Tswana. In this study, concentrating on one ward or social unit of a settlement, the historical development of the ward was investigated, and the physical changes or continuities were described. The physical organization of space was then used as an armature for investigating the values inherent in the form and whether the values were maintained or changed over time. Both studies, therefore, focused on what was culturally important and how, over time, these values have affected the organization of space.

Two researchers (Muller, 1984; Smith, 1981) have looked at the cultural values of the inhabitants of urban housing and the impact they have had on different housing projects in Nairobi. They particularly emphasized different behavior norms as they affected the sleeping quarters and the cooking of food. Smith, an architect, focused on developer-built housing and the changes that homeowners had made or desired to make in order for the houses to be more satisfactory. These included separate rooms from the rest of the family for visitors or parents, a master bedroom large enough to allow for children's beds and thus to allow a high degree of social intimacy for the nuclear family, and a large kitchen with direct access to a private yard. Muller, a sociologist, bemoaned the fact that social scientists, on the whole, are not trained to deal with environmental issues. She showed different kinds of dysfunctions that exist between behavior and the environment on a Kenyan housing estate. For example, she described how the occupants of one-room dwellings struggle to create spatial separation in order to maintain important symbolic divisions inherent in the culture of the household.

These two researchers have raised important issues of privacy, which have received relatively little attention in spite of the crowded housing conditions existing in developing countries. How do families moving from a rural homestead, where space allows them to have separate sleeping quarters based on various criteria, such as gender or age, manage to cope in an urban environment? Moving into the city, such a family will be required to share one room in a house that may accommodate a number of households. Few studies have considered the resulting impact on the behavior of the family, the accompanying stress, and the family's means of dealing with these issues spatially.

By contrast with research in other developing countries, a number of researchers in Latin America have looked at environmental stress in the residential environment, partly because of the emphasis on psychological issues of environment and behavior research in this part of the world. In Mexico, Ribes, Galvan, and Villanueva (1973) registered the "density of use" of different areas of residences, finding that frequency of use is not related to size. In Venezuela, studies have been made of the relationship between vandalism and housing (Montero, 1976, 1979), neighbor relations (Curtis, Garcia, & Naveda, 1978), and the perception of crowding (Castro, Guerrero, & Hernandes, 1978; Wiesenfeld, 1984; Wisenfeld & Sanchez, 1983).

Housing Technology and Infrastructure

In the literature, there has been much discussion of how to foster methods of house construction that are appropriate to the situation. This appropriateness implies that the technology would be low-cost, would use local materials, would mobilize local labor, and would be climatically appropriate. Seldom is there discussion of the values held by the people and how these might influence the acceptance of new technology.

By way of example, it was found in Botswana (Hardie, 1980) that the people had a specific conception of what they considered a good wall. In recent times, the concrete block has been introduced, is now widely used, and has become a material that exemplifies both higher status, because of the cost, and lower maintenance than the traditional hand-molded mud blocks. However, the hollow cores of the blocks had been filled with concrete, a costly addition to the construction costs and a feature that the design of the block specifically attempts to avoid. When one asks why this was done, the people say that it is because a good wall is a solid wall. The people have adopted a new technology, but in fact, they have not adopted the values inherent in it and instead have used it according to those values that seem more appropriate to them.

There are countless examples of this kind; yet, few have found their way into the literature, and more important, few have influenced technologists' developing appropriate technology. Bein (1981) investigated the system of domestic water supply in the Sudan. He discovered that the inability to further improve health conditions in rural Sudan is the result of cultural value differences between decision makers and the people who must live with the decisions. Such cultural value differences, he maintained, may be attributed to differences in education, in access to information, and in physical contact with the resource. Often, water development efforts are made without knowledge of the user's habits and perceptions of the drinking-water source. Bein believed that this results in government projects' being implanted in a situation without any understanding of how the new facility will be accepted by the local people. Often, water facilities have been installed at great expense, meeting world health standards and population requirements, only to be used in such a way that the water becomes contaminated or the facility breaks down and becomes nonfunctional.

There is also little reporting of the involvement of communities in the creation of their own infrastructure as a means of stimulating community development. Communities that I have visited have dug trenches for their water pipes in days rather than having contractors do the same task in weeks, and this approach had had a major impact on the community's sense of control over the resulting water provided. However, these approaches have not been well documented nor has the way in which they can positively influence a community been studied. Infrastructure is all too often seen as a technical intervention and thus as presenting limited options, not only in terms of what is appropriate but also in how it is administered.

Another personal example comes to mind. A squatter community consisting mainly of women had water supplied to their area for which they had to pay. But there was no billing system in place, and so to pay the bill once a year, the community organized a meeting at which all were expected to contribute. "Of course," the chairwoman of the neighborhood told me, "we could not expect unemployed or the sick to pay," but everyone else would gather in a circle and, passing by a collection plate, contribute a required amount. This procedure would continue until all the money required to pay the bill was collected. This would not necessarily have been the procedure that a planner would adopt, but it was an approach that worked for that community, and it maintained local control. This example highlights informal mechanisms that could have great impact on management of the environment but that require understanding and insight. It goes counter to the rational, Western, and often technologically based solutions that most planners would adopt. Investigating appropriate approaches is an area of research to which those interested in people's behavior and its impact on the environment in terms of technology and health could bring new understanding.

Participation and Housing

A number of Third World governments have recently withdrawn from building housing for low-income urban dwellers and have encouraged self-help and self-build approaches, which enlist household initiative on land either provided or commandeered by the state. Some of the emphasis on governments' building houses has fallen away, although where it remains, it is usually for middle-income urban dwellers. In these new approaches, it is the people who control the process and therefore make decisions as best they can about the environments in which they live. It is the government that usually provides services.

The early self-help literature (Turner, 1976) extolled the positive social virtues of self-built housing, such as community cooperation, self-esteem, potential for self-improvement, and local control. This thesis was generalized to operate universally and was based on Turner's experience in Peru. More recent research by Lobo (1982) gives insight into the context of Lima and makes it understandable why Turner came to his particular conclusions. As an anthropologist, Lobo followed a number of families through the process of securing an urban foothold, documenting their successes and failures. She found that the situation allowed the home builders to use their landholdings as a basis for upward mobility because there was flexibility in the system, which is not usually an option open to dwellers in similar locations elsewhere. The group commandeering of land was supported, in part, by the government, which then compensated the landowners for their land once it had been settled on.

It cannot be denied that the writings of Turner and others have made many governments, faced with large squatter settlements, rethink their approaches to self-help housing processes, which in many instances had been seen as a negative development, uncontrolled, and a blight on the landscape. Some governments did so, however, believing that the positive attributes of self-help would—almost magically—result in their particular countries if they accepted this procedure as an alternative to government built low-cost housing. But the results were not as advantageous as Turner had projected because the total economic and political system inhibited other essential freedoms that are necessary if the full social benefits of self-help are to be realized. What was in place often totally inhibited these benefits from happening and even worked against the self-help home-builder. For example, if the access to land was limited, there could be no upward social mobility. In some situations, the land made available was so removed from work and other revenue-producing activities that it actually made the life of the self-help builder impossible. On the other hand, if self-building households could not operate small, informal-sector businesses from their homes, this would also inhibit their progress. A major incentive to governments to adopt self-help housing procedures, was that it often appeared to relieve them of their financial responsibilities with regard to housing.

As a result, the focus of researchers in this area has shifted to understanding what supports or hinders the self-help home-builder's willingness to take action (Fernandez-White, 1983; Hardie & Hart, 1984; Hart, 1983; Hart & Hardie, 1985, 1987; Nell, Sifrin, Hardie, Strelitz, & Sturgeon, 1983; Sanchez *et al.*, 1987c; Ward, 1983; Wegge, 1983). How much pressure should exist and what kind of support is most appropriate in forwarding behaviors that will create an environment that is supportive and helpful to the occupants are issues that need to be investigated.

Some of the few researchers to document the advantage of the participatory process itself are Sanchez *et al.* (1987c) in Venezuela. They compared two self-help projects being organized in different communities. They documented the impact of differing levels of community participation in the process and how this impact affected the outcome of the projects. They focused on the experiential impact of participation, an aspect which has not usually been expounded in the literature. They stated that, in the literature, attention is focused on certain key moments of the experience, which are, in a way, achievements, such as planning or decision making. The literature, they suggested, puts little emphasis on the internal workings of the group, such as its organizational aspects or its experience of solidarity. "Power" is considered mostly in terms that are vague, such as "giving power to the citizens," rather than in terms of its nature, or the internal workings of organization, or particular aspects of manipulating power when dealing with interorganizational considerations. These reseachers have considered behavior not usually associated within environmental concerns, and yet, if understood, this behavior turns the building of houses into a community development exercise. More work needs to be undertaken in this area if we are to understand human processes that would be more supportive of community action in the self-build housing-process.

By contrast, an article describing the importance of community organization in the development of a squatter settlement in Turkey focuses on community action leading to physical development. Aksoylu (1986) described the establishment of a school and a water system more than the behavioral aspects that related to the dynamics of the community development process. This is perhaps a more standard approach to physical interventions; community development is seen only in the products generated and not in the process itself. This approach does not present the dynamic relationship, seen in the work of Sanchez *et al.*, between the people, their participation, and what they produce.

Housing Policy

Enough housing policies have been written around the world to sink a battleship. On the whole, they remain standardized, yet idealistic, economic proposals. Seldom do policymakers consider how acceptable their ideas will be to their clients. Tipple (1987) condemned most housing policies, which, although they are replete with excellent ideas, fail in implementation because of some incongruence between the planners' concept and the clients' willingness to cooperate. He believes that any strategy based on a wrong assumption about the clients' view of their housing is likely to be a weak link in the chain of implementation.

Tipple (1983b) proposed a reexamination of policies that ignore the dichotomy between the local culture and low-income housing policies transferred from elsewhere. "The gap between policy and implementation," he stated "has been very wide and may be at least partly explicable in cultural terms" (p. 24). Tipple is one of the few researchers in the environment and behavior field who has confronted this issue. It has taken a long while for housing policies to be questioned from this standpoint. Most developing countries can elaborate great endeavors and intentions incorporated into housing policies that are never realized or implemented or, if they have been acted upon, may even have backfired.

The thesis developed by Tipple is based on this research work in Ghana, where he investigated the land ethic of the Asante. He discovered that, because of their religious and cultural orientations toward the land, property and housing were most profoundly and deeply embedded in their understanding of the world. Tipple (1983a) documented the various ways in which the Asante view land-based issues and that thus affect their actions and run counter to their ever finding acceptable the government's housing policy. Tipple believes that it is necessary to face the dichotomy and make a careful choice between the importance of continuing urban development in the face of modern rates of migration and constraints on the use of foreign exchange, on one side, and the desire to keep all, or reluctance to change any, of the tenets of Asante culture, on the other. He concluded that "rapid urbanization and Asante culture as it stands are irreconcilable and this dichotomy must be grasped by policy makers or else the current ineffectiveness of housing policy cannot be overcome" (1983a, p. 17).

Recently, a similar misunderstanding of a people's values was exemplified in South Africa. The government made 500,000 low-cost houses that it owned available for purchase to the occupants at very nominal costs. The sale progressed very slowly indeed and was unsuccessful in the short run. An investigation (Hart & Hardie, 1987) showed that the people had valid reasons for not wanting to purchase the houses, but that their reasoning had not been considered before the launching of the scheme. They resisted, and even their awareness that profits could be made on the house seldom acted as an incentive to purchase. Their most universal experience was that owning a house was more expensive than renting. They were also suspicious of the state's motives. These attitudes were not helped by state policy vacillation, and a perceived lack of clarity and consistency in administrative costs and procedures did little to reverse the endemic distrust. Much research remains to be undertaken in this area, so that policy issues related to housing are investigated and understood from the perspective of the people whom the policy is intended to support.

FIELDWORK TECHNIQUES

Carrying out research in developing countries calls into question many research methods that are used elsewhere. Not only what is being investigated but also the context of the research should define the most appropriate methodology. Standard investigation procedures may not be appropriate. What follows is a description of various methods that have been used in undertaking research in developing countries.

In the same way that it takes the yin to understand the yang, the method by which attitudes in one area are compared to attitudes in another can be used to great effect to show what is special and significant in each place. This approach can be a most powerful tool in environment–behavior research, particularly in situations where everything is exotic, because comparison allows the particular to stand out. For instance, in considering self-help procedures, Hart and Hardie (1987) showed that the comparative approach can give insight into the context-boundedness of issues. Two housing schemes were compared in which the social and political history, the location, and the prior housing experience were all shown to influence how each had developed. If only one example had been considered, the generalized perceptions of self-help schemes would have been completely skewed by what was discovered from that one example.

What will be realized from the different methods presented below is that flexibility is needed in the way research is undertaken. The context, the ability of the people being interviewed to disclose information, and the issues being investigated all need to be thoroughly reviewed before one decides on an appropriate method of investigation. In conclusion, much caution should be exercised in transferring techniques used in more literate and perhaps less verbal societies to situations where illiteracy is high. The use of photographic or outline drawings, for example, can cause a problem because the viewers may not be able to discern the third dimension that most Westerners, through familiarity, take for granted.

Physical Baseline Surveys

Often, the claim is made that the data of social scientists—and more particularly, of anthropologists—are "soft" in nature, too descriptive, and lacking in sufficient quantitative assessment to be useful. Physical surveys can help to allay some of these criticisms and to tap into the methods used by human geographers in their surveys of the landscape.

This data-gathering technique is often the first step in research. Hart and Hardie (1987) used this approach on a number of occasions, particularly in trying to assess the progress of home builders involved in self-built housing. A survey of houses was undertaken on a house-by-house, street-by-street basis, the degree of construction being noted on each site. Repeating this exercise regularly over a number of years enables one to make a quantitative statement about progress in an area. If this information can then be matched with socioeconomic data, it is possible to make hypotheses about who is making progress and why.

Likewise, photographs can be used comparatively for good effect

FIGURE 4. Photography can record the growth of a new town: The "tent town" Inanda Newtown, in Durban, South Africa, in February 1982.

FIGURE 5. Inanda Newtown in Durban, South Africa, in August 1982.

FIGURE 6. Inanda Newtown in Durban, South Africa, in January 1983.

FIGURE 7. Inanda Newtown in Durban, South Africa, in September 1983.

FIGURE 8. Inanda Newtown in Durban, South Africa, in February 1984.

and impact. These can be taken at different times from one viewpoint, either from the ground or from the air. Often, air photographs are publicly available from government survey offices. Aerial photographs taken at significant intervals, such as 5 years apart, can enable a researcher to gain an historical overview of an area from which the rate of development can be assessed. This information can then be combined with qualitative data, so that the physical survey can be analyzed and fully understood. The method may even define a significant pattern of development that can then serve as the basis for selecting case studies representative of a particular category. Photographic studies can also be used to map behavioral data, such as human action in a plaza or other social spaces, which can then be used as a basis for investigating the behavior noted.

Physical baseline studies can lead one to wrong conclusions if taken in isolation. I discovered this recently firsthand when I conducted a survey in a town by counting the number of houses that required thatch roofing. I discovered that there were a great many in need of reroofing. My immediate conclusion was that these houses remained unthatched because they were traditional houses and that, therefore, the state of disrepair was an expression of the people's lack of concern about houses of the past. However, it was drawn to my attention that there had been a drought for 6 years, and that the drought had substantially decreased the supply of thatching grasses; until this situation returned to normal and the availability of grasses changed, it would not be possible to validate my original conclusion that traditional houses were being disregarded. So researchers should be cautious in taking the physical too literally, without substantiating it with other information.

Structured Interviews

In eliciting broad-based opinions, structured questionnaires have proved to be a useful means of investigation. However, this method also poses some difficulties. The respondents, often illiterate, do not understand the significance of the questions and withdraw at an early stage, leaving the interviewer either to forgo that interview or to complete the schedule by inventing what would seem to be appropriate responses. This has happened even in situations where the interviewers are well trained and well supervised (through personal communication).

A further difficulty with structured interviews is dealing with different languages. It is possible that the researcher, the interviewer, and those interviewed may each be of a different language group. Language difficulties can, in turn, create difficulties in interpreting data and in ascertaining whether the questions have been correctly understood or, conversely, whether the interviewer has clearly interpreted the response.

Another problem relates to what people are willing to tell when confronted by an interviewer. In one large survey of a self-help project, it was discovered that the income of female-headed households averaged only about half the incomes of male-headed households (Hart & Hardie, 1987). Yet, surprisingly, the survey showed that the size of the house built was about the same, regardless of whether it was constructed by a female- or male-headed household. The results, on the surface, made one marvel at the ingenuity of the female-headed households, which managed to build as much as the male-headed households on approximately half the income. On reinterviewing a number of women in an unstructured format, different issues emerged that clarified the results. The formal incomes were different, as had been recorded, but many of the women had some form of small, informal-sector enterprise, whether it be brewing and selling beer; selling meat, vegetables, or other supplies; or making clothes. Because these activities were considered illegal, they had not been declared to the interviewers, who were strangers. This experience revealed the potential openness to misinterpretation of information gained through surveys. The reality remained hidden because it was not in the interests of those being interviewed to declare themselves fully.

From personal experience, I have found that very often the kind of results described above, which seem anomalous (the 10% that doesn't fit the overall pattern), when investigated often lead to crucial insights into the dynamics of a situation and thus should not be ignored and cast aside because of seeming lack of significance.

After working in Tunisia, Enders (1981) supported these difficulties with traditional survey methods. He stated that this methodology can only rarely identify the values and desires of a target population when the group is poorly educated. Illiterate and poorly educated persons often have great difficulty in responding to questions that ask them to make choices on subjects that are beyond their range of experience. Enders believes not only that such people have trouble expressing their values and desires but, in addition, that their participation in such surveys often causes them to raise their expectations beyond what they can realistically hope to obtain from their government.

A structured questionnaire gives the researcher a snapshot view of what prevails in a particular situation. Interviewing in environments over an extended period has shown me that the time dimension is a most important variable (Hardie, 1983). Events can affect the attitudes in a place and can cloud issues and perspectives. Consistently, where interviews have been repeated and contacts made at regular intervals, a much better insight is gained into a situation and the prevailing attitudes. Similarly, random sampling techniques, although appropriate, may not be possible. In some situations, where the environment is racked with conflict and violence, it may be impossible to go from door to door. Interviewing in self-help schemes where the houses are not numbered adds to the complication of setting up a sampling strategy. It may also be found that interviewing at the residence is considered an invasion of privacy or a show of disregard for the status of women. In such circumstances, alternate ways of reaching the sample have to be considered and used. From personal experience in Islamic countries, Alnowaiser (personal communication) proposed that a quota-sampling method be used where appropriate. In his own research in Saudi Arabia, he found this to be especially applicable; because of the social conditions, he was unable to randomly interview people at their domiciles and had to carry out interviews at the respondents' place of work, at school, and in the market.

Open-Ended In-Depth Interviews

Where possible, in-depth interviews, with the researcher directly involved in the data collection rather than receiving data from secondhand sources, such as interviewers, result in a better informed researcher. From personal experience, this is my preferred approach, which I have adopted in the course of over 20 contract projects that involved interviewing more than 1,000 households over a span of 4½ years. Many of the households were revisited, and each visit focused on different issues. With time, rapport develops between the researcher and the respondents; those being interviewed gain confidence in the researcher, and a greater openness results, which increases the quality of the data received. This experience has shown that issues related to housing are always open to discussion and that, in connection with this topic, it is possible to discuss a great many others; a discussion of the physical features of the house can get one in the door. Seldom has a household refused entry, but then, my investigations have never been in situations where there were taboos about men interviewing women. The fact that someone is interested in what a householder has created is taken positively, and the pride felt in such accomplishments acts as a way of establishing contact. Taking a photograph of a family and either leaving it with them or sending it to them can also establish a continuing dialogue and involvement.

Repeat visits are found to be essential. In one instance, a woman, a retired domestic, stated that she was managing to build a house on her pension. This seemed impossible, and although the fact that she had a large kitchen was noted, it was only after many interviews that I discovered that she cooked lunches for local building contractors. However, because she did not have a health permit to cook food commercially in a domestic kitchen and her operation was, in fact, illegal, she had never disclosed this information. On another occasion, an elderly couple proudly displayed their plans for their house. They had doors and window frames stored on their roof, but they had no savings and lived on their government pension. It was impossible to comprehend how they would ever manage to complete their home. One day, on a chance return to the site, the reseacher met the couple's daughter, who was home from a neighboring town where she was employed as a nurse. It was then discovered that it was she who was building the house, which eventually would be used by her. The couple had never indicated this, and the deceptive hopelessness of their situation had left a deep impression on the researcher, affecting his overall perspective and evaluation of the project. Only by returning, was it possible to discover that their building was in hand and progressing.

Open-ended interviews are crucial as a first step in developing a structured questionnaire and, similarly, as a follow-up to a large survey as check not only on the interviewer, but also on the data that have emerged from the survey. At times, open-ended interviews can stand on their own as a means not only of gathering generalizable information or broad-based opinions, but of giving greater insight. Because they are open-ended, these interviews are highly qualitative and rich and, as a result, can serve designers well, who often seem to appreciate the nuances of people's needs rather than bald opinions. Obviously, the approach adopted needs to be appropriate to both the situation and the way in which the data will be used.

At times, investigations are undertaken in which no random sampling methods can be applied. The approach called for may best be described as journalistic rather than scientific, a methodology that many researchers may not believe to be appropriate to the field of environment and behavior. However, some issues may call for a response that cannot be handled rigorously.

With the task of assessing the abilities of and the constraints acting on black building contractors in South Africa and the impact that these would have on the home-building industry, it was impossible to establish normal sampling procedures (Hart & Hardie, 1982). Legislation, in fact, made some of the work undertaken by black builders illegal, so that, in some areas of the country, they were under threat even of criminal conviction. The situation was quite variable around the country. What, then, could be described as being representative? Contractors were interviewed in different towns around the country. As pertinent issues emerged in each place, these were pursued and investigated further in order to clarify the particular context and problem. When presented, the final report, although giving an overview, also dealt with specific incidents of exploitation and thus was not, after cabinet review, allowed to be publicly released—yet, it later brought about legislative changes.

This is only one example of an approach that may be appropriate in particular circumstances in which a topic is pursued qualitatively, so that a whole spectrum of views may be presented and investigated in greater depth, and so that, with time, a global picture of the issue may be established.

Group Discussions

This is a method that can also provide qualitatively rich results and can expose what are often deep and emotionally charged issues that may not come through with the use of the methods discussed above. Mutual support can bias the response; although conversely, it may also enable participants to be more outspoken.

This approach was used in an investigation into the reasons that the occupants of government-owned houses in South Africa had not found it advantageous to purchase their houses even when the houses were offered at very low prices (Hardie & Hart, 1986). Following a series of pilot interviews, it was decided that a research procedure should be developed that would stimulate black respondents to freely discuss their attitudes toward the housing sale, to reveal their fears and the current rumors, and to recount experiences related to the sale or to homeownership in general. To meet these requirements, a group discussion procedure was adopted. These discussion were arranged by companies, which assembled, on average, 15 employees, representing various ranks, who were willing to discuss the issue with us. Names were not recorded, and company officials were asked not to attend.

Because the group discussions were based on formally employed black people, the group of respondents was deliberately biased to include those with a steady income, and therefore, they were those people most likely to consider the purchase of state-owned houses as a serious option. Because of this rather crude stratification, and against the background of the relatively informal criteria for company and group selection, it cannot be claimed that the responses elicited in the group discussions were representative of any specific black population. It is, however, valid to analyze the qualitatively rich results of the group discussions within the context of the group of respondents. Nor is it unreasonable to propose that some of the deep and often emotional issues exposed were felt by other urban blacks in South Africa as well.

The researchers believed that no other method would have given such wide-ranging insight. The method of discussion revolved around recording publicly every idea presented, meaning that there were no hidden minutes. All the participants saw their ideas appear before the group, so that they felt their contribution was valuable; this feeling then reinforced the cooperation and discussion within the group. This method can also be useful in situations where it may be impossible to interview people in their homes.

Walk-About Verbal Descriptions

Another approach to the group discussion is to have the group walk through a building or a particular environment and to elicit their opinions. By questions and discussion, the group can be encouraged to describe what is of value to them in a particular space. This description can lead to a very free and open discussion, which may reveal a whole range of unexpected issues and aspects of value. Many researchers have used this method to assist those under examination to present their environmental preferences.

Working among the indigenous populations of Canada, Simon (1983) evaluated proposals for a housing development for an Ojibway community. Walking over the land together with those to be affected by the housing development, Simon uncovered their concepts of land and ownership, whether it be to obtain individual parcels of land or how to use a lake that was located on the site. On the latter issue, the planners had seen the lake-front property as a prime residential location. However, the community saw it as a place not to be touched by anyone's house but to be available to all. Likewise, the layout of roads leading to individual house sites with their own boundaries was found, by walking on the site, to be in conflict with the prevailing approach of the community to ownership, in which all the land was held in common.

This method of using walk-abouts has been developed by researchers in New Zealand (Daish, Gray, Kernohan, & Salmond, 1982) among Maori communities and others. These researchers have used this technique a great deal and have much insight to offer into how it may best be undertaken.

Simulation and Projective Techniques

Simulations, used along with a number of the interview techniques covered above, can greatly facilitate discussion. These methods may be particularly helpful in the presentation of abstract ideas. Models and other graphic means of presentation can help a great deal in involving the interviewee and also in stimulating more open dialogue.

When investigating attitudes toward house and settlement patterns in Botswana, I turned to the use of three-dimensional models. This approach (Hardie, 1981) showed that there was a high degree of environmental cognition on the part of people being studied. They had no difficulty at all in modeling their homes exactly, even though the actual dimensions were not known. They were able to do so with accuracy, simply relating doors and windows and wall height to the overall scale of the rooms. De Long's experiments (1976) with perception and the use of models confirm this correlation between models and reality in North American participants.

The modeling technique used in Botswana for investigating alternative house forms was highly flexible, and individuals could with ease model both their present house and the house they desired to build in the future (Figure 9). As a research tool, modeling was remarkable for the way in which it generated discussion on the use of domestic spaces,

FIGURE 9. A house-modeling apparatus.

as well as for the accuracy attained by people modeling their own home, even when doing so far from their house. In this connection, plans had little meaning, as the people could not understand what was represented. One woman who had bought a plan for her future house expressed total disbelief when she saw it modeled. "Is that my house? What are those spaces for?" she asked, referring to leftover passage space. She had agreed to the design on the basis of the elevations and the number of rooms the house offered, but she had been unable to follow the intricacies of the design in the architect's drawing. The model, however, made the plan understandable.

The modeling apparatus has gone beyond being used only as a research tool, for it is aiding homeowners and self-builders in the design of their own houses. A number of self-help housing projects in operation in Southern Africa use the model as a basis for counseling people designing their houses, such as in the Ikaya Lethu and Katlehong Housing Project. It has also been used in housing projects in Dar es Salaam, Tanzania, and Nairobi, Kenya. Some observations can be made on the use of modeling systems. Users are unable to go beyond their spatial experience. For instance, during my early research with models in rural areas, I discovered that, in some cases, participants had never been in a house with a passage: the traditional house comprised a single space with multiple units on each site (Hardie, 1983). A four-room house

FIGURE 10. A householder models his future home in the yard of his temporary structure.

presented no problem. However, a five-room house was confusing, until it was suggested that a passage could be used to connect the fifth room. The use of models seems to be no problem for participants in terms of scale. It would appear that people who have built their own house, shaping it with their own hands, have little difficulty in dealing with scale. On the other hand, two-dimensional drawings are not understood and are of little use. In terms of satisfaction, the involvement of the household in the design of the home, using the model, means that, from the outset, they feel that the house they are designing, and will later be building through their own efforts, is their own creation. In other schemes, set plans are made available, but this approach inhibits the sense of identification with the house. Using the modeling method, the participants are able to make adjustments to the design according to their own preferences and what they can afford. Should they make a mistake, they appear to be more accepting of what they have planned and built than if the plan had been developed by someone else.

Another technique that I have used on a number of occasions was developed to discover the preferences of people in terms of the future design of their town. This procedure consists of an evaluation of four different street patterns presented in model form. The interviewees were asked their reasons for liking or disliking each alternative. The models displayed a street with a circular dead end (Figure 11), one with

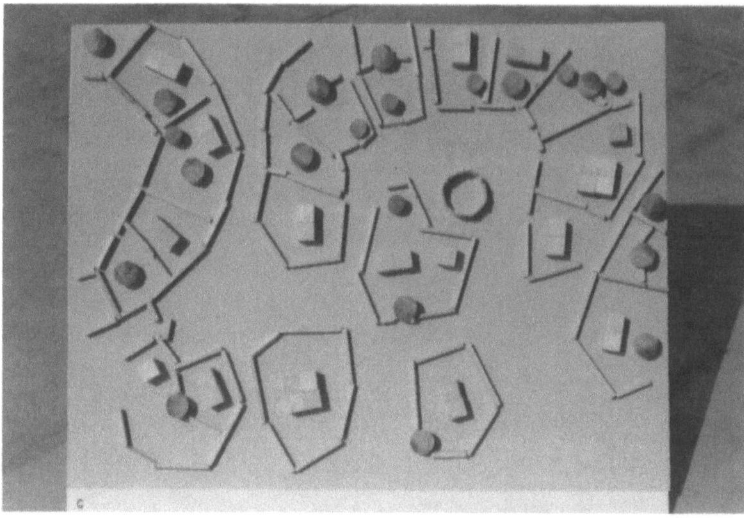

FIGURE 11. A street with a circular dead end.

a random layout (Figure 12), a traditional layout of a typical ward (Figure 13), and a straight street (Figure 14). The amount of response elicited was extremely rich and varied. The participants often discussed the models in their own language and could voice their opinions (Figure 15). Often, they were elderly and illiterate, and if a questionnaire-type approach had been used, they would have been unable to express their views. With the modeling approach, they could freely contribute their ideas.

The following is a selection of the comments that were obtained and will give an idea of the variety of reactions that the model stimulated (for a full description and analysis, see Hardie, 1980, 1988). On describing the traditional street (Figure 13), respondents said:

> "I prefer this because it is good for a traditional Motswana."
> "This is good since the children grow up knowing their relatives, and we should not desert our traditional ways."
> "Good, typical Tswana style."
> "This is an old type of settlement. Land is wasted. The kraal must not exist because it is dirty."
> "With this you can't fence the plot easily because of the peculiar plot shapes."
> "Good in that in emergencies you can quickly summon your neighbor."
> "There is no outlet at the end. If I am attacked, where will I get out with my vehicle?"

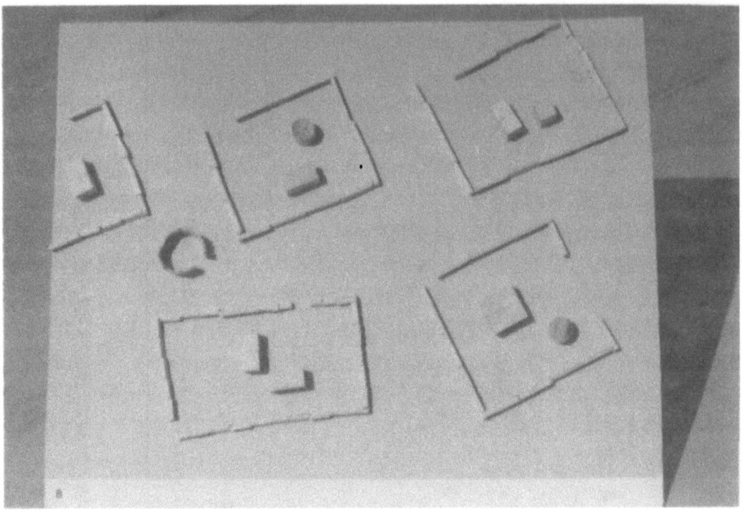

FIGURE 12. An area with a random pattern.

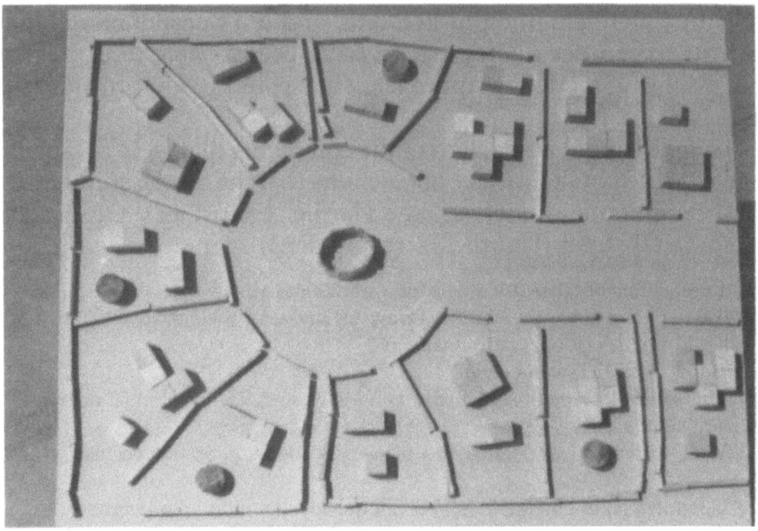

FIGURE 13. A traditional layout of a ward of a Tswana town.

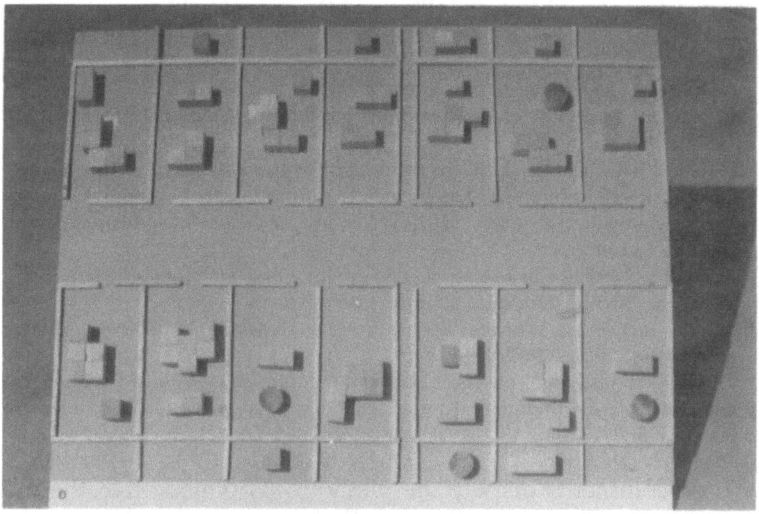

FIGURE 14. A straight street with rectangular plots.

FIGURE 15. Townsfolk of Mochudi, Botswana, gather around the models to express their opinions.

About the straight street (Figure 14), people said:

> "This is beautiful."
> "First-class."
> "These are moden times, and we must build for development."
> "It is a civilized type."
> "I'm not used to this; I still have to learn to live this way." "This is the European way of building. Let us stick to our customary procedures."
> "This belongs to the whites. However, if anyone wishes to settle in it, they shouldn't be stopped."
> "I don't want to live in a location. You don't know your neighbors there and life in the town is very difficult and oppressive."
> "Only suitable for vehicles. People using ox wagons should be isolated; otherwise they face difficulties here."
> "I would choose more passages between the houses, because if I settle first and build a fence, then the next man will not use any money to build a fence on his side."

These comments display the plethora of opinions that the models generated and touch on implications that can be explored further by the planner. Take, for instance, the comment by the person looking at the dead-end street and asking, "If I am attacked where will I get out?" My immediate reaction was "Attacked by whom?" There had been no wars for eons. But the man was, in fact, referring to spiritual attack, which could be initiated by a traditional doctor who could place his potions on the road very effectively, if there was only one way to leave the house. The desire for many footpaths was based on the same belief, with the people wanting many routes to their homes, and feeling it wise to use a different route each day.

It is obviously difficult to decide on the right sampling procedure in the use of these techniques. With the models, it was also difficult to sample on a random basis because the models were not that maneuverable. In the end, we used a self-selected opportunity sample. The models were placed at various places around the towns, such as on street fences and in shops. The displayed models drew crowds of people, from which individuals were invited to be interviewed.

As a result of this selection procedure, the sample cannot be claimed to be representative of the communities in which the interviewing took place, nor of the potential residents. Obviously, this research procedure trades off information that is qualitative with a more structured and abstract approach to planning issues that may not elicit the important attitudes that the models stimulate. In the instances where this procedure has been used, it has been decided to make the interviewing process as dynamic as possible so as to seek the greatest insight into the attitudes of the potential residents. The planners agreed that, for their purposes, this approach should be adopted because it would give them greater insight how the potential residents viewed their world.

Numerous researchers have developed gaming methods for stimulating discussion on planning issues. There is an extensive literature on this subject, and it is appropriate to mention here the methods that have been used in developing countries. Rodman (1986) used photographs as a way of discovering what people had to say about their houses and of interpreting what their comments revealed about the culturally constructed meanings they assigned to house form and to the domestic space that is part of, yet goes beyond, the house itself. Stea (1980) used models in many different contexts, and Sanoff (1979) focused on particular programmic aspects.

Participant Observation

Participant observation is a standard anthropological method and can be most effective in revealing how shared meanings and values are publicly acted out in space and thus are observable in people's behavior. It is this behavior that conceptually and practically organizes space because most behavioral acts presuppose a specific area for their performance. In the environment and behavior field, it is this interaction that is our concern. It is not surprising that anthropologists gave the title *participant observation* to this method because in the study of a community with values different from their own, they could assume nothing and much seemed strange, so they simply observed.

This method can be used most effectively to discover how particular functions, social networks, status, and other variables affect the organization of space. Lobo (1982) used this approach to great effect in her study of self-help housing in Peru. Low (personal communication) described how watching a plaza in Costa Rica, and particularly using her camera to record the action, introduced her to the workings of the area. She came to identify the various community subgroups that, at different times of the day, occupied and used the square. Through observing the action in the plaza and how the plaza was used, she was able to unravel the panoply of social interactions in the town.

IMPLICATIONS FOR FUTURE RESEARCH AND APPLICATION

In this chapter, four issues have been raised that require further investigation: (1) the role of participation and its effectiveness, particularly in self-help schemes; (2) the ways in which researchers in the environment and behavior field can help to develop housing policy and planning that recognize the people involved and their value systems; (3) the need to understand new technologies and the introduction of household services according to local values; and (4) the need to develop appropriate ways of communicating cross-culturally on environmental issues. The chapter has also attempted to show some practical ways in which this communication may be achieved by discussing practical field techniques.

But what are the real lessons that can be learned from this interchange between researchers in the developed world and the developing world? One lesson, perhaps the most profound, is that our approaches and even our "science" are ethnocentric and, when transferred to other contexts, are found wanting. Ahrentzen (1987) made this point most poignantly when she described the difficulties she had in teaching environment–behavior studies in the People's Republic of China. She described how the pursuit of privacy and individualism dominant in our American thinking and even in our American research was a foreign concept to the Chinese, who do not consider privacy but rather believe that the built form must be viewed in terms of what is best for the public good and what the State considers best for the people.

New, insightful studies are beginning to appear that are cross-culturally comparative. For example, Zube, Vining, Law, and Bechtel (1985) investigated and compared perceived urban residential quality in relation to the acoustic sensitivity of Arab and American respondents. The research was undertaken by means of both on-site and simulation procedures. The results showed that the Arabs had a higher noise tolerance than the Americans. Churchman (1986) considered differences between environment and behavior findings in the United States and is Israel. She compared Israeli expectations about the availability and form of urban housing with attitudes found in housing projects in the United States. This work highlights unique and perhaps anomalous conditions in each country that would otherwise have been overlooked by researchers. Stea (1980) creatively compared the sense of place of nomads and squatters from an environmental psychology perspective. Kimura and Sime (1986) compared Japanese and English homes and the differences in their use of domestic space. These studies show that comparison leads to greater understanding of what is important in each context in the differing relations between behavior and the environment. This direction of research could be most fruitful in the future and could illuminate the meeting point between the developing and the developed world. It may prove that developed countries have a great deal to learn from the so-called developing countries about the best ways to integrate behavior sensitively and positively with the environment.

REFERENCES

Ahrentzen, S. (1987). Reflections on teaching environment-behavior relations in the People's Republic of China. *Design Research News, 18*(2), 1, 7–13.

Ahuja, S. (1980). Socio-cultural dimensions in some Indian villages. *Ekistics, 283*, 289–295.

Aksoylu, Y. (1986). Community organization: The Pinar experience in Turkey. *Open House International, 11*(3), 44–48.

Alexander, C., with Davis, H., Martinez, J., & Corner, D. (1985). *The production of houses*. New York: Oxford University Press.

Alnowaiser, M. A. (1985). Traditional and Modern Settlements in Saudi Arabia. *Habitat International 9.1*, 5–17.

Bein, F. L. (1981). Domestic water resource use in rural Sudan. *Ekistics, 291*, 430–433.

Bentz, B. (1981). User participation in the design of housing. *Open House International, 6*(2), 13–18.

Blitzer, S., Hardoy, J. E., & Satterthwaite, D., (1981). Shelter: People's needs and government response. *Ekistics, 286*, 4–13.

Bor, W. (1982). Planning and development in developing countries. *Ekistics, 92*, 28–31.

Boukraa, R. (1981). The features of Islamic urbanism and social change in Hammamet, Tunisia. *Ekistics, 291*, 426–429.

Centre for Architectural Research and Development Overseas. (1987). Unpublished poster. School of Architecture, University of Newcastle-Upon-Tyne.

Castro, I., Guerrero, R., & Hernandes H. (1978). *Estudio exploratorid sobre la influencia de la densidad residencial y el nivel socio-econ6mico en la cohesi6n de group vecinal*. Unpublished manuscript. Universidad Centrale de Venezuela, Escuela de Psycologia, Caracas, Venezuela.

Chandhoke, S. K. (1980). Human habitations in India: An anthropological approach. *Ekistics, 283*, 296–300.

Churchman, A. (1986). *What is or is not special about the Israeli context*. Unpublished paper presented at the Conference on Built Form and Culture Research, Lawrence, Kansas.

Curtis, I., Garcia, B., & Naveda E. (1978). *Influencia del diseño arquitect6nica de las veviendas multifamiliares sobre las relaciones vecinales*. Unpublished manuscript. Universidad Central de Venezuela, Escuelo de Psicologia, Caracas, Venezuela.

Daish, J., Gray, J., Kernohan, D., & Salmond, A., (1982). Post occupancy evaluation in New Zealand. *Design Studies, 3*(2), 77–83.

Dandekar, H. (1981). Social and spatial constraints on rural women's sexuality: observations from an Indian village. *Ekistics, 291*, 422–426.

De Long, (1976). The use of scale models in spatial-behaviorial research. *Man-Environment Systems, 6*, 179–182.

Duncan, J. (Ed.) (1981). *Culture and identity: Cross-cultural perspectives*. New York: Holmes and Meier.

Duncan, J. (1986). Symbolic mediation: Ideology, landscape, and social being. In J. W. Carswell & D. G. Saile (Eds.), *Purposes in built form and culture research*. University of Kansas, Lawrence: School of Architecture and Urban Design.

Enders, M. J. (1981). Housing and settlement patterns in Central Tunisia: An example of the conflict between the individual and the organization. *Ekistics, 287*, 137–144.

Fathy, H. (1973). *Architecture for the poor.* Chicago: University of Chicago Press.

Fernandez-White, J. (1983). Organizing squatters as training in participation. In UN Centre for Human Settlements (Ed.), *Community participation for improving human settlements.* Nairobi: United Nations Centre for Human Settlements.

Guinness, P. (1986). *Harmony and hierarchy in a javanese kampung.* New York: Oxford University Press.

Hardie, G. J. (1980). *Tswana design of house and settlement—continuity and change in expressive space.* Boston University (Unpublished thesis).

Hardie, G. J. (1981). *Tswana design of house and settlement: Continuity and change in expressive space.* Doctoral dissertation, Boston University, 1981. Ann Arbor, MI: University Microfilms, #81:12201.

Hardie, G. J. (1983). Stimulating user participation in the design process. In D. Joiner, G. Brimilcombe, J. Daish, J. Gray, & D. Kernohan (Eds.), *Conference on people and physical environment research.* Wellington: New Zealand Ministry of Works and Development.

Hardie, G. J. (1986). Continuity and change in Tswana's house and settlement form. In I. Altman & C. Werner (Eds.), *Home environments.* New York: Plenum.

Hardie, G. J. (1988). Community participation based on three-dimensional simulation models. *Design Studies, 9,* 56–61.

Hardie, G. J., & Hart, T. (1984). The willingness to act: Self-helps vital factor. *Housing in South Africa,* May, 18–19.

Hardie, G. J., & Hart, T. (1986). A black perspective on the sale of state-owned houses. *Planning and Building Developments* (Modern Mass Housing Supplement), March, 17–20.

Hart, T. (1983). "The public face of black self-help housing: Policy and practice post Soweto 1976." Presented at the conference of the South African Geographical Society, Pietermaritzburg.

Hart, T., & Hardie, G. J. (1982). *Factors inhibiting the effectiveness of black building contractors in South Africa.* CSIR Confidential Contract Report C/Pers 333. Johannesburg: National Institute for Personnel Research.

Hart, T., & Hardie, G. J. (1987). State sanctioned self-help and self-help homebuilders in South Africa. *Environment and Behavior, 19,* 353–370.

Kimura, M., & Sime, J. D. (1986). Spatial boundaries in Japanese and English family homes. In J. W. Carswell & D. G. Saile (Eds.), *Purposes in built form and culture research.* Lawrence, University of Kansas, School of Architecture and Urban Design.

Lobo, S. (1982). *A house of my own.* Tucson: University of Arizona Press.

Montero, M. (1976). Estudio psico social de la propiedad. *Psicologiá, 3,* 229–235.

Montero, M. (1979). Relación entre tipo de vivienda, deteriore y clase social percibida. *Psicologiá, 7,* 125–132.

Muller, M. S. (1984). Traditional cultural identity in new dwellings of urban Africa. *Ekistics, 307,* 359–365.

Nell, M., Sifrin, G., Hardie, G. J., Strelitz, J., & Sturgeon, K. (1983). *Guidelines for self-help housing: Principles, practical issues, and case studies.* Johannesburg: Urban Foundation and the South Africa Institute of Housing.

Pellow, D. (1986). African renewal: Continuity and change. In J. W. Carswell & D. G. Saile (Eds.), *Purposes in built form and culture research.* Lawrence: University of Kansas, School of Architecture and Urban Design.

Prussin, L. (1969). *Architecture in northern Ghana.* Berkeley: University of California Press.

Rapoport, A. (1983). Development, culture change, and supportive design. *Habitat International, 7*(5/6), 249–268.

Rastorfer, D. 1985. The man and his work. In J. M. Richards, I. Serageldin, & D. Rastorfer (Eds.), *Hasan Fathy.* Singapore, Mimar.

Ribes, E., Galvan, E., & Villaneuva L. (1973). *Estudio piloto de evaluacuión conductual de interieres de vivienda.* Unpublished manuscript, Mexico.Instituto para el Fondo Nacional de la Vivienda para los Trabajadores.

Rodman, M. C. Why do we need to know what makes a house "good to live in" in Vanuatu? In J. W. Carswell & D. G. Saile (Eds.), *Purposes in built form and culture research.* University of Kansas, Lawrence: School of Architecture and Urban Design.

Sanchez, E., Cronick, K., & Wiesenfeld, E. (1987a). Environmental psychology from a Latin American perspective. In D. Stokols & I. Altman (Eds.), *Handbook of environmental psychology.* New York: Wiley.

Sanchez, E., Cronick, K., & Wiesenfeld, E. (1987b). Group research synopses. *Architectural Psychology Newsletter, 16*(4), 2–5.

Sanchez, E., Cronick, K., & Wiesenfeld, E. (1987c). Psychosocial variables in participation: A case study. In D. Canter, M. Krampen, & D. Stea (Eds.), *Ethnoscapes.* London: Gower.

Sanoff, H. (1979). *Design games.* New York: Kaufman.

Sanoff, H. (1988). Participatory design in focus. *Architecture and Comportement/Architecture and Behaviour, 4.1,* 27–42.

Setiawan, B., Subanu, L. P., & Hermanislamet, B. (1987). *Code River Settlement: A preliminary study of the community potential and environmental development of an urban kampung in Yogyakarta.* Yogyakarta, Indonesia: Gadjah Mada University, Department of Architecture.

Simon, J. (1983). Employing socio-cultural factors in environmental design. In D. Amedeo, J. B. Griffin, J. J. Potter (Eds.), *EDRA 1983: Proceedings of the Fourteenth International Conference of the Environmental Design Research Association.* Washington, DC: Environmental Design Research Association.

Smith, D. L. (1981). Household design and family needs in Nairobi, Kenya. *Ekistics, 287,* 145–151.

Stea, D. (1980). Psychosocial studies of nomads and squatters: Two applications of cross-cultural environmental psychology. *Interamerican Journal of Psychology, 14,* 1–20.

Stea, D. (1981). *Environmental modelling as participatory planning.* Monograph FW #5, Fourth World Studies in Planning, School of Architecture and Urban Planning, University of California, Los Angeles.

Tipple, A. G. (1983a). *Asante culture and low-income housing policies: An examination of antithesis.* Washington, DC: Water Supply and Urban Development Department, World Bank UDD-45.

Tipple, A. G. (1983b). Housing policy and culture in Kumasi. *African Urban Studies, 15,* 17–30.

Tipple, A. G. (1987). Housing policy and culture in Kumasi, Ghana. *Environment and Behavior, 19,* 331–352.

Turner, J. (1976). *Housing by people.* London: Marion Boyars.

van Staden, F. (1986). 'N oorsig van die ontwikkeling van omgewingsielkunde in Súid-Afrika. In C. Plug (Ed.), *Reports of the Department of Psychology,* No 16. Pretoria: University of South Africa.

Ward, P. (1982). *Self-help housing: A critique.* London: Mansell.

Wegge, J. (1983). Modes of participation by the urban poor in the construction, maintenance, and improvement of their homes and neighbourhoods. In UN Centre for Human Settlements (Ed.), *Community participation for improving human settlements.* Nairobi: United Nations Centre for Human Settlements.

Wiesenfeld, E. (1984). *Relación entre densidad residencial, locus de control y percepcion de aglomeración en la vivienda.* Unpublished master's thesis, Universidad Central de Venezuela, Caracas, Venezuela.

Wiesenfeld, E., & Sanchez, E. (1983). *Creencias y actitudes hacia la aglomeración en al hogar.* Unpublished manuscript, Universidad Central de Venezuela, Instituto de Psicologia, Caracas, Venezuela.

Zube, E. H., Vining, J., Law, C. S., & Bechtel, R. B. (1985). Perceived urban residential quality: A cross-cultural bimodal study. *Environment and Behavior, 17,* 327–350.

III

ADVANCES IN USER GROUP RESEARCH

6

Hospital Design, Health Providers, and the Delivery of Effective Health Care

SALLY A. SHUMAKER and WILLO PEQUEGNAT

The acute-care hospital is among the most complex of institutions, comprised of diverse user groups that often have competing needs, an intricate organizational structure, varied settings, and a mandate to be responsive in design to a rapidly changing medical technology. All of these factors are embedded within one of the highest and perhaps most stress-evoking purposes for the existence of any institution: the preservation of life. The degree to which this purpose may be thwarted by the sociophysical environment has been the theme of previous articles and chapters (Moos & Schaefer, 1987; Newman, 1984; Rachman & Philips, 1978; Shumaker & Reizenstein, 1982; Sommer & DeWar, 1963; Taylor, 1979; Winkel & Holahan, 1985), and recommendations have been made about ways in which the environment can be improved. Here, we reevaluate these recommendations and consider the reasons that so few have been adopted and tested.

In this chapter, we also discuss two implicit models that link sociophysical factors to the well-being of health providers and patients in

Sally A. Shumaker • Behavioral Medicine Branch, National Heart, Lung and Blood Institute, National Institutes of Health, Bethesda, Maryland 20892. **Willo Pequegnat** • Environmental Design Research Association, Bethesda, Maryland 20814.

hospitals. We consider potential obstacles to empirically based hospital research and, within this context, the complexity of the hospital design process and the efficacy of patient-based interventions. Next, we examine health-provider-based interventions. We explore issues of the work environment and the provision of care in a hospital setting. The relationship between nurses and patients and the differences between the two groups in their interactions with the designed hospital environment are discussed. Then, we elaborate on the two *nonexclusive* models of hospital design and consider how patient care may be negatively influenced by the impact of the sociophysical environment on health providers. We track a "typical" myocardial infarction (heart attack) patient through a hospital to illustrate the variety of settings in which health providers work. We close the chapter with explicit suggestions for future research on the hospital environment as it relates to the delivery of effective health care.

THE SOCIOPHYSICAL ENVIRONMENT AND PATIENT CARE

Implicit Models

Scientists argue that the organization and design of the acute-care hospital may undermine the health and well-being of its various user groups in two nonexclusive ways. The unstated assumptions in this literature are that poor design and a cumbersome organization may (1) *directly* affect health by impeding the effective and timely delivery and receipt of health care and (2) *indirectly* affect health by creating a stressful milieu for most, if not all, user groups.

The assumptions underlying the first model are straightforward and derive from human factors and organizational management theories. From the patients' perspective, design features that result in difficult way-finding, for example, can directly hinder their timely *receipt* of health care (Carpman, Grant, & Simmons, 1986; Shumaker & Reizenstein, 1982). Design features such as poor lighting, inappropriate placement of equipment, long distances among related functions, and inadequately designed equipment directly impede the *delivery* of health care.

The premise behind the second assumption derives from a stressor–coping–strain theory. That is, stressors challenge the adaptive abilities of individuals, and long-term exposure to stressors can deplete physical and psychological coping resources (Selye, 1974). An extensive research literature has documented that repeated exposure to aversive stimuli (stressors) can lead to greater physiological arousal, which can contribute to psychological problems such as apathy or poor judgment (Seligman, 1975), to physical disorders such as brain lesions (Walsh & Greenough, 1976), to the etiology of certain diseases such as coronary heart disease (Califano, 1979), and even to death (Selye, 1956).

The basic argument linking the stress model to the hospital environment is that sociophysical settings that impede rather than facilitate goal attainment may be stressors (Stokols, 1979). In addition to the stress resulting from poor design, other sources of stress are factors related to the sociphysical environment that contribute to perceptions of inadequate control, to lack of access to sufficient privacy and restorative settings, to multiple and competing stimuli, and to competing role demands.

Patients are vulnerable at the time of hospitalization. Stressors in the hospital setting can further threaten their health by diverting their physiological and psychological resources from recovery. Also, these relentless and often subtle stressors can impair the ability of the health provider to give efficacious treatment and can thereby have an indirect effect on the patient.

As noted, a number of scientists have explored how the environment of the acute-care hospital may actually thwart the hospital's primary objective, the preservation of life. Investigators have considered the implications of this problem for staff (Moos & Schaeffer, 1987), patients (Shumaker & Reizenstein, 1982), and visitors (Carpman *et al.*, 1986). Specific suggestions for design and organizational interventions have been made, and to a limited degree, some of these have been implemented and evaluated. However, there continues to be a noteworthy lack of empirical investigations in each of the following areas: (1) the *sources* of stress and degree of inefficient design within the hospital environment; (2) the direct and indirect impact of stress and inefficient design on the various hospital user-groups; and (3) the possible limits to intervening effectively in hospital design in order to attenuate stress and increase efficient health care. Thus, in spite of compelling arguments regarding the potentially negative impact of the hospital environment on patients, visitors, and health-care providers, direct tests of the effects of these settings on the different user groups and empirically sophisticated evaluations of the impact of design-based interventions are seriously lacking (see Jenna, 1986b, for a recent exception).

Limits of Empirical Studies of Hospital Settings

Without a clear test of either the inefficient-design (direct) or the environmental-stress (indirect) hypothesis, it is impossible to demonstrate unequivocally their impact on users or to target effective design interventions. A closer examination of these issues suggested to us that the lack of empirical research in this area may be due to two fundamental research problems: (1) the complexity and duration of the hospital design process may mitigate against empirical research, and (2) the efficacy of patient-oriented design changes may be limited.

THE EVOLUTION OF HOSPITAL DESIGN

Although writers have often considered the complex organizational structure of the hospital, they have rarely explored the process by which hospitals are planned, designed, constructed, and evaluated. The building of today's acute-care hospital is embedded in a history of medical folklore and technology, and in the complexity of current regulatory and fiscal policies. This design process helps to explain the decisions and strategies that lead to an apparently nonoptimal and stressful environment. It also helps to identify points at which changes are most feasible in order to balance the needs among the competing user groups, that is, health providers, patients, and visitors.

Historical theories of medicine play a major role in the design of medical facilities. More recently, treatment, medical technology, the changing profiles of patients, and the diversification of health professionals' roles, along with economic and regulatory policies, have combined to influence the organization and design of the acute-care hospital.

Historical Origins of Hospitals

There is an intricate relationship between the theory of medicine and the evolution of hospital design (see Rosenfield, 1969, for a detailed discussion). In the Middle Ages, disease was thought to consist only of observable symptoms. Medicine and healing were linked with the supernatural and treatment areas resembled temples. Interestingly, during this period, the "Hôtel-Dieu," the forerunner of today's hospital, was run by nursing sisters, who were often responsible for the design, construction, and maintenance of the structures (Canter, 1983; Kenny & Canter, 1983).

Careful observation of a disease's course and the information gained from autopsies led to the more sophisticated view of disease as an internal malfunction. Miasma, an early (ca. 1850) theory about the causes of these malfunctions, attributed the spread of infection to gases generated by organic filth. In response to this theory, hospitals were built with separate pavillions, each serving a different function (e.g., a pavilion for baths and one for operating rooms).

As early as 1863, Florence Nightingale questioned the miasma theory and argued that problems in hospitals were due to inadequate standards, which resulted in poor ventilation, the crowding together of large numbers of patients in one building, and a reliance on natural lighting (Seymer, 1954). She established some of the first criteria for the design of a hospital, which led to the development of the Nightingale ward: long narrow rooms with beds perpendicular to the exterior wall and a nursing station from which the nurses could oversee the entire ward. These wards, which persisted into modern times, housed about 30 patients. Although patient surveillance was enhanced at the expense of personal privacy, the areas were less clearly demarcated than today, and there were opportunities for "spillover" care because nurses attended to more than one patient in the same space (Canter, 1983). The trade-offs between privacy and ease of surveillance and specific versus incidental care persist into today's modern hospital and exemplify two recurring issues in hospital design: (1) the gains of one user group may negatively affect another group, and (2) the gains of one user group in one domain (e.g., increased privacy in a single room) may cost that group in another domain (e.g., loss of incidental care).

Louis Pasteur established the relationship between disease and bacteria (and soon thereafter, viruses), giving rise to the germ theory of disease. Wilheim Roentgen (in 1895) and Pierre and Marie Curie (in 1898) published their scientific workers on X-ray and radium emanations, which were developed as both diagnostic and therapeutic tools. External symptoms were determined to result directly from infections of the body. These discoveries spawned the general medical model of disease, in which diagnosis precedes treatment and is a scientific endeavor. The linking of diagnostics to technology, followed by treatment and recovery, broadened the functions of hospitals and the types of spaces required.

In the late 1800s, the trustees of Johns Hopkins University commissioned recommendations for a new hospital in which "light and air" were to be of primary importance (Rosenfield, 1969). It opened in 1890 and comprised a series of one-story pavilions bounded by a one-third-mile-long corridor to the administration and service buildings (see Figure 1). This design set the pattern for American hospitals for the next 25 years.

Several factors eventually converged to make the pavilion-style hospital unwieldy. As hospitals grew, the connecting corridors became longer. Patients needed to be moved more frequently for diagnostic and treatment procedures, and additional supplies and equipment were required for treatment; long corridors made such movements inefficient. Finally, the development of building technology—steel construction,

FIGURE 1. Typical corridor of a hospital shared by medical personnel and equipment, patients, and visitors. (Courtesy of Mimi Levine, The Washington Hospital Center.)

elevators, and air-conditioning—permitted the new upward architecture of hospitals, and hospitals adopted a vertical "monoblock" form (known as the *American hospital*).

Paralleling the changes in building technology were the upgrading of medical education and services, initiated by the Flexner Report after world War I, and the continued gain in medical knowledge and technology. Medicine was becoming more of a science, and physicians gained status and became the purveyors of special knowledge (Canter, 1983). Hospitals went from being places to avoid to becoming the only places in which to avail oneself of the new diagnostic and treatment procedures that were significantly alleviating suffering and prolonging lives.

In response to these changes, the United States went from 3 general hospitals in 1821 to over 6,000 in 1920 (Vogel, 1980). The rapid growth in hospitals continued, especially after 1946, when the Hill-Burton Act led to the federal funding of hospitals in underserved areas. By the 1970s, the number of hospital beds in the United States was increasing at a rate 50% faster than the population (Easterbrook, 1987).

In 1948, the U.S. Public Health Services published *Elements of the General Hospital*, which set the standards of safety, construction, and performance of materials. These standards reinforced the new, vertical, monoblock design, which came to be known as a "matchbox on a

FIGURE 2. An American hospital whose design is known as a "matchbox on a muffin." (Courtesy of the Virginia Hospital Association.)

muffin" (see Figure 2). The vertical element (the matchbox) accommodated patient rooms, and the horizontal element (the muffin) contained the services, such as the X-ray department, the laboratories, and the clinics. Although this division of functions has changed over the years, the basic design persists.

A number of factors continue to influence the interior layout of hospitals. After World War II, the Nuffield Provincial Hospitals Trust (1955) in England established a multidisciplinary team of research workers to study hospital function and design. By systematically documenting the activities that occur in hospitals and their implications for the staff, this team developed annotated designs for specific wards. One result was the Riggs ward, adopted by most hospitals in the United States since the 1940s, which includes smaller wards with nursing stations placed within access of both equipment and patients. Beds are turned parallel to the outside wall, anticipating the development of the sophisticated Springfield wall which conveniently places behind the patients' beds the support systems for resuscitation and ventilation (see Figure 3).

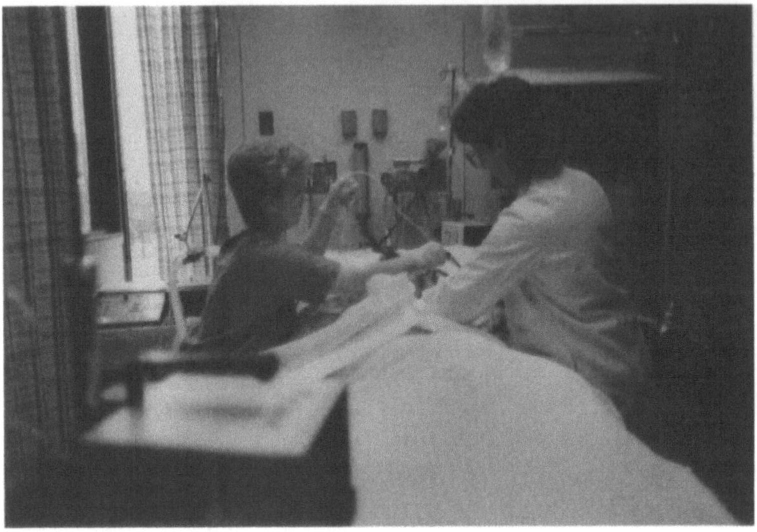

FIGURE 3. A patient's room, showing the "Springfield wall" that supports the treatment equipment being used by nurses. (Courtesy of The Washington Hospital Center.)

Traditionally, the hospital departments were defined by the type of intervention required for the patient. There were a medical ward, a medical team, and a medical intensive-care unit, or a surgical ward, a surgical team, and a surgical intensive-care unit (Canter, 1983). As treatments became more complex, departmentalization occurred with classifications such as laboratories, radiology, medical records, outpatient, rehabilitation, coronary-care unit, and pediatrics (Clibbon & Sachs, 1970). Designing hospitals to accommodate these departments, however, has become increasingly difficult.

Researchers have argued for flexible hospital designs that respond to changing needs. For example, Weeks and Best (1970) suggested that (1) hospitals should never be complete, and (2) "an envelope fitting particular functions rather loosely should be standardized for use as widely as possible" (p. 264). Few people would argue with the need for flexible designs; the problem is how to implement such a strategy while making space attractive and usable and still controlling costs. Breakout walls for new modules have been used (Bobrow & Van Gelder, 1980), as well as interstitial spaces that allow major changes without a serious interruption in hospital functions (Weeks, 1980). However, flexibility in design does not address how hospitals should be optimally organized. There continues to be a struggle over whether it is better to design a hospital in which the administrative units or departments represent a clustering of similar *services*, regardless of who will receive them, or whether the *people* who provide the services should be clustered together. In recent years, doctors have been pressing for increasing services (e.g., X-ray machines) within their departments, rather than relying on more centralized services.

Current Factors Influencing Hospital Design

Medical Technology. With the application of technology to medicine, hospitals are continually reshaped and reorganized to meet the requirements of new equipment, specialized personnel, and new interventions. The changes in technology associated with diagnosis and treatment continue to be major and to directly influence patient care and hospital design. Most equipment is expensive and demands large spaces and the specialized skills of both health professionals and technicians. Although some argue that medical technology dehumanizes patients (Leventhal, Nerenz, & Leventhal, 1982), many of the more recent advances (e.g., laser cauterizations, computed axial tomography, and magnetic resonance imagers) allow minimal, noninvasive treatment and diagnosis and promote an increase in outpatient care. It may be that the most negative aspects of increased technology are not patient dehumanization but the *overcrowding* of the already limited hospital space, an increase in the inequitable utilization of technology (Faden, 1986), and a "hardening" of the primary work spaces of health professionals (Sommer, 1974). The current instrument panel at the nurses' station in an intensive care unit (ICU; see Figure 4) and the amount of equipment required in the operating room for open heart surgery (see Figure 5) exemplify the space demands and the complexity associated with medical technology.

Most researchers agree that technology will continue to have a major impact on the design of medical settings. Panther (1984) predicted that, by the year 2015, there will be a greater use of artificial and transplant organs, of less invasive surgical techniques, of new imaging techniques, and of more efficient emergency transportation. With the exception of electronic communication and medical record-keeping, these changes will continue to exert pressure for flexibility and more space.

The Changing Patient Profile. At the turn of this century, the leading causes of death in the United States were acute diseases (e.g., influenza and pneumonia) that are now largely controlled by medical interventions. Chronic illnesses have emerged as America's primary health problem. The burden of illness, death, and disability has shifted to heart

FIGURE 4. A nurse monitoring a number of patients at an instrument panel in an intensive-care unit. (Courtesy of Steve Spartana, The Washington Hospital Center.)

disease, cancer, and stroke, and to accidents and violence. Whereas only 1% of the population dies from infectious diseases before the age of 75 (Califano, 1979), 46% of the population is limited by chronic disease.

Chronic illnesses tend to ebb and flow over a long period of time, and a "cure" is often impossible. A health-care system that is structured for acute care and end-stage intervention is not the best model for treating chronic disease, and more imaginative approaches to prevention and health care are needed. Hospital organization and design will need to be more responsive to these issues. Against the backdrop of a change from acute to chronic disesases, as well as an aging patient population, there have been major public-policy initiatives that dramatically affect the environment in which acute-care hospitals[1] are designed and function.

Public Policy and the Cost of Health Care. The United States spends per capita more money than any other country on health care; yet, we do not have the longest life expectancy or the lowest infant mortality rate. In the early 1900s, the philosophy of health care was that individuals

[1]Although the majority of patients treated in this country are suffering from chronic diseases, people experience *episodic* crises requiring hospitalization. Hospitals are therefore still referred to as *acute-care hospitals*.

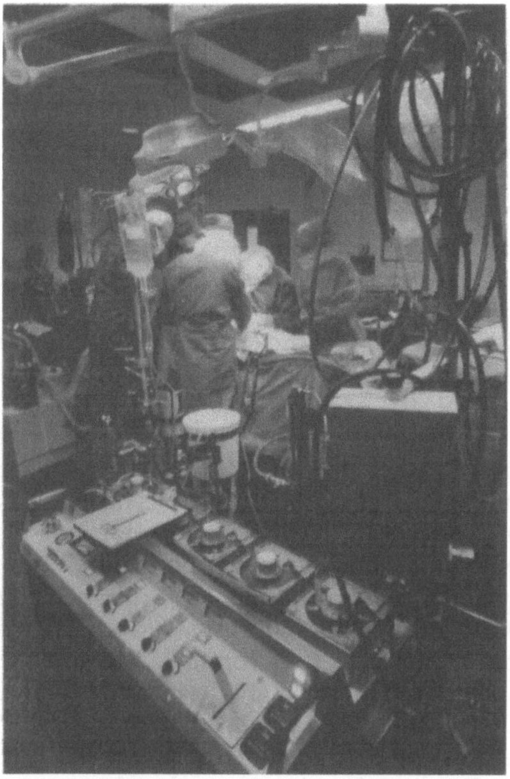

FIGURE 5. A surgical theater showing the complicated equipment (heart–lung machine in foreground) and the number of personnel required during cardiac surgery. (Courtesy of Ira Wexler, The Washington Hospital Center.)

were entitled to only those services they could afford (Letourneau, 1965). Poor people were treated in almshouses, which evolved into what we recognize as hospitals, and wealthy people were treated in hotels or in their homes. A decision to build a hospital that would care primarily for the poor was based on the finances available rather than on the needs of the community. Philanthropists provided some contributions for those who could not afford their own care, but these donations did not always meet the needs of the public. During the 1930s, Franklin D. Roosevelt argued that the health of citizens was a *responsibility* of the federal government, and that every citizen had a right to benefit from developments in medical science, whether or not they could personally afford them. This stance, and the concomitant funding, altered the status of public hospitals in the United States.

The percentage of the cost of health care paid for by the government has gone from 22% in 1950 to 40% in 1985. By 1985, the total cost of health care in the United States was in excess of $400 billion (U.S. Department of Health and Human Services—USDHHS, 1986). (It should be noted, however, that a relatively small proportion of this money goes *directly* to the design of *public* hospitals.) Beginning in the late 1970s, measures for "cost containment" of treatment emerged. The latest measure is a prospective fee structure based on diagnostic related groups (DRGs; Fetter, Thompson, & Mills, 1976; Thompson & Golden, 1975). Patients in a DRG category are assumed to represent a homogeneous group that is similar in terms of the clinical description of the illness, the expected length of the hospital stay for treatment, the types and amount of resources consumed, and, ultimately, the cost of the treatment (Ziegenfus, 1985). Reimbursement is based on a set rate *determined in advance of treatment,* rather than the traditional fee for service.

As a result of DRGs, since the early 1980s hospitals have gone from overcrowded to underused, having vacancy rates as high as 69%. In addition, hospitalization may no longer be required in order to trigger third-party payments (by private medical-insurance companies). This has contributed to the increase in outpatient services (e.g., "surgical centers"), where minor surgery is conducted in small clinics. Medical insurance companies are increasingly designating the hospital or treatment facility that is the *preferred provider organization* (PPO), where a specific service must be performed because it is the least expensive in the community. Patients choosing to go elsewhere must pay the additional fees.

The long-term positive and negative impact of these changes has yet to be evaluated. Smaller hospitals are turning to education and prevention programs to attract patients who are suffering from a myriad of chronic illnesses, for which the most potent intervention may be lifestyle changes. Depending on the sophistication and appropriateness of these approaches, they may fill an important gap in medical treatment. In addition, some researchers argue that the larger spaces made available in hospitals because of vacancy rates may improve patient care by providing more private rooms and a greater nurse to patient ratio. However, with more nurses leaving than entering the profession, this expectation may not be realized.

DRGs represent an attempt to make hospitals function in the free market economy. Patient satisfaction with the hospitals and their care should therefore become more important. The full effect of DRGs is not yet known, and well-designed assessments of their impact are needed. In addition to an evaluation of patients' satisfaction with care and of total cost changes, it is critical that changes in the allocation of space within existing hospitals, as well as the design strategies for the few new facilities, be carefully examined.

The DRG-based method of reimbursement is related to another important concept in health care: quality assurance. The "Principle of Quality Assurance Standard" (Grimeldi & Micheletti, 1983) which a hospital must meet in order to achieve accreditation, states that "the hospital should demonstrate a consistent endeavor to deliver patient care that is optimal within available resources and consistent with achievable goals. A major component in the application of this principle is the operation of a quality assurance program" (p. 23). Implicit in the switch to the prospective fee structure (DRGs) is the assumption that quality can be maintained by improving the effectiveness and the efficiency of the service delivery system. These new health policy measures, together with changes in the profile of patients and in medical technology, exert a major influence on the design of today's acute-care hospital.

The Hospital Design Process

It is beyond the scope of this chapter to provide a detailed description of the multiple decision-making relationships that exist in the planning, design, construction, and evaluation of an acute-care hospital. Our intent is to provide simply a flavor of the complexity of this process in order to identify the points at which the various user groups may be able to influence the final product.

Constituents. For a variety of reasons, the number of groups that participate in the design of a modern health-care facility has escalated substantially in the past 30–40 years. Medical care regulations, construction and design codes, and the rapid growth of medical technology all influence hospital design. The problems to be solved by a building, because of the broader range of patients and chronic diseases that are treated, have further increased the complexity of the hospital milieu. Finally, the "client" for whom the modern hospital is built is no longer a single, private patron, but a constellation of public and private groups. In addition to the primary users (e.g., health professionals, support staff, and patients), an additional range of interested parties is involved in the design of the modern health-care facility.

In Figure 6, we present the three broad groups of constituents that participate in hospital design. To varying degrees, representatives from each of these groups participate in the planning of a hospital. In addition to the constraints of as many as 200 or more building codes and rules that apply to hospital design (Price, 1984), planners may have to con-

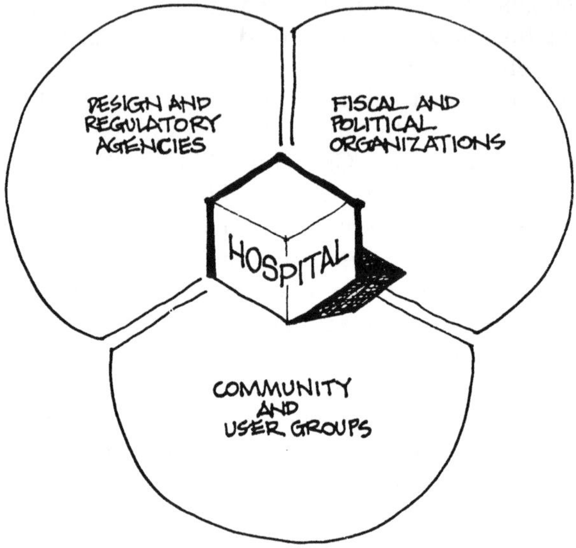

FIGURE 6. Constituents in the design process. (Drawing courtesy of Michael Brill.)

tend with groups like the local businesspeople pushing for a large facility to symbolize the area's economic viability, construction trade unions fighting against prefabricated units and other labor-saving building techniques, neighborhood organizations fighting for and against locations proposed for the hospital, and banks and other fiscal agencies oscillating on financial commitments. At the center of these constituents is the architectural team and its various consultant members, who must sort through competing needs and seemingly unreconcilable goals (see Figure 7).

Design and Building Phases. In Tables 1 and 2, we present the major phases in the hospital building process, as well as the degree to which the various constituents participate in each phase. The architectural team that will eventually orchestrate the design and construction of the hospital does not enter the process until after some strategic planning has occurred within the community. The degree of participation for each constituency varies considerably at each phase. In addition, the influence of the members of a particular constituency depends on four major factors: (1) their interest in and knowledge of the process; (2) their status within the community and their professional field; (3) their organizational strength; and (4) their long-term proximity to the planning and design process. User groups are at a disadvantage on all four of these

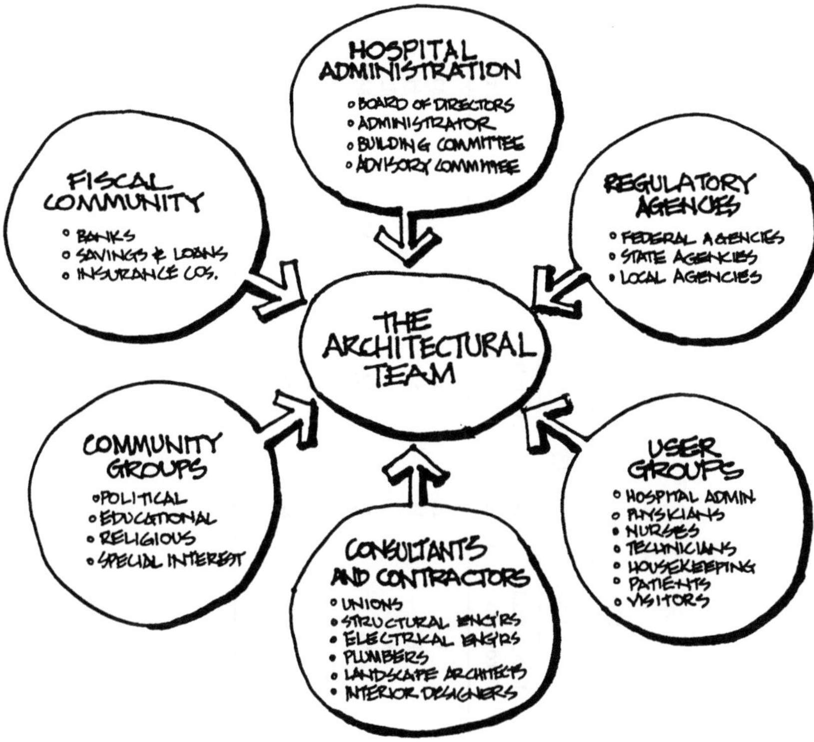

FIGURE 7. Role of constituents in the design process. (Drawing courtesy of Michael Brill.)

factors (i.e., knowledge, status, strength, and proximity) and therefore have a difficult time participating effectively.

Improving the Design Process. In order to resolve competing needs and to facilitate the hospital building process, the architectural team has a number of techniques at its disposal. Kirk, Eichinger, and Spreckelmeyer (1984) suggested three strategies that the team can use to improve its decision making. The first strategy, *multiobjective analysis,* is used to delineate the stated goals of the building and the design alternatives and then to select the design alternative that most closely meets these goals. The second technique, *life-cycle costing,* assesses competitive design alternatives with respect to the life of the building (e.g., maintenance, energy, and staffing costs). This technique provides investors with an estimate of the return-on-investment by evaluating the utility of the options (e.g., high-rise versus low-rise buildings, initial costs versus future costs). Finally, *postoccupancy evaluation* (POE) is used to study how the building

TABLE 1. Participation of Constituents in Phases of Hospital Building Process

Constituents	Major phases in building process								
	Strategic planning	Regulatory planning	Financial planning	Functional program	Master site plan	Architect. design	Construction	Opening	POE
Architectural team		M	M	H	H	H	H	M	L
Hospital administration	H[a]	H	H	H	H	M	L	H	H
Fiscal community	M	L	H	M				L	
Regulatory agencies		H		M	L	M	H	M	
Community groups	H				M			M	
User groups	H		L	H	M	H		H	H
Consultants and sub-contractors		L	M	H	M	H	H	M	L

[a]Denotes the level of participation (H = high; M = medium; and L = low)

TABLE 2. Description of Building Phases

Strategic planning
Perception of need for building program; organizing for planning feasibility assessments; data gathering and forecasting for community.
Regulatory planning
Developing relationships with Health Systems Agency (HSA) and State Health Planning and Development Agency (SHPDA); obtaining Certificate of Need (CON) from state agency.
Financial planning
Review methods of financing and scheduling process.
Functional programs
Develop framework for defining the problem to be solved by the design; conduct physical and functional evaluations; prepare workload projections; develop spacing program; determine equipment needs.
Master site plan
Prepare plan for site selection and evaluation; traffic planning.
Architectural design
Design physical form which is interpretation of requirements of various operational constituents of the facility in the Functional Program; prepare block, schematic, and design development documents.
Construction
Obtain bids; define construction manager role; break ground.
Opening
Building occupied and reprogramming of the designed space by user groups begins.
Postoccupancy evaluation (POE)
Evaluation of the performance of the building and useful behavior.

performs and how users behave in it (see Wener, this volume). Though there are a number of inherent problems in conducting a POE, this particular strategy holds a great deal of promise for improving future hospital design and its products. The firms of Caudill, Rowlett, and Scott and of Kaplan, McLaughlin, and Diaz (McLaughlin, 1975) are notable for developing programming strategies that incorporate POE data from the firms' previous projects. There is also a group of POEs on Veterans Administration Medical Centers (Kirk *et al.*, 1984). In addition, Kenny and Canter (1981) evaluated British hospitals, using a complicated methodology called *facet structure*.

In spite of these exceptions, well-designed hospital POEs are infrequent (Bobrow & Van Gelder, 1980). In order to change this situation, the problems underlying hospital POEs must be addressed. First, because of the multiple users and their competing needs, it is difficult to develop clear goal definitions for the POE. The conflict between patients' desire for privacy and health professionals' need for easy patient surveillance exemplifies this issue.

The modern hospital is an evolving system rather than a finite architectural work; thus, the *timing* of the POE is important. As noted, planners argue for hospital buildings that have structural stability, while allowing for the flexibility to meet changing needs. A POE conducted at a specific time will not capture the interplay between stability and flexibility. In order for POEs to influence future hospital design, multiple assessments must be made at different times.

The cost of hospital POEs is a major obstacle to better research in this area. The same complexity that underlies the design process (see Figure 7) exacerbates the difficulties in the evaluation process. Many of the constituencies involved in the building of a specific hospital are not likely to benefit directly from the POE. Therefore, it is not clear which constituent is primarily responsible for the funding of an acute-care hospital POE. The potential benefit of a centralized data bank of hospital POEs has yet to be recognized. Such a resource could include data on measures taken at different points and could emphasize the needs of the various user groups. Thus, designers would not be dependent on a single POE in evaluating the potential effects of various design alternatives. Although federal or state agencies appear to be the most appropriate sources of funds for such a data bank, current fiscal trends in health care make such funding unlikely in the near future. Thus, it becomes even more critical for interested groups to provide well-considered input for the design of medical facilities, so that these groups represent the needs of the multiple user groups.

Humanizing Hospital Space

Several recent trends provide hope that hospital space can be humanized. Studies conducted on the use of color and lighting (Weeks, 1970) suggest a growing sensitivity to the ambient environment and a willingness to go beyond a superficial approach to interior design, and to implement positive changes. Bobrow and Van Gelder (1980) argued that there is a clear trend toward the use of light, materials, and color to "enliven" the human qualities of the environment. And although there has been an increase in windowless units in health-care environments because windows are expensive, research demonstrating the importance of access to the natural environment for patients and staff (Kaplan & Kaplan, 1982; Ulrich, 1984; Verderber, 1982) may reverse this trend.

Perhaps the most promising sign that hospitals can become more humane settings is exemplified by the Planetree experiment (Jenna, 1986a,b). Planetree is a model hospital unit oriented to the user groups and designed to bring together humanistic medicine and technology. It was designed by the late Roslyn Lindheim, an architect who attempted to go beyond cosmetic changes (e.g., the huge hospital graphics of the 1970s) and to implement, *through design,* organizational changes that would fundamentally alter the setting (Lindheim, Gloser, & Coffin, 1972). To this end, the ward includes a communal kitchen for staff, patients, and visitors; the use of attractive, "homelike" furniture and designs; a lounge adjacent to the nursing station for staff, patients, and visitors; and an open, low-fronted nursing station that promotes easy access for patients and visitors. This innovative design has produced mixed responses from the staff. It is currently being rigorously evaluated, however, and its impact on the health and well-being of both the patients and the staff are among the outcome criteria.

Planetree is not representative of hospital design, but it does demonstrate the potential for positive changes. As noted by Bobrow and Van Gelder (1980), however, we have a long way to go:

> For all the absolute knowledge and power that hospitals symbolize and shelter, for all the organizational, technology, and mechanical modes that have been developed these past 30 years to expedite the delivery of health care services—for all of this, the average health care environment has yet to "touch" people. (p. 108)

The complexity and duration of the design process may provide a partial explanation of why the hospital milieu does not "touch people." Perhaps another problem in this field is the group traditionally targeted for design-based interventions: the patients.

CHOOSING THE USER GROUP TO OPTIMIZE CHANGE

Patient-Based Design Interventions

Most design-based recommendations for hospitals have focused on the patients. Despite our best intentions, we may be unable to *significantly* reduce the stress-inducing aspects of the hospital environment for patients.[2] Many of the factors that influence people's negative associations with hospitals, like the pain and discomfort that occur with disease

[2]The new labor–delivery–recovery suites (LDRs), outfitted with all the amenities of a hotel while maintaining the technology of a delivery room, represent one exception. As noted by Roland Binker, AIA, (personal communication, 1987), one out of four hospitals is updating its LDR suites as a marketing tool for the hospital.

and with diagnostic and treatment processes, are unalterable. Evaluations of the efficacy of design-oriented interventions for the patient show that they may not be potent enough to contribute significantly to improved recovery. In addition, even under the most optimal design intervention circumstances, designs that meet the needs and goals of *all* hospital user groups may not exist; compromises are inevitable in the design of complex institutions.

We are not suggesting that the attempts to humanize hospitals for patients should not continue. However, these attempts should be made with a recognition of the cost of intervention versus the obtained benefits. Relatively low-cost design changes have been suggested in the literature (Shumaker & Reizenstein, 1982; Sommer & DeWar, 1963) and remain to be evaluated. Focusing more expensive design changes directly on patients may not be the most cost-effective way to improve this environment for *all* user groups. It may be worthwhile to turn the current emphasis in this literature away from patients and to health-care providers, a group which has been underrepresented with respect to suggestions for *design-based* changes.

Health-Provider-Based Design Interventions

The average individual is hospitalized very few times during his or her lifetime, and the current average length of stay in acute-care hospitals is 6.6 days (USDHSS, 1986). Thus, as negative as the setting may be for the patients, its impact is short-lived. In contrast, the acute-care hospital is the primary workplace for most health professionals in the United States. Although they may not be ill and therefore as immediately vulnerable to the setting as are patients, for health professionals the hospital remains a stressful environment. Recognizing the potential sources of design-related stress for health professionals and identifying methods for minimizing or eliminating the stressors may enhance health benefits for both the staff and the patients.

Nurses as the Primary User Group. To varying degrees, the hospital environment is stressful for all health-care providers. Rather than using examples across different groups, we focus our discussion on a single user group, nurses, with the expectation that readers will extrapolate the issues to other health-care professionals. There are several reasons that nurses are an ideal target for hospital-design-related strategies. First, nurses fit the profile of a user group that is vulnerable to the effects of stress, that is, persons with low control and high responsibility. There is an extensive literature on occupational stress that documents the negative effects of these characteristics in work settings (House, 1981; Karacek, 1979; Stokols, 1986).

Nurses work in a variety of physical settings, allowing us to track the relationship between patients and nurses in different behavior settings, as well as the degree of fit between nurses and their environment. Also, there is a body of research on hospital design and nurses' satisfaction with the environment (Jaco, 1979); historically, their professional roles have been tied to the design of hospitals (Canter, 1983; Seymer, 1954).

Nurses have the most contact with patients, often know the patients and their conditions better than anyone else in the hospital (Theis, 1986), and, from the patients' perspective, are (along with other patients) the most important hospital-user group (Kenny & Canter, 1981). Nurses are the primary patient advocates (Theis, 1986; Jenna, 1986a). According to the most recent American Nurses' Association code, "respect for human dignity and support of the patients' rights to self-determination are an integral part of nursing practice" (Theis, 1986, p. 1223). Because of their advocacy roles, their more holistic views of patients, and the fact that nurses often carry out and experience the effects of medical decisions "at the point of impact" (Theis, 1986), the aspects of the hospital environment that impact on nurses have major implications for patients' welfare.

HOSPITAL DESIGN AND NURSING CARE

The Sociophysical Environment and Nurses

The small body of empirical research that addresses the effect of settings on health professionals focuses on two issues that are consistently separated in the research: (1) the effect of the social and organizational climate on staff satisfaction and turnover and (2) the effect of the physical design and layout on staff satisfaction.

Organizational Climate. There is extensive research on the negative effects of the hospital milieu on nurses. Interest in this issue has escalated in recent years as nursing shortages have become a crisis. Hospitals are vulnerable to nursing shortages, having unfilled position rates of over 20%, which in some cases lead to double shifts (Curran, 1987).

Several aspects of the nursing role are associated with stress. Nurses have a high level of responsibility with limited control. They receive multiple and competing demands from hospital administrators, other health professionals, and patients and visitors. They are often placed in a mediating role between physicians and patients, or between patients and technicians, with limited power to resolve the conflicts among these groups. They remain in low-status roles within the health professionals' hierarchy, and this low status is reflected in both their low salaries and their limited authority.

The research focus has been on role ambiguity, role conflict, status differentials among health professionals, job demand versus job discretion, interpersonal support among the staff, and the supervisor's relationship to the staff (Cartwright, 1979; Moos & Schaefer, 1987). For example, Gray-Toft and Anderson (1985) found that organizational climate, supervisory style, and work-group relations were primary determinants of stress for nurses and a direct cause of job dissatisfaction and absenteeism. In an elegant path analysis, Decker (1985) determined that, although job satisfaction was directly related to the social and organizational work environment, other factors external to the setting (e.g., education and income) directly affected nurses' propensity to leave a particular hospital or even the profession.

There is little doubt that components of the nursing role and the hospital's social environment contribute to stress and the high rate of nurse burnout and attrition. External factors such as the broadening role opportunities for women, coupled with the comparatively low salary and status of nurses, also contribute to the high drop-out rates in this profession, as well as to the significant decrease in new students entering formal nursing programs. Thus, steps to alleviate this national crisis in nursing care must be taken that attend to both the social milieu in the hospital and the external status-related issues.

As overwhelming as the organizational climate may be in contributing to nurses' stress and burnout, it is inextricably embedded within and influenced by the physical setting. Limited research has addressed the relationship between the physical environment and nurses' well-being.

Physical Setting. Hospital settings vary widely on a number of dimensions (see Table 3). The *pattern* of these dimensions within a particular setting may directly influence nurses' satisfaction and their delivery of health care or may indirectly affect nurses by the impact of the design on the organizational structure. Studies have not explicitly identified the dimensions of a particular hospital setting and have not examined the multiple effects of altering these dimensions on nurses' satisfaction or delivery of care. There are a few studies, however, in which broad aspects of a setting have been manipulated, and nurses' satisfaction has been assessed[3] (Jaco, 1979; Jenna, 1986b; Kenny & Canter, 1981; Winkel & Holahan, 1985). Data from these studies are promising in that they

[3]There is an extensive literature on user-group satisfaction with the organizational and physical environments of the hospital. However, satisfaction is not the best direct test of either the direct (inefficient-design) or the indirect (environmental-stress) hypothesis.

Table 3. Dimensions of Settings Used by Myocardial Infarction Patients That Influence Patterns of Use and Potential for Change

	Settings							
Dimensions	Medical emergency helicopter	Emergency room	Diagnostics	Operating room	SICU	Patient's room	Corridors	Nurses' station
Health providers								
Numbers	2 to 3	variable	3 to 4	5 to 8	3 to 4	1 to 2 at time	variable	variable 1 to 10
Roles	2 to 3	3 to 4	3 to 4	5 to 8	2 to 3	3 to 4	all roles	all roles
Interdependency	high	high	high	high	moderate	low	low	low
Hierarchy	high	dur.emer.,hi.	high	high	moderate	low	low	low
Setting controller	clear	variable	clear	clear	clear	ambiguous	ambiguous	ambiguous
Patients								
Number	1	variable	1	1	variable	var.; 1 to 2	variable	usu. low
Medical status	critical	critical	var.,usu.stab.	variable	crit.to stab.	var.,usu.stab.	stable	stable
Designated space use								
Number of functions	1	1	1	1	1	multiple	multiple	multiple
Complementary versus antagonistic features of functions	comple.	dur. emer. compl; between emer. antagonistic	comple.	comple.	comple.	moderately antagonistic	antag.	highly antagonstic
Clarity versus ambiguity of space use	clear	dur.emer. clear; bet. ambig.	clear	clear	clear	moderate ambiguity	ambig.	high ambiguity
Equipment								
Size		variable	variable	variable	variable	moderate	variable	variable
Amount		variable	high	high	high	low	high	moderate
Complexity	high	high	high	high	high	low	low	low to mod.
Fixed versus flexible function	fixed	both	fixed	fixed	fixed	both	flex.	flex.
Fixed versus flexible placement	fixed	both; mostly flexible	fixed	fixed	fixed	both	flex.	mod. flex.

demonstrate a clear effect of environmental manipulation on satisfaction.

In research that lends support to the environmental-stress (indirect) hypothesis, Trites, Dalbraith, Sturdavant, and Leckwart (1970) studied the effect of three different ward designs on health care. They found that the radial design—shorter distances between the nurses' station and patients' rooms—led to the nurses' spending more time with patients. In the often-cited Jaco (1979) study contrasting two ward designs (radial and corridor), there was an interaction between design preference and the number of patients on the ward, nurses' preferring a radial design with high occupancy rates over the corridor style with low occupancy rates. This phenomenon underscores the inextricable relationship between social and physical environments, as well as the complexity of the research in this field, in addition highlighting the need for explicit models of the setting's dimensions that may affect its users.

Investigations of the influence of design on organizational structure within *health facilities* are limited. Coser (1958) evaluated communication patterns between physicians and nurses in different settings in the hospital. He found that communication in the surgical theater was extremely formal, whereas in the nursing station and the hallways, nurses felt freer to venture opinions about patient care. However, the literature on occupational stress and work settings clearly demonstrates a relationship between work-space arrangement and such factors as worker interaction patterns, perceived privacy and control, and job efficiency and productivity (House, 1981; Sundstrom, 1987). Components of workplace environments that are stressful, such as noise, crowding, HVAC, and lighting, have also been investigated, and clear influences of these and other ambient stressors on job satisfaction and health have been demonstrated (Evans, Cohen, Stokols, & Krantz, 1986). Health settings are optimal for testing the models and hypotheses regarding work-setting stress that have been developed and tested in other institutions.

In addition to research on other work settings, there are some data on how the hospital environment affects patients that may provide insights into the impact of hospital settings on health professionals. In order to extrapolate from these studies, however, it is important to understand the different ways in which nurses and patients interact with hospital settings.

Nurses and Patients in Hospital Settings

There are several key ways in which nurses and patients differ in their interactions with the hospital environment, and each of these has implications for design-based interventions. A review of these differences in setting use highlights the ways in which the goals and needs of user groups may be in conflict. Even though nurses are identified as patient advocates, the diverse circumstances in a hospital often put these two groups in competition for scarce space and resources. Equitable compromises may not be possible, and a system that rates the value of certain design decisions for the various users is needed.

Patients and nurses differ in their *familiarity* with the hospital environment. Depending on their length of residency within a given institution, nurses are familiar with the hospital layout, with the meaning of medical terms and signage systems, and with how to access and use various facilities. In sharp contrast, for most patients the hospital is a strange environment, where the descriptors for various spaces sound like a foreign language, and wayfinding is difficult. The various types of equipment used daily by the medical staff are unfamiliar and often frightening to the average patient. The impact of this strange environment on patients has been considered elsewhere (Carpman *et al.*, 1986).

The nurses' familiarity with the medical environment may also affect them and their patients negatively. For example, nurses may become insensitive to how the setting and equipment affect others. In addition, familiarity may cause nurses to adapt to the negative components of settings, and over time, it may become increasingly difficult for them to identify irritants within the hospital environment. For example, they may adjust to poorly organized equipment or convoluted pathways between hospital functions without being aware of the extra time it takes them to perform a specific job or the amount of unnecessary walking they are forced to do on a typical shift.

Nurses are the active providers of care, whereas patients are often placed in the passive, recipient role. Patients are more vulnerable in that they are in the midst of a health crisis, whereas the nurses' role is crisis intervention. As providers, nurses have more responsibility in hospitals, and their behavior can have a major impact on others.

In terms of environmental control, patients often have a specific space (their room) that is identified as theirs during their stay. A comparable space that "belongs" to nurses does not exist. Even the nurses' station (see Figure 8) is used by all hospital user groups, and because of the competition for space within hospitals, adequate nurses' lounges are a rare phenomenon. Thus, in terms of environmental control, patients may actually be in a better position than nurses, and design interventions that enhance patient space at the cost of nurses' space may be misguided. In the Planetree experiment discussed earlier (Jenna, 1986b), the interventions included a merging of patients' and nurses' spaces.

FIGURE 8. A nursing station showing the number of health personnel occupying the same space while involved in different activities (patients' medical charts are on the carousel in the foreground). (Courtesy of The Washington Hospital Center.)

Such an intervention forces patients and nurses to stay "in role." Yet, one could speculate that, as nurses are responsible for the well-being of others and are often in a crisis intervention role, they may need to periodically escape role demands and have access to private space that they exclusively control.

A third difference between patients' and nurses' interactions with the hospital environment is their *occupancy patterns*. Most patients are in the hospital for short periods of time (less than 7 days) and only two to three times throughout their lives. In contrast, the hospital is the nurses' workplace, and they are there for the tenure of their career. The intensity of the stay is also varied. Whereas patients remain in the setting 24 hours a day, nurses are free to leave. The differential effect of intensity and length of stay on patients and nurses is not known. The degree to which access to nonhospital settings compensates for the stress of the medical setting has not been investigated, although some theorists (e.g., Price, 1982) have speculated on this issue in the context of other work-versus-home settings.

A better understanding of the interdependency of the nurses' and the patients' roles may provide research guidelines to space allocation decisions. To that end, we present two models of how the sociophysical environment may affect nurses and patients.

Modeling the Effects of the Sociophysical Environment on the Delivery of Health Care

In a review of the design and behavior research on health providers and patients, two nonexclusive models emerge. These models provide a framework for the interrelationship of the sociophysical environment, and nurses and patients.

Direct-Effect Model. The first and simplest model assumes that the physical layout and organization of a workspace and the layout of related settings within an institution directly influence employee efficiency and productivity. Even within this apparently simple model, however, there are important trade-offs and complications that have implications for the well-being of nurses and patients. First, depending on the number of *functions* within a given setting, design efficiency for one function may decrease efficiency in another (see Table 3). Similarly, designs may not be equally efficient for all user groups occupying a particular setting (e.g., a nurses' station; see Figure 8). Thus, depending on the dimensions of a particular setting, a complex weighting scheme of goals and functions is needed that compares the relative efficiency of different design arrangements. The more specific the goal (e.g., a complex surgical procedure), and the more interdependent the users with respect to achieving that goal, the simpler the model. In contrast, spaces with multiple goals and independent users (e.g., a hospital corridor) are difficult to model in terms of efficiency.

One problem with the direct-effects (inefficient-design) model is the underlying assumption that efficiency benefits all user groups. There are a number of design strategies that can enhance efficient patient care at some cost to the patient. The early Nightingale wards, for example, allowed nurses to survey many patients at one time; patients' privacy, however, was limited. More recently, having windows in "private" rooms or leaving patients' room doors ajar achieves the same mixed benefits for nurses and patients. The advent of monitoring techniques (e.g., telemetry, such as EKGs being visible in the nursing station) may attenuate this problem, but nurses still prefer to *see* the patient.

Indirect-Effect Model. The second model of the sociophysical environment derives from the stress–coping–strain theory. Aspects of the designed environment may stress health professionals and may thus affect their ability to deliver effective health care. There are two broad categories of stressors in medical settings: (1) those that derive from inefficient design and that, by impeding goal-attainment, stress the health professionals and (2) ambient stressors, such as noise, inadequate privacy or control, uncomfortable temperature oscillations, and crowding.[4]

There are several ways in which a stressful environment can effect a degradation in the provision of health care. First, exposure to stress impairs an individual's ability to attend to stimuli within the environment and leads to fatigue in the short run and to burnout in the long run. Thus, in stressful settings, health-care providers may not respond to important cues regarding patient needs, or they may be less able to concentrate during the delivery of care and may be more likely to make errors when following doctors' orders.

When exposed to a stressful environment over time, individuals may become emotionally labile and physically ill. The relationship between stress and absenteeism in nurses has been documented (Pardes, 1982). In addition, high vacancy rates and staff turnover have been attributed in part to stress (Rigger, 1985). Lack of continuity in care and inadequate staffing reduce the quality (e.g., length of experience) and the amount of care of patients (Pardes, 1982).

As in the direct-effects model, there are a number of complexities inherent in the indirect model of the sociophysical environment and patient care. The complexity of the settings and the competing needs of the users are issues that must always be addressed in research in this area.

THE INTERFACE BETWEEN NURSES AND SPECIFIC SETTINGS

To explore the complexity of medical settings, we track a typical myocardial infarction patient through an acute care hospital, using the two models and the setting dimensions outlined in Table 1.

Helicopter Rescue

Initially, our "typical" patient experiences a myocardial infarction (heart attack) on the highway and is evacuated by a medical emergency helicopter, designed to efficiently treat medical emergencies (see Figure 9). The space is well defined and inflexible, and the equipment is well-secured. An emergency helicopter is often staffed by a nurse and an assistant trained in emergency medicine. Only one or two people are

[4]We should note that the difference between the first category of stressors and the ones in the indirect-effects model is that the direct-effects model does not assume a mediating role of stress on patient outcomes.

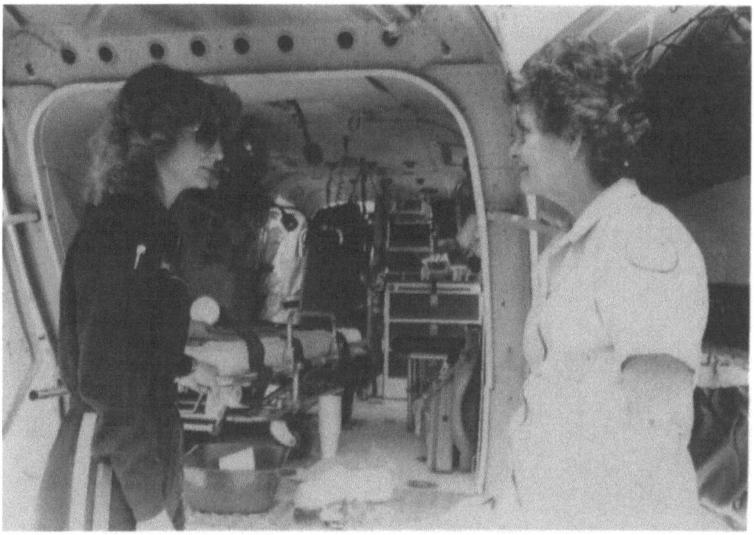

FIGURE 9. A helicopter designed to evacuate patients experiencing severe trauma to a designated hospital. (Courtesy of The Washington Hospital Center.)

working on the patient, who is in extreme distress and is often unconscious. While the patient is in the helicopter, the work is continuous, often involving a tense life-and-death struggle. In this setting, the criteria for design efficiency are unambiguous, and the goals of all the users are complementary and highly interdependent. Given the condition of the patient and the intensity of the work, where seconds count, efficient design is salient to all users. Thus, identification of the targets for change is straightforward.

THE EMERGENCY ROOM

The helicopter delivers the patient to a shock trauma unit or an emergency room of the hospital, where the patient is stabilized by a medical team using highly technical equipment (Figure 10). This setting is designed to provide immediate and continuous treatment. In contrast to the medical emergency helicopter, however, this space is very flexible; individual areas may be demarcated only by curtains. The equipment is on wheels so that it can be quickly taken to where it is needed. The nurses work closely with the physicians in providing multiple interventions. The roles are highly interdependent, and the nurses follow the doctors' immediate verbal orders. The patient is usually in great distress

FIGURE 10. A medical team treating a patient in an emergency room. (Courtesy of Cathaleen Curtiss, *The Washington Times*, 1987. Reprinted with permission.)

and is not able to do much on his or her own behalf. The interdependency of the users, the shared goals, and the clear hierarchy of the roles in the emergency room make the performance criteria comparatively clear *during an emergency*. However, the intervals between emergencies may be stressful, in that the space and the equipment are designated for unpredictable, episodic events. During these intervals, role hierarchies, goals, and "ownership" of space may be more ambiguous. Design methods for alleviating the potential strain of these intervals are difficult, in that the integrity of the setting must be maintained for emergencies.

DIAGNOSTICS

When the patient is well enough, an angiogram (a diagnostic test) is conducted to determine the extent of the damage from the heart attack and its cause (Figure 11). Diagnostic spaces are frequently redesigned to accommodate technological advances. Nurses function as assistants to physicians and are part of a two- or three-person team. It is careful, deliberate work that is rarely conducted in an atmosphere of emergency. The patient is passive but awake.

The catheterization laboratory provides an excellent example of the

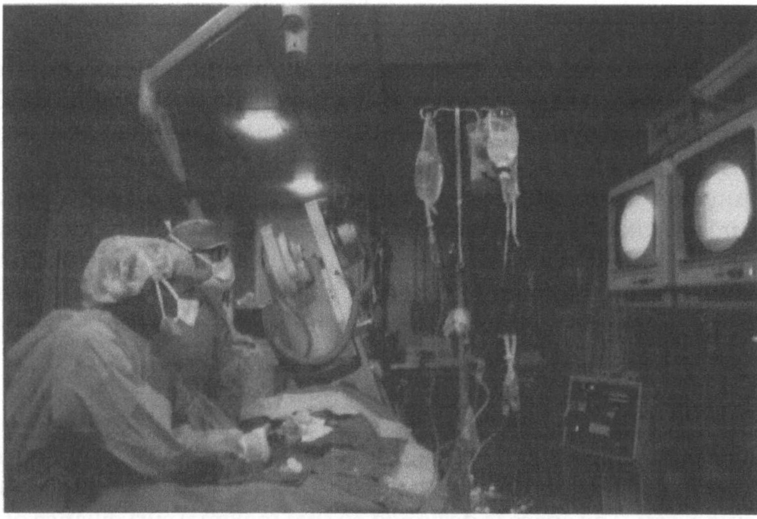

FIGURE 11. A doctor and nurse viewing the monitor while the patient is having an angiogram (a diagnostic cardiac exam). (Courtesy of Greg Pease, The Washington Hospital Center.)

encroachment of technology on hospital space. As this technology advances, larger spaces with multiple pieces of equipment have become necessary. Because most hospitals were built before this technology, catheterization labs have taken over spaces allocated to other purposes. Labs often "feel" crowded and disorganized. The roles of each health professional are explicit, and again, there is a great deal of interdependency among the participants. Design efficiency is important, and the identification of the design goals is unambiguous.

Catheterization laboratories are "hard" settings (Sommer, 1974) with minimal (if any) softening touches. They are usually placed in interior spaces with no window access. Personalization of this space is not possible. Thus, relief from this setting means leaving it. Design interventions such as adjacent lounges or nearby access to exterior views could counter the harshness of the laboratory by providing the health professional with intermittent periods of relief. One problem with such interventions is the fact that catheterization (and other highly technical) laboratories are expanding, further encroaching on hospital space. Unfortunately, space allocation for lounges is low in priority, as there are no direct links between these spaces and economic gain. This point underscores the need to empirically demonstrate the cost-benefit ratio of restorative spaces for health professionals.

The Operating Room

Based on the results from the angiogram, the patient may be scheduled for coronary-artery-bypass graft surgery. The operating room (OR) has a hierarchical system presided over by a cardiac surgeon (see Figure 5). Each person has a specific role and a specific place, defined by status and function (Goffman, 1969). The patient is encircled by people and equipment. Like the catheterization laboratory, the OR is a crowded, hard environment with minimal opportunities for personalization or softening because of the need for efficiency and sterilization. Thus, easy access to restorative settings is important to the well-being of health providers.

The Surgical Intensive-Care Unit

The patient spends the initial 24–48 hours in the surgical intensive-care unit (SICU; Figure 12). This is a high-tech environment, functioning 24 hours a day. The space is designed so that maximal surveillance of the patient is possible both visually and instrumentally. The nurse is one-on-one with the patient and follows an intricate set of written orders provided by the attending physician. This setting involves repetitive tasks interspersed with intense episodic events (Hay & Oken, 1986).

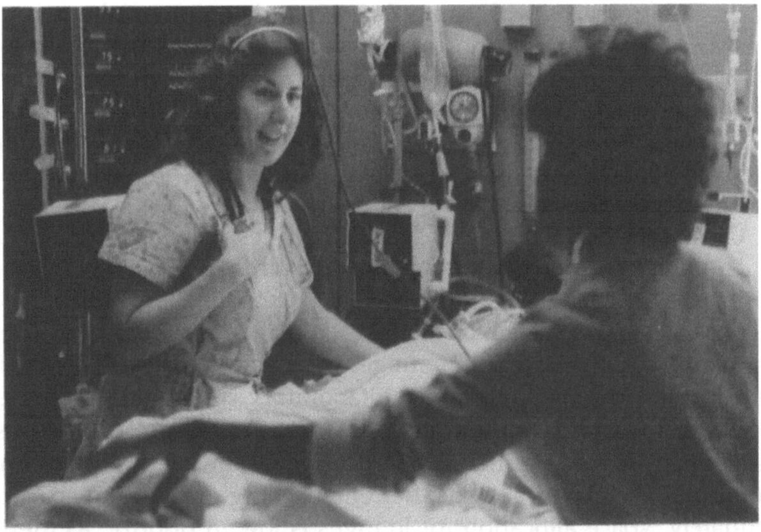

Figure 12. A patient being treated by two nurses in a surgical intensive care unit; the monitoring equipment is shown. (Courtesy of Susan Fitzhugh, The Washington Hospital Center.)

Even the episodic events require a specific set of routine responses that can produce boredom and fatigue. Nurses are continuously administering to patients who are extremely sick and who will either shortly move out to a stepdown unit or die. In addition, continual patient assessments are required, and nurses are called on to interact with the families and friends of the patients. Breaks from these settings are critical in maintaining the concentration of health providers.

The Stepdown Unit or Acute-Care Ward

From the SICU, the patient is taken to a stepdown unit or an acute-care ward for final convalescing before going home. In this area, there are clearly designated spaces: the patient's room (Figure 13), the ward corridors (Figure 14), and the nurses' station (Figure 8). Although the last is called the nurses' station, it is really the command center for the unit and is used by virtually everyone on the health-care team, including patients and visitors. It is usually a place where some staff personaliza-

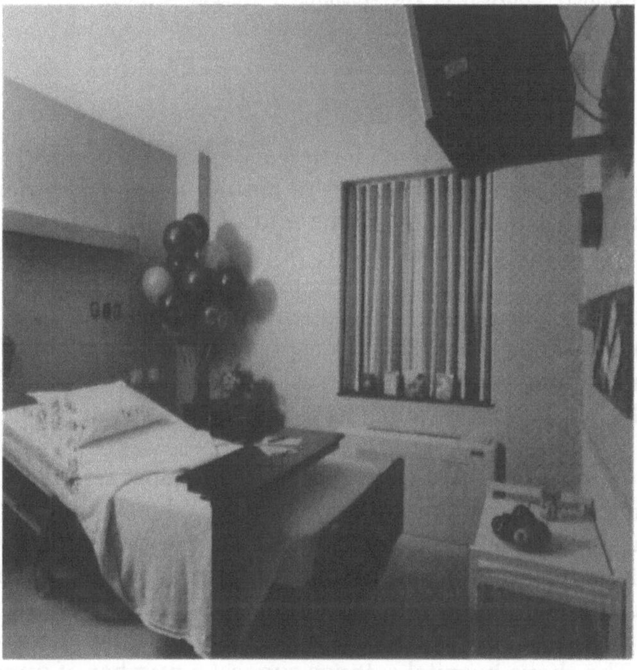

Figure 13. A typical private room for patients in a modern hospital. (Courtesy of The Washington Hospital Center.)

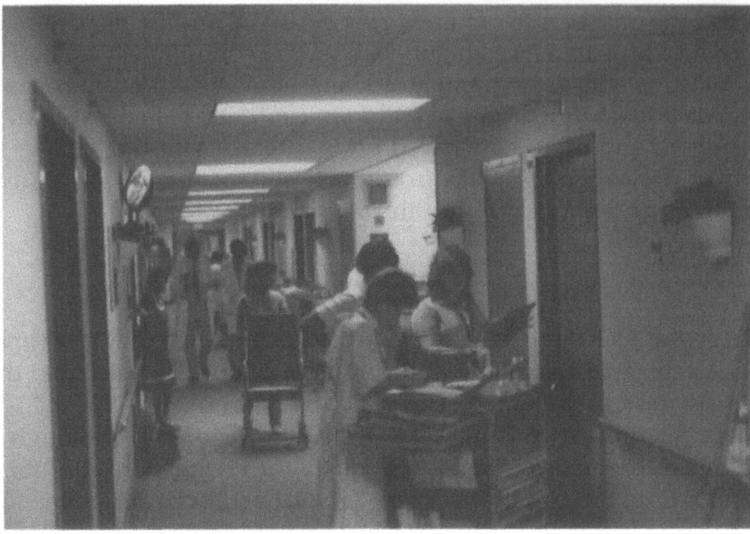

FIGURE 14. A hallway of a patient ward showing the multiple use by patients, visitors, and medical personnel. (Courtesy of The Washington Hospital Center.)

tion and changes have been made; however, because of poor design, the staff often must lean over counters or around poles to do their work, a requirement that takes a toll at the end of a long shift. On this unit, nurses are autonomous and follow physicians' written orders without direct supervision. The patients' conditions can vary dramatically from day to day, but in general, the patients are stable and do not require continuous bedside care. The atmosphere from the patients' point of view can be quite relaxed; the nurses, however, are continually on the move. They may walk as much as 3–5 miles on a shift in a unit that is only 150 feet long.

The various acute care ward spaces are the most amenable to design change in that the functions are flexible and fixed technology is minimal. However, these spaces exemplify many of the problems inherent in design-based interventions. Multiple user-groups with conflicting needs dominate the ward spaces. In addition, ambiguity in the functions of some spaces (e.g., the corridors) makes setting criteria for efficiency difficult. The above hospital settings suggest a number of opportunities for research-based design interventions and evaluations.

IMPLICATIONS FOR FUTURE RESEARCH AND APPLICATION

Research and application in hospital design break down into three broad categories: (1) investigations designed to assess the impact of current settings on health care; (2) design interventions and evaluations; and (3) the involvement of health-care providers in the design process.

Well designed studies are needed that provide clear tests of the models discussed in this chapter, as well as other models linking environment to behavior. Such studies need to identify clearly the initial sociophysical dimensions of a particular setting (see Table 3), and to incorporate a broad range of outcome variables that link design to behavior and health. Thus, for example, measures of satisfaction with a setting should be augmented with indices such as staff turnover; the satisfaction of different user groups; the time needed to deliver various services; and direct health related measures of stress (e.g., blood pressure, staff sickness records, and staff absenteeism). The data sources should include self report, more objective measures of particular issues, and unobtrusive assessments of such variables as intergroup conflict and space utilization patterns of staff and patients.

Table 3 provides information on the settings most likely to be a problem and the areas amenable to change. Interventions based on data obtained from the types of investigations suggested above should further identify specific targets for interventions. All interventions should be designed so that their impact can be unambiguiously interpreted. That is, experimental or quasi-experimental designs are critical, and multiple outcome criteria should be collected from all impacted user groups.

For the most part, health-care providers are excluded from the hospital design process. To some degree, this is an inevitable aspect of the profession. As discussed in this chapter, hospital design and construction represent a complex process that may take several years, from inception to a fully operating building. Few health professionals will remain in the specific setting throughout this process. Furthermore, most health professionals are unfamiliar with the design process and their opportunities to implement positive changes. However, neither of these obstacles to involvement is insurmountable. To minimize the effect of turnover, health professional *organizations,* not specific individuals, should initiate and maintain an active role in the design process. Furthermore, it is in the best interest of these organizations to become familiar with the design process and to maintain an active staff liaison.

Health professionals are directly in touch with the building and therefore can provide useful data about both the performance of the building and the behavior of its users that can contribute to improvement in the design process for another hospital. We do not know of any ethnographic studies in which the decisions that have been made as part of the design process are mapped, and certainly each situation is different. Yet, there are specific points at which health professionals could become actively involved in the hospital design (see Tables 1 and 2).

First, professional organizations representing health-care providers (e.g., the American Medical Association) can be participants in the policymaking within their community. The federal and state codes and regulations are important, and feedback from the health-care community in their development is essential. Next, within the area of strategic planning, health-care providers may perceive the need for a new facility and may be a part of the group that develops the feasibility assessment. Through the director of their hospital department, health-care providers can have a direct voice in the development of a functional program by providing data for the physical and functional evaluations and by making recommendations about the development of the space program.

The directors of hospital departments are usually involved in the review of the block plan drawings, which test internal building arrangements to ensure the best functional allocation of space. The staff becomes central when the new facility is occupied. In large part, it will be left to the health professionals to reprogram some of the space so that it will work for the variety of user groups that it serves. Thus, their involvement throughout the design process forms a natural feedback loop that could enhance future hospital designs.

ACKNOWLEDGEMENTS

We would like to thank Roland Binker, American Institute of Architects, for discussing the hospital design process in detail early in the development of this chapter and for reviewing it after it was written. We would like to thank Susan Boehm, Judith Kulick, and Eleanor Schron for their critical comments. Special thanks are due to Barbara B. Oliver and the Washington Hospital Center, which provided the photographs that illustrate the text.

REFERENCES

Bobrow, M., & Van Gelder, P. (1980). The well-being of design quality in the health-care world. *Architectural Record*, May, 102–126.

Califano, J. A., Jr. (1979). *Healthy people.* Washington, DC: U.S. Government Printing Office, U.S. Public Health Services Publication No. 79-55071.

Canter, D. (1983). The environmental context of nursing: Loading beyond the ward. In S. Stevington (Ed.), *Social psychology and nursing.* Chichester, England: Wiley.

Carpman, J. R., Grant, M. A., & Simmons, P. A. (1986). *Design that cares.* Chicago: American Hospital Association Publishing.

Cartwright, L. K. (1979). Sources and effects of stress in health careers. In G. C. Stone, F. Cohen, N. E. Adler & Associates (Eds.), *Health psychology: A handbook.* San Francisco: Jossey-Bass.

Clibbon, S., & Sachs, M. L. (1970). Like-spaces versus bailiwick approach to the design of health care facilities. *Health Services Research,* Fall, 172–185.

Cohen, S., Evans, G. W., Stokols, D., & Krantz, D. S. (1986). *Behavior, health, and environmental stress.* New York: Plenum.

Coser, R. (1958). Authority and decision-making in a hospital. *American Sociology Review, 23,* 5763.

Decker, F. H. (1985). Socialization and interprofessional environment in nurses: Affective reactions to work. *Social Science Medicine, 20,* 499–509.

Easterbrook, G. (1987). The revolution in medicine. *Newsweek,* January 26, 4065.

Faden, R. (1986). *Presymptomatic screening in fetuses and adults: Moral and psychological issues.* Invited address presented at the annual meeting of the American Psychological Association, Washington, DC.

Fetter, R. B., Thompson, J. D., & Mills, R. E. (1976). A system for cost and reimbursement in hospitals. *Yale Biology Journal, 49,* 123.

Lindheim, R., Glaser, H. H., & Coffin, C. (1972). *Changing hospital environments for children.* Cambridge: Harvard University Press.

Goffman, E. (1969). *Asylums.* New York: Doubleday.

Gray-Toft, P. A., & Anderson, J. G. (1985). Organizational stress in the hospital. *Health Services Research, 19,* 753–774.

Grimaldi, P. L., & Micheletti, J. A. (1983). *Diagnosis related groups.* Chicago: Pluribus.

Hay, D., & Oken, D. (1986). The psychological stress of intensive care nursing. *Intensive Care Unit Nursing,* 118–131.

House, J. (1981). *Work stress and social support.* Reading, MA: Addison-Wesley.

Jaco, E. G. (1979). Ecological aspects of hospital patient care: An experimental study. In E. G. Jaco (Ed.), *Patients, physicians, and illness.* New York: Free Press.

Jenna, J. K. (1986a). Toward the patient-driven hospital: 1. *Healthcare Forum,* May–June, 9–18.

Jenna, J. K. (1986b). Toward the patient-driven hospital: 2. *Healthcare Forum,* July–August, 53–57.

Kaplan, S., & Kaplan, R. (Eds.). (1982). *Humanscape environments for people* (2nd ed.). Belmont, CA: Duxbury-Wadsworth.

Karacek, R. (1979). Job demands, job decision latitude, and mental strain. *Administrative Science Quarterly, 24,* 285–308.

Kenny, C., & Canter, D. (1981). A facet structure for nurses' evaluations of ward design. *Journal of Occupational Psychology, 54,* 92–108.

Kirk, S., Eichinger, C., & Spreckelmeyer, K. (1984). Decision methods for evaluating design. In G. S. Lasdon & J. S. Gann, (Eds.), *The future of hospital design.* Washington, DC: U.S. Department of Health and Human Services.

Letourneau, C. V. (1965). Regional hospital planning. *Hospital Management,* May, 41–50.

Leventhal, H., Nerenz, D. R., & Leventhal, E. (1982). Feelings of threat and private views

of illness. In J. Singer & A. Baum (Eds.), *Environment and health*, Vol. 4. Hillsdale, NJ: Lawrence Erlbaum.
McLaughlin, H. P. (1975). Post-occupancy evaluation of hospitals. *American Institute of Architect Journal*, 30–34.
Moos, R. H., & Schaefer, J. A. (1987). Evaluating health care work settings. *Psychology and Health*, *1*, 97122.
Nuffield Provincial Hospitals Trust. (1955). *Studies in the function and design of hospitals*. London: Oxford University Press.
Panther, R. E. (1984). Hospital design in the year 2015. In G. S. Lasdon & J. S. Gann (Eds.), *The future of hospital design*. Washington, DC: U.S. Department of Health and Human Services.
Pardes, K. R. (1982). Occupational stress among student nurses: A national experiment. *Journal of Applied Psychology*, *67*, 784–796.
Price, D. (1972). Fast track and new methodology in hospital planning and construction. *Hospital Progress*, June, 50–68.
Price, R. (1982). *Work and community*. Presidential address for division 27, American Psychological Association, Toronto, Canada.
Rachman, S. I., & Philips, C. (1978). *Psychology and medicine*. New York: Penguin.
Rosenfield, I. (1969). *Hospital architecture and beyond*. New York: Van Nostrand Reinhold.
Seligman, M. E. P. (1975). *Helplessness: On depression, development and death*. San Francisco: Freeman.
Selye, H. (1956). *The stress of life*. New York: McGraw-Hill.
Selye, H. (1974). *Stress without distress*. Philadelphia: Lippincott.
Seymer, L. R. (1954). *Selected writings of Florence Nightengale*. New York: Macmillan.
Shumaker, S. A., & Reizenstein, J. E. (1982). Environmental factors affecting inpatient stress in acute care hospitals. In G. Evans (Ed.), *Environmental stress*. New York: Cambridge University Press.
Sommer, R. (1974). *Tight spaces*. Englewood Cliffs, NJ: Prentice-Hall.
Sommer, R., & DeWar, R. (1963). The physical environment of the ward. In E. Friedson (Ed.). *The hospital in modern society*. London: Free Press.
Stokols, D. (1979). A congruence analysis of human stress. In I. G. Sarason & C. D. Spielberger (Eds.), *Stress and anxiety*, Vol. 6) New York: Wiley.
Stokols, D. (1986). Developing contextual theories of environment and behavior. In M. E. Dolden & R. Ward (Eds.), *The impact of work environments on productivity*. Washington, DC: Architectural Research Centers Consortium.
Sundstrom, E. (1987). Work environment. In D. Stokols & I. Altman (Eds.), Handbook of environment psychology. New York: Wiley.
Taylor, S. (1979). Hospital patient behavior. *Journal of Social Issues*, *35*, 156–184.
Theis, (1986).
Thompson, J. D., & Golden, C. (1975). *The hospital*. New Haven: Yale University Press.
Trites, D. K., Dalbraith, F. O., Sturdavard, M., & Leckwart, J. F. (1970). Influences of nursing unit design on the activities and subjective feelings of nursing personnel. *Environment and Behavior*, *2*, 303–334.
Ulrich, R. (1984). View from a window may influence recovery from surgery. *Science*, *223*, 420–444.
U.S. Department of Health and Human Services (USDHHS). (1986). *Utilization of short-stay hospital*. Washington, DC: U.S. Printing Office, U.S. Department of Health and Human Services Publication No. 86-1745.
Verderber, S. F. (1982). Designing for the therapeutic function of windows in the hospital

rehabilitation environment. In P. Bart, A. Chen, & G. Francescato (Eds.), *Knowledge for design*. Washington, DC: Environmental Design Research Association.

Vogel, M. J. (1980). *The intervention of the modern hospital*. Chicago: University of Chicago Press.

Walsh, R. N., & Grenough, W. T. (1976). *Environments as therapy for brain dysfunction*. New York: Plenum.

Weeks, J., & Best, G. (1970). Design strategy for flexible health sciences facilities. *Health Services Research*, Fall, 263–281

Weeks, R. (1980). Interior design: Hospitals. *Architectural Review*, July, 46–56.

Wener, R. (1989). Advances in evaluation of the built environment. In E. H. Zube & G. T. Moore (Eds.), *Advances in environment, behavior, and design*. New York: Plenum.

Winkel, G. H., & Holahan, C. J. (1985). The environmental psychology of the hospital: Is the cure worse than the illness? In A. Wandersman & R. Hess (Eds.), *Beyond the individual: Environment approaches and problems*. New York: Haworth.

Ziegenfus, J. T., Jr. (1985). *DRGs and hospital impact*. New York: McGraw-Hill.

IV

ADVANCES IN SOCIOBEHAVIORAL RESEARCH

7

Environmental Perception and Cognition

TOMMY GÄRLING and
REGINALD G. GOLLEDGE

Environment and behavior research investigates the mutual relations between the sociophysical environment at all scales and human behavior at all levels of analysis (Moore, 1987). It is generally assumed that there are an "actor," an external environment or situation, an input from situation to actor resulting in a "psychological response," and, on the basis of this response, an output from actor to situation termed *action* (Golledge, 1987). In studies of environmental perception and cognition, the existence of motives, goals, and attitudes toward action alternatives are usually taken for granted, and the psychological responses or processes mediating between the environment and actions are made a primary focus. These processes include the picking up of information from, and about, the environment; the internal, perceptual, and cognitive representation of the information; and judgments, decisions, and choices made on the basis of the represented information. Knowledge gained about perceptual–cognitive processes may improve the quality of human environments through policy, planning, and design, to the extent that it tells how to plan and design environments that do not interfere with the proper functioning of these processes. Examples are the work of Ap-

Tommy Gärling • Environmental Psychology Research Unit, Department of Psychology, University of Umeå, S-90187 Umeå, Sweden. **Reginald G. Golledge** • Department of Geography, University of California, Santa Barbara, California 93106.

pleyard (1976) and Lynch (1960), resulting in suggested guidelines for how to make cities legible.

Research on environmental perception and cognition has been concerned with environmental settings at different spatial scales. Some studies have focused on perceptions of particular attributes of settings or of the ambient environment; others have tried to assess the nature of the complex, multifaceted perceptual-cognitive response to settings. This chapter aims primarily at reviewing studies of the perception and cognition of the physical-spatial properties of settings, studies of the perception and cognition of the properties of places, and studies of spatial preferences, decisions, and choices based on information picked up from the environment. A secondary aim is to try to integrate these three research traditions. Excluded from the review are studies of the perception of properties of the ambient environment, like light, sound, air, temperature, and density of people. Such perceptions are often studied in relation to performance and in relation to psychological and physiological stress responses, which are reviewed elsewhere in this series (Weiss & Baum, 1987). Furthermore, because research on the psychological response to the environment of special populations is intended to be reviewed in later volumes, studies comparing different populations, or studies focusing on a particular special population, are reviewed selectively.

A primary goal of much research on environmental perception and cognition is the development of theory. The review in this chapter reflects this goal. The first section describes how major theoretical perspectives have emerged. Such theoretical perspectives, or orientations, indicate how to frame problems, what to look for, and in what ways to look. Low-level explanatory theories within these theoretical perspectives (which may themselves be explanatory theories) as well as empirical findings with bearings on the explanatory theories are then reviewed and evaluated. Research on the perception and the cognition of the physical-spatial properties of environments at different spatial scales is reviewed in two sections. Previous such reviews are found in Moore (1979), in Evans (1980), and, most recently, in Golledge (1987) and, confined to children, in Hart and Moore (1973), in E. J. Gibson and Spelke (1983), and in Heft and Wohlwill (1987). Research on complex perceptual-cognitive response to places is reviewed next in this chapter. The most recent reviews are by Holohan (1982, 1986). A focused review appeared in Kameron (1973), and more general reviews appear in Stokols (1978) and in Russell and Ward (1982). Methods used in such research were reviewed and discussed by Golledge (1976) and more recently by Canter, Brown, and Groat (1985). Research on spatial preferences, decisions, and choices is much too extensive to be reviewed comprehensively here (see Halperin, Gale, & Gorenflo, 1988, for a recent review). The final section is therefore limited to a discussion of the relationships that exist between the perceptual-cognitive processes that pick up information from the environment and spatial decision and choice processes.

THEORETICAL PERSPECTIVES

Despite 100 years of perception research by psychologists, Ittelson (1973) saw reasons to question the validity of this research for theories of the perception of environments mainly because of its focus on phenomenal experiences of abstract objects and space. Ittelson argued that not only is the perception of environments more complex, but the quality of the complexity also differs. First, environments provide information conveyed by patterns of sensory inputs that are not specific to particular sensory channels. Second, environments are unbounded and surround the person, forcing him or her to explore rather than to observe the environment. Thus, the perception of environments at almost any scale usually extends over long time spans and entails transient and permanent storage in memory and the integration of information inputs from various sources. Third, both central and peripheral information are present in environments; at any moment, the information is more than can possibly be processed; and the information may simultaneously be redundant, inadequate, and ambiguous. Thus, information in environments needs to be selectively attended to, and the relevance and validity of the information need to be judged. Fourth, environments have atmospheres, difficult to define, but overriding in importance. Environments are encountered as part of social activities, they have aesthetic and affective qualities, and they are perceived holistically, as consisting of parts and components interrelated in meaningful units. Fifth and finally, perception is primarily directed at purposeful action. Rather than being observed and experienced, information that can be brought to bear on actions is sought out in environments. No direct, invariable correspondence between properties of the environment and a person's perception of it can be expected. By taking the person's current intentions and plans into account, a better understanding of why the environment is perceived in certain ways may be attained.

According to Ittelson (1973), it is possible to make no sharp distinctions between perceptual and cognitive processes. In this vein, information-processing theories in cognitive psychology (Lachman, Lachman, &

Butterfield, 1979) view perception as being determined by a sequence of internal processes, corresponding to a series of transformations of information selectively attended to. Cognitive factors such as past experiences stored permanently in memory, as well as the temporary storage of information in memory, play a crucial role, and so do judgments, decisions, and choices. In environmental perception and cognition research, information-processing theories constitute a dominant theoretical perspective and have done so for a long time. For instance, Downs and Stea (1973) and Siegel and White (1975) outlined such a perspective in their studies of environmental images, and more recently, Leff (1974) and Kaplan and Kaplan (1982) introduced a broader such perspective on the perception and cognition of the global properties of environments.

Information-processing theories have been challenged, more in cognitive than in environmental psychology (Heft, 1981), first by the ecological theory of perception proposed by Gibson (1966, 1979; see Landwehr, 1988; Ohno, 1986) and later by Neisser's modification (1976) of Gibson's theory. Gibson's analysis and arguments were similar to Ittelson's (1973), but Gibson reached another conclusion. In his theory, the units of perception are events that are spatiotemporally extended. No mediating cognitive processes need be assumed because a skilled person is, over time, capable of directly perceiving invariants of events that are specified in, for instance, the ambient light reflected from surfaces. Such invariants that are important in the environment are the spatial layout and moving and stationary animate and inanimate objects. Gibson (1979) also proposed that species' perceptual systems have evolved to enable them to perceive functionally significant conditions in the environment, called affordances, or utilities.

Neisser (1976) believes that information-processing theories of perception and cognition need to be revised in the direction of the Gibsonian perspective. According to Neisser's revision, information is processed not linearly from input to response, as is assumed by information-processing theories, but in cycles. The concept of schema is proposed as an internal information structure that accepts certain information and directs a search for more information. In a perceptual cycle, some available information initiates a schema, which then directs perceptual exploration, locomotion, and action leading to the picking up of more information, which modifies the schema, and so on. A person does not construct the world in perception as implied by information-processing theories; only general hypotheses or directions for further information search are activated, strengthened, or possibly modified, on the basis of information sampled from the environment. Some schemata are tuned to pick up information about the physical-spatial properties of environments, objects, and events; some are tuned to their affordances.

Ittelson (1973) pointed out how important it is to investigate the role of perception for the overall functioning of the person in the environment, for instance, by studying the extent to which perception directs purposeful actions. Ittelson's arguments were followed up by Moore (1979) and, more recently, by Russell and Ward (1982), who noted a need for more research on how perception and cognition of the environment are related to actions. Action theory has recently emerged within cognitive and cognitive social psychology in response to similar criticisms of information-processing theories. As Frese and Sabini (1985, p. xviii) stated:

> Modern action theory starts where Tolman (1932)[1] left off: There is a goal, and there is a map. But are the map and the goal related to behavior? How is the gap between cognition and action to be bridged? . . . And finally, how can one bridge the gap between the real world outside and cognition inside?

The emphasize of modern action theory is on the objective environment, not on the individual's life space or subjective environment. Human behavior is conceptualized as being directed toward the accomplishments of goals, and as being controlled by hierarchically organized plans and by feedback from the environment. Action theory stresses how people go about creating specific plans in specific environments to reach specific goals. The theory is an approach to perception and cognition that emphasizes how knowledge is acquired, represented, and accessed so that it is brought to bear on actions.

This section has indicated that theoretical perspectives in research on environmental perception and cognition have undergone changes in the direction that Ittelson saw as desirable in 1973. "Elementary" perceptions of color and form and other sensory qualities are no longer thought to be the blocks of which complex perceptions are built, and therefore, knowledge about such perceptions is of less value for an understanding of the perception and cognition of environments. Information-processing theoretical perspectives were adopted early in environmental perception and cognition research. In such theoretical perspectives, a person's knowledge structures are assumed to bear on the information picked up from the environment and to determine his or her processing of the information and, consequently, the end product,

[1]Tolman (1932) introduced in psychology the concept of a cognitive map as referring to an acquired internal model of the environment, including its spatial layout, its causal structure, and its affordances.

the perceptual response. This view is being challenged. In particular, the assumption that the perceptual response is determined by organismic factors is being questioned as being too strong when carried to the extreme. Furthermore, the relative neglect in such theoretical perspectives of the relationship between the internal processing of information, the environment, and a person's actions in the environment has been noted and criticized. Actions are part of the perceptual process, and specific purposes and plans for actions in specific environments determine what information will be processed.

EMPIRICAL RESEARCH IN SMALL-SCALE AND MEDIUM-SCALE ENVIRONMENTS

Perception and Cognitive Representation of Physical-Spatial Properties

Movement and locomotion are a basic general characteristic of human actions in the environment. Gibson (1966, 1979) proposed that our perceptual systems, therefore, are tuned to directly perceiving the permanent spatial layout of the environment, consisting of surfaces and their affordances for locomotion. Another important movement-related function of perception is to make possible spatial orientation and navigation in the environment.

Recent research in environmental perception and cognition has focused almost exclusively on the navigation function of spatial perception described by Gibson, but most researchers have turned to an information-processing theoretical perspective rather than to Gibson's theory of direct perception. The ability to navigate has then been assumed to rely on the availability of acquired mental models, or cognitive maps, of the environment. The important role thus ascribed to cognitive maps makes it appropriate first to review research investigating the content and properties of cognitive maps, then to review research on how they are acquired, and finally, to review research on their use in perception, spatial orientation, and navigation.

Content and Properties of Cognitive Maps. Although there has been much debate about what kinds of knowledge about the environment is represented in cognitive maps, recent reviews (Evans, 1980; Golledge, 1987) suggest that there is agreement about the distinction between knowledge about places and knowledge about spatial relationships. Furthermore, it is generally assumed that spatial knowledge consists of knowledge of routes and of the locations in the environment of at least some salient places. A question of debate still remaining centers on the viability of the map metaphor (Kuipers, 1982). This question entails two other questions: Is the information represented in cognitive maps of the same type as information represented in cartographic maps, and do cognitive maps have a maplike (or picturelike) representational format?

A cartographic map is a transformation of the environment into an image that preserves some of its contents and properties, disregards others, and distorts still others (Downs, 1981). Maps are, by this definition, models of the environment, and the question is what people's mental models of the environment—that is, their cognitive maps—have in common with cartographic maps. One may, for instance, ask whether cognitive maps, like cartographic maps, preserve the Euclidean-geometrical properties of the real world. A substantial amount of research in which people's cognitive maps of familiar environments have been assessed has tried to answer this question. Baird, Wagner, & Noma (1982) were inclined to conclude, on the basis of a review of some of this research, that both directly perceived distances and memorized distances are nonlinearly related to the real-world distances. Directions have been found to be more accurately represented than distances (Gärling, Böök, Lindberg, & Nilsson, 1981), but the non-Euclidean character even of representations of directions was demonstrated by Moar and Bowers (1983).

An affirmative answer to the question of whether cognitive maps are Euclidean has been given in studies of how spatial cognition develops (e.g., Biel, 1983; Hazen, 1982; Hazen, Lockman, & Pick, 1978; Herman, Shiraki, & Miller, 1985; Spencer & Darvizeh, 1981). For instance, Conning and Byrne (1984) showed that children 3–4 years old are able to acquire Euclidean mental representations of familiar environments, especially when familiarity is brought about through self-exploration. In studies of children, but less frequently in studies of adults, subjects have been tested in small-scale spaces, like rooms, and the acquisition of the representation has been carefully controlled. Such differences in procedure may be important, but the results of a series of well-controlled experiments (Sadalla & Staplin, 1980), in which adult subjects walked various distances in a room, suggested that real-world distances may be inaccurately represented because they are influenced by the number of the segments of traversed paths. In a more recent study (Herman, Norton, & Klein, 1986), however, these results were not possible to replicate with schoolchildren. That distance estimates are distorted was also found by Allen and Kirasic (1985) in another controlled study, in which sequences of slides depicting segments of paths in a larger scale environment were shown. A direct comparison, using this task with adults and children, aged 7–11 in this case, revealed worse performance by the children (Allen, 1981). All this seems to suggest that people have the competence to represent distances and directions in the ways in which they are represented on cartographic maps, but that many factors may cause performance to be suboptimal in actual environments.

On cartographic maps, directions are encoded in relation to a common reference system, like the cardinal compass directions. The question has been raised whether directions are encoded in cognitive maps in relation to an objective, nonegocentric reference system, and if so, what properties this objective reference system has. Much of the related research has been carried out with infants and was reviewed by Acredolo (1985). The results of such studies indicate that, after displacement, prelocomotor 4-month-old infants respond to locations as originally encountered, that is, egocentrically, but by 8–9 months of age, when the infants are becoming mobile, they react to objective locations if the test conditions are optimal, and at 11 months of age, they react predominantly in this way. The evidence thus suggests that the competence to encode directions relative to objective reference systems, presumably landmark-based, is developed very early. However, in any unfamiliar environment, such reference systems must be encoded, and if the environment is of a larger scale, this encoding may take some time. For instance, Kirasic, Allen, and Siegel (1984) found that the degree of familiarity with an environment affected adult subjects' accuracy in judging distances and directions in that environment if the judgments were made in the context of a perspective-taking task. Those subjects with more experience were assumed to have acquired an abstract, coordinated reference system. In another study of adults in a familiar, larger-scale environment (Gärling, Lindberg, Carreiras, & Böök, 1986b), directions to occluded places were found to have been encoded in relation to place-based reference systems, which, furthermore, were in better agreement in different, spatially separated places if the places were closer, and if a regular street grid system connected them. No research seems to have explicitly addressed the question of what features in the environment are encoded as reference systems. In outdoor environments, both local features, such as the main directions of buildings, and more global features, such as streets and rivers, are possible candidates.

The physical realization of cartographic maps as plane projections may have precluded reseachers from considering the possibility that the cognitive map is three-dimensional. This conception may explain the scarcity of studies of whether elevation differences in the environment are mentally represented and, if so, how they are represented in cognitive maps. Introspective observations suggest, however, that retrieving information from the cognitive map of a small-scale or medium-scale environment is like the reinstatement of the original perception of it, and that the environment is never perceived as two-dimensional, viewed from above. A single study (Foley & Cohen, 1984) speaks directly to this point. In that study, subjects judged the distances between targets located in a building, both on the same floor and on different floors. Multidimensional scaling (MDS) (see Marans & Ahrentzen, 1987) of the distance judgments yielded three-dimensional solutions, indicating that the vertical dimension of the building was taken into account by the subjects. The existence of three-dimensional visual imagery was also documented.

The question has been asked: What representational format do cognitive maps have? Research by Kosslyn (1980) and others on visual imagery has indicated that information may be internally represented in both analogue and propositional formats.[2] In a study of spatial problem-solving, Levine, Jankovic, and Palij (1982) tested some implications of the hypothesis that cognitive maps are analogue, maplike structures. One implication, which these authors verified in a series of experiments in which subjects walked various patterns in a room, was the principle of the equiavailability of spatial relations; that is, as in maps, all spatial relations are equally easy to recall whether they have been directly perceived or not. Another principle, also empirically verified (Levine *et al.*, 1982; Levine, Marchon, & Hanley, 1984), was orientation specificity: If a map of an environment is acquired in a certain orientation, then its use in another orientation will be impaired. Evans and Pezdek (1980) obtained results not congruent with this principle when they tested subjects for their knowledge of campus locations. This difference in results may be related to whether the cognitive map is acquired from a cartographic map or by direct experience (Presson & Hazelrigg, 1984).

Research on the representational formats of cognitive maps have, in recent years, benefited greatly from the use of the technique of computer simulation. This technique is being used in artificial intelligence and cognitive psychology to model internal information structures and processing (Smith, Pellegrino, & Golledge, 1982). The technique is not a substitute for theory but may help in its development. It has, for instance, been instrumental in developing the information-processing the-

[2]An internal representation in memory has a format distinguishable from what it represents. It is generally thought that facts expressed in ordinary language are represented as propositions that may be true or false. This format is abstract in the sense that it has an arbitrary relationship to what it represents. In contrast, an analogue representation preserves some of the properties of what it represents. For instance, Kooslyn (1980) argued that visual images occur in a spatial medium, and that they have an abstract spatial and surface (texture and color) property isomorphism.

oretical perspective by introducing concepts like schema and frame as structures that represent declarative knowledge and by modeling actions as procedural knowledge structures.[3] By 1978, Kuipers had published a description of a computer simulation of a cognitive map, how it is acquired in the course of interactions with the environment, and how it is used for different purposes like spatial orientation and navigation.

More recently, Golledge, Smith, Pellegrino, Doherty, and Marshall (1985) described a computer simulation of how an individual's cognitive processes relating to perception, inferences, storage, and retrieval interact with short-term and long-term memory structures in the acquisition of a knowledge representation of a route that is subsequently accessed during spatial orientation and navigation. The results of perceiving a single object in the environment (e.g., a choice point) is in the working computer program, temporarily stored in a perceptual buffer; salient features are extracted and stored for some longer time in a limited-capacity short-term memory; and some features are finally passed on to long-term memory, where they are permanently stored. The feature-extraction process is done through a pattern-matching mechanism controlled by frames (that is, general descriptions common to all instances of the frame), and the features are stored in long-term memory as particular instances of frames (e.g., a large blue house if the frame is "house"). Each object in long-term memory is also gradually incorporated into a structure of linear links connecting choice points in the order in which they are encountered. Thus, salient objects or places in the environment become represented, in the form of declarative knowledge, making it possible for the individual to know their existence and to recognize them when they are encountered. Routes between places are stored mainly in the form of procedural knowledge, that is, as decisions at choice points. Several aspects of a child's navigational performance along a route were possible to successfully simulate as acquisition proceeded (Golledge *et al.*, 1985), but as the authors themselves pointed out, how a geometrical representation of an area—that is, a cognitive map—is acquired and stored on the basis of episodic encounters with the environment has yet to be worked out. Nevertheless, the richness and precision of the representational medium that computer programs provide seem promising.

Acquisition of Cognitive Maps. Research questions concerning how cognitive maps are acquired have been asked, partly independently of

[3]Declarative knowledge refers to a representation of facts about the world. Actions may be based on such facts, but the facts do not by themselves indicate how to act, which is the domain of procedural knowledge representations.

questions asked about their content and properties. The assumption behind most of these studies is that cognitive maps of small-scale and medium-scale environments are acquired predominantly through direct experience. In a walk from one place to another, knowledge is gained about the perceptual characteristics of the places. Furthermore, the spatial relationships between such places are learned when one moves from one to another.

Even though the acquisition of cognitive maps may often by very fast and resistant to forgetting (Gärling *et al.*, 1981; Gärling, Lindberg, & Mäntylä, 1983), conditions may exist in which acquisition is less fast. Furthermore, different types of information have been found to be acquired at different rates (Evans, 1980; Golledge, 1987). This finding has been interpreted as reflecting that microgenesis is similar to ontogenesis (Hart & Moore, 1973; Moore, 1979; Siegel & White, 1975). However, more recent evidence has tended not to support this assertion (Evans, Marrero, & Butler, 1981; Spencer & Weetman, 1981). At least, there are other possibilities, such as the hypothesis that the rate depends on how much processing it takes to acquire the information (Lindberg & Gärling, 1983). Places are perceived directly, and information about places is therefore more easily acquired than information about routes and spatial configurations of places, which, under ordinary circumstances, cannot be directly perceived. Of the latter types of information, spatial configurations are, in general, found to be more difficult to acquire. Both route knowledge and the spatial configurations of places need to be encoded piecemeal, but in the latter case, inferences are usually made on the basis of route knowledge, and such inferences have been found to require more processing (Lindberg & Gärling, 1982).

Consistent with the assumption of different processing requirements, factors that can be assumed to reduce these requirements have been shown to facilitate acquisition. For instance, a direct perception of spatial configurations, as well as access to a map representation before experiencing the environment, has been found to make acquisition faster (Gärling *et al.*, 1983; Thorndyke & Hayes-Roth, 1982). Furthermore, a pathway grid has been found to enhance the acquisition of both route knowledge and spatial configurations (Evans, Skorpanich, Gärling, Bryant, & Bresolin, 1984; Gärling *et al.*, 1986b). One disadvantage of visual impairment is probably that the direct perception of spatial configurations is precluded (Rieser, Lockman, & Pick, 1980), although more recent evidence suggests that visual impairment may be compensated for by a movement-based acquisition of knowledge of spatial configurations (Rieser, Guth, & Hill, 1988).

Gärling, Böök, and Lindberg (1984, 1986a) proposed that cognitive maps are acquired in connection with the formation and execution of travel plans. The formation of travel plans in unfamiliar and partly familiar environments is not possible unless maps, other people, and so on are consulted. Spatial configurations of places may be learned mainly in this phase, whereas the perceptual characteristics of places and detailed information about particular routes may be learned in connection with the execution of travel plans. How this learning is accomplished was addressed by Golledge *et al.* (1985) in their computer simulation. Furthermore, both Golledge *et al.* and Doherty and Pellegrino (1985) empirically demonstrated the importance of navigational decisions for memory of places along a route. Places at choice points were found to be more easily recognized than places at nonchoice points. Other factors making these places distinctive, such as uniqueness in form, color, and size, may also enhance acquisition, as is suggested in studies by, among others, Appleyard (1976), Chalmers and Knight (1985), and Evans *et al.* (1984).

Spatial Orientation and Navigation. In Neisser's theory (1976), a cognitive map is a schema that embeds other schemata directing the picking up of immediately available information in the environment. As Figure 1 illustrates, a cognitive map of the environment is thus both influenced by and influences the information pick-up from the environment. This mutual dependency of direct perception of the physical-spatial properties of the immediate environment and the cognitive map of the environment, which cannot be directly perceived, has been noted by others. For instance, Hunt (1985a) identified two interrelated component processes of spatial orientation and navigation: recognizing perceptually available places (current location) and localizing perceptually nonavailable places through a process that keys the cognitive map to the perceptual input from the environment.

There is an extensive literature on human recognition memory for nonverbal visual information, which was recently reviewed by Böök (1986a). Obviously, the frequently mentioned notion of an extremely good capacity for the recognition of visual scenes, based on the results of a few early studies, has had to be revised in light of much further evidence. Whether recognition is good or not depends mainly on how similar a particular scene is to other encountered scenes; this similarity may vary along both perceptual and conceptual dimensions. However, the stimuli used in this research have almost exclusively been pictures. More detailed knowledge about the process of the recognition of places in actual environments is needed. A point of departure for such research is provided by hypotheses about and studies of how places are represented in cognitive maps. Golledge *et al.* (1985) proposed a conceptually

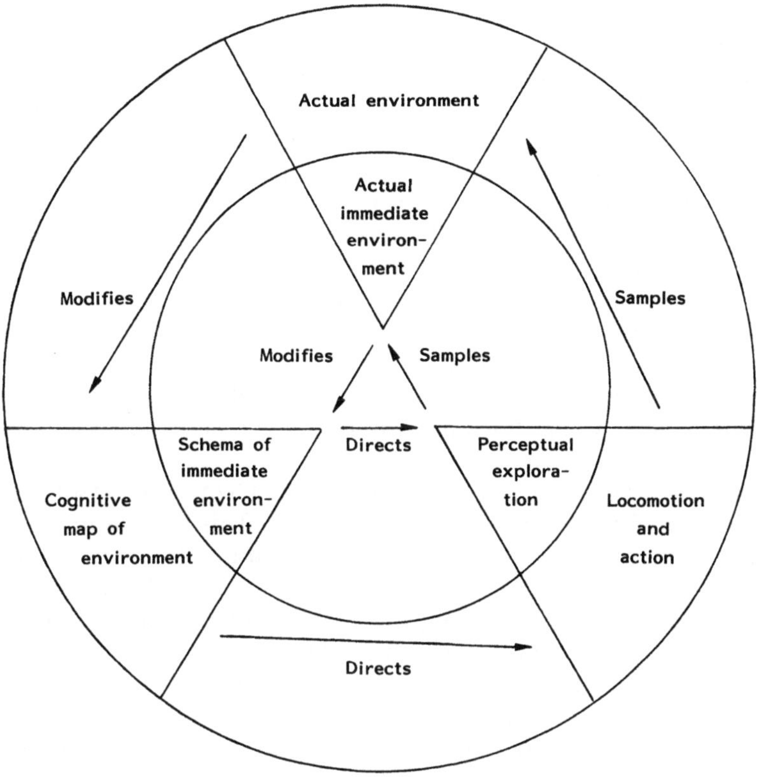

FIGURE 1. An illustration of how a cognitive map of the environment that cannot be directly perceived embeds a schemata that directs the picking up of immediately available information in the environment. (After Neisser, 1976.)

based frame representation of the perceptual features of places, whereas Gärling *et al.* (1984) suggested that places are represented in an analogue format, to explain the capacity for visual imagery of the places in an environment (Foley & Cohen, 1984). Other work has attempted to reveal the basic perceptual dimensions that account for the similarity between places (Ward & Russell, 1981), perceptual categories (Groat, 1982; Tversky & Hemenway, 1983), and the properties of place schemata (Mainardi-Peron, Baroni, Job, & Salmaso, 1985).

In unfamiliar environments, spatial orientation may rely on two different, complementary methods. As shown by, for instance, Böök and Gärling (1981) and, more recently, by Rieser *et al.* (1988), one is able to keep track of one's movement even though no visual information about the environment is available. A more reliable method in most ordinary environments requires, as input, visual information about perspective transformations. Böök (1986b) and Péruch, Pailhous, and Deutsch (1985) found that perspective-dependent information can be retained in memory. Such perspective transformations may subsequently be used (even if they are noncontinuous because of occluding objects, during navigation in an environment) to infer a current location relative to previously occupied locations. This method of spatial orientation appears to develop earlier than the method relying on the ability to keep track of one's movement (Acredolo, 1985). Acredolo, Adams, and Goodwyn (1984) found that 12-month-old children responded nonegocentrically—that is, in a place-based way—after active or passive displacement if they were allowed to track the displacement visually. At 16 months of age, visual tracking was no longer necessary for nonegocentric responding.

Spatial orientation refers to a process by which a person knows where he or she is relative to something else. This process informs the person neither about where he or she is heading, nor about how to get there. Gärling *et al.* (1984, 1986a) therefore argued that spatial orientation should be viewed as being interfaced with the navigational decisions entailed by a travel plan. Such decisions have been studied in depth by Passini (1984). Subjects were taken on tours in an unfamiliar environment (for instance, a parking structure), were asked to find a destination, and were then observed by an accompanying experimenter. The decisions that the subjects made were inferred from analyses of think-aloud protocols obtained by tape-recording the subjects' verbal reports of what they thought during the tours. A decision-tree analysis was used to decompose higher-level decisions (e.g., route choices) into decisions about what information to look for in the environment, what conditions should be fulfilled before executing the decision, and what to expect. The empirical data analyzed in this way vividly illustrated the complexity of the decision process, suggesting that, in many environments, navigation is no simple task. Moreover, most of the subjects were found to rely on those methods of orientation discussed above, whereas some subjects appeared to rely predominantly on signs and messages.

A question of obvious practical interest is whether people are better at spatial orientation and navigation in some environments than others and, if so, in what ways those environments differ. Practical problems, unavoidable confounding of different factors, and an unknown degree of generalizability of the results to other environments, to other users, and across time prohibit the direct answering of this question. Some of these problems may be solved by employing environmental simulation techniques, as indicated by some promising attempts in that direction (Evans *et al.*, 1984; Hunt, 1985b; Weisman, 1981). It may also be possible to find an answer by means of a theoretically and empirically based model that identifies the general properties of environments that are likely to affect spatial orientation and navigation. Recently, Gärling *et al.* (1986a) tentatively proposed such a model. As Figure 2 shows, the degree of direct or indirect visual access to destinations in a particular environment is assumed to be one factor that facilitates spatial orientation and navigation. However, visual access is useful only if the parts of the environment are differentiated (e.g., by form and color) so that they can be recognized. Thus, the degree of differentiation is a second factor. The complexity of the path system is a third factor. Visual access and differentiation may reduce the negative effect of a complex path system. However, if the path system is simple, differentiation and good visual access may not further enhance the ability to maintain one's orientation and to find one's way.

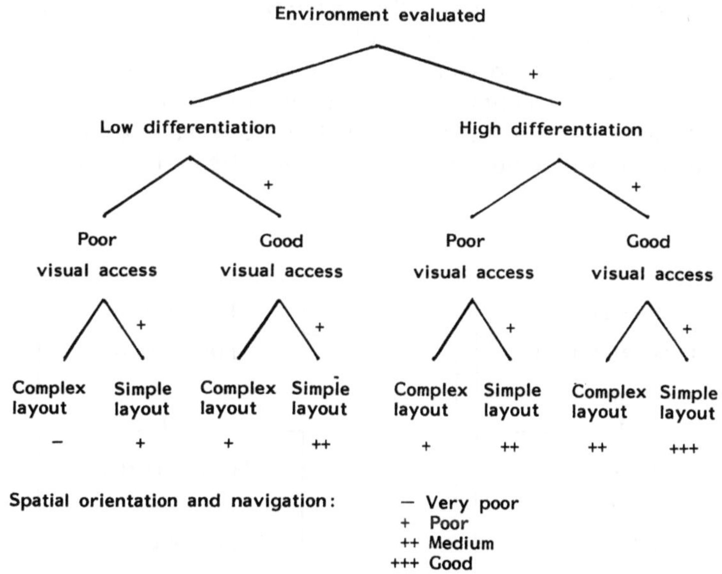

FIGURE 2. A model that identifies factors affecting spatial orientation and navigation in small-scale and medium-scale environments. (After Gärling *et al.*, 1986b.)

Perception and Cognitive Representation of Properties of Places

The concept of place experience has emerged to denote a holistic, affectively toned perception of the environment (Canter, 1977, 1983; Sime, 1986), similar in meaning to what Ittelson (1973) referred to as "atmosphere." In recent conceptualizations in personality and social psychology, physical environments are considered features of more global units of analysis, like situations (Magnusson, 1981), scripts (Schank, 1982), or behavior settings (Wicker, 1987). However, place may be a more fundamental unit of experience. People probably think of places as the major contexts for actions and interactions, within which a number of situations, scripts, and behavior settings may occur.

Some of the relevant research on place experiences was cited in the preceding section in connection with place recognition, which is considered a component of navigation and spatial orientation. However, place recognition in the context of the execution of travel plans does not serve only the purpose of making navigation possible. A person has purposes and intentions of acting in certain ways in places, and as proposed by Genereux, Ward, and Russell (1983), he or she is likely to imagine whether a place is suitable to choose to go to and, once there, whether to stay and engage in the desired activities (Figure 3). The criteria for suitability may be pragmatic (e.g., costs), perceptual–cognitive, and/or affective. Research on the experience of places while one is there is reviewed first, followed by research on how places are imagined.

Place Experiences. The pragmatic aim of environmental assessment is to predict how properties of places relate to perceptual–cognitive–affective responses and to actions (Craik & Feimar, 1987). Investigations of

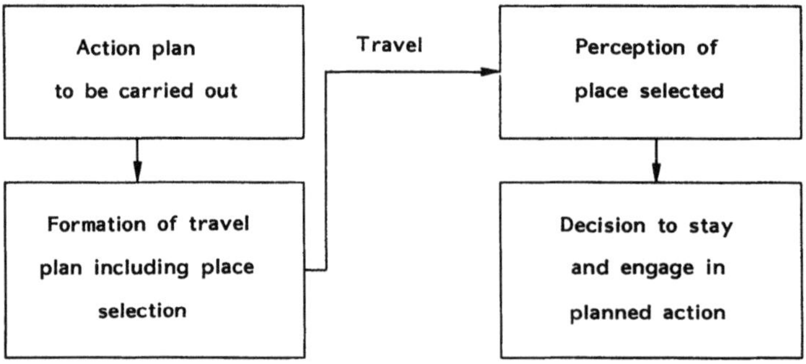

FIGURE 3. An illustration of how place experiences and place images are interrelated. (After Genereaux *et al.*, 1983.)

such relationships face decisions concerning what responses to measure. Thus, some research, following a psychometric tradition, has tried to reveal the basic components of place experiences. In an extensive, representative study, Ward and Russell (1981) presented subjects with color slides of a set of diverse places. In this single study, these researchers used the major techniques of analysis employed in a large number of other studies (e.g., Gärling, 1976; Hall, Purcell, Thorne, & Metcalfe, 1976; Kasmar, 1970): MDS of judged dissimilarities between pairs of places, MDS of the amount of overlap of the adjectival descriptions checked for each place, and factor analysis of ratings of the settings on semantic-differential types of adjectival scales. Ratings of anticipated actions in each place were also collected and were subjected to similar analyses. The main findings of the study were that the different methods yielded similar results, and that the action ratings were possible to predict from the underlying dimensions of similarity among the places. These dimensions were not only perceptual (for instance, scale, built versus natural, and vertical versus horizontal), but also affective (such as pleasant and arousing; see also Russell, Ward, & Pratt, 1981). After reanalyses of the action ratings, along with some cross-validation data, Genereux *et al.* (1983) reached the conclusion that people distinguish between places on the basis of suitable actions. Another finding was that some actions (e.g., walking) were perceived as suitable in most places; other actions (e.g., watching TV), only in a few places.

Ward and Russell (1981) noted that a basic tenet of their study and of similar studies—namely, that responses to places can be represented by a small set of uncorrelated dimensions—may not be tenable. A different but complementary approach takes as its point of departure the work by Rosch on so-called natural categories (Mervis & Rosch, 1981). Instead of focusing on underlying dimensions, this research aims at revealing a perceptual taxonomy. Tversky and Hemenway (1983) asked subjects to name slides of ordinary places (e.g., stores, restaurants, and beaches). As in other stimulus domains, such as situations, scripts, everyday objects, plants, and animals, the results could be described as a hierarchy of categories with a basic, preferred intermediate level of abstraction, as well as superordinate and subordinate levels. The superordinate level (e.g., built-natural environments) coincided with dimensions revealed by factor analysis in some of the studies referred to above; the subordinate levels coincided with results in such studies of other types of analyses (cluster analyses) that do not presuppose uncorrelated dimensions. Thus, despite the different assumptions of this study, as compared to previous studies (e.g., Ward & Russell, 1981), commonalities in results were obtained.

In similar but more inclusive studies, concepts from semiotics and substantive propositions based on architectural theories of meaning have been introduced as a framework for the study of place experiences (Groat, 1981). Although even taxonomic studies need to be guided by theory, one obvious problem, documented by Groat (1982) in a study of categorizations of building façades, is that general populations may not respond in the same way as architects and designers. As Gibson (1966, 1979) argued, with experience people increase their sensitivity to invariants in the environment. The fact that experts are able to make a certain kind of discrimination therefore tells us that this is something nonexperts can learn to do. However, the very fact that nonexperts do not make this discrimination may mean that it has little utility to them. It is of value to demonstrate such differences between different populations but a rationale other than experts' own experiences is clearly needed for the derivation of theories about laypeople's place experiences.

Recent advances in theories about place evaluations contribute to the further understanding of place experiences. Canter (1983) proposed that places are evaluated with respect to whether they facilitate a person's objectives. Such objectives, or referents of the place experience, may be spatial access, convenience, comfort, security, and satisfying social interactions. Different aspects of places vary in their degree of focus, depending on their relation to the person's objectives. A third factor is spatial scale. An individual's purposes may be facilitated at any level of interaction with a place. No particular scale can be identified *a priori* as being more important than any other. A schema for the internal representation and organization of place experiences was illustrated in Canter (1983), and Canter and Rees (1982) provided an empirical example of people's satisfaction with their housing. In this study, the schema guided the selection of questionnaire items. MDS of the questionnaire responses provided support for the structure entailed by the schema.

In a study by Craik and Appleyard (1980), the question of the different foci of the perceptual response was raised. Taking as their point of departure the theory proposed by Brunswik (1956), these authors conceived of the environment as organized in layers at various "depths" from the observer. For instance, security (crime rate) cannot be directly perceived; it has to be inferred from "cues" that, in a certain ecology, are more-or-less reliable indications of the crime rate. More than in many other studies, the inferential nature of environmental perception and cognition is highlighted in this research. Furthermore, the foci may, as in Canter's theory, vary, depending on the person's purposes and intentions. Physical-spatial properties may sometimes be perceived as such, and sometimes as overt cues related to covert factors. To what extent and in what respects the environment is probabilistic have been debated (e.g., Gibson, 1966, 1979). It seems clear, however, that many perceptual-cognitive responses are predictive in nature, and that many contingencies in human environments cannot be predicted with certainty.

Another general approach was taken by S. Kaplan (1983), who started with the notion of fit or congruence between people and environments. Perceptual-cognitive processes contribute to such a congruence and, to the extent that the congruence is subjectively achieved, to positive feelings toward the environment. It is further argued that congruence is experienced when a place facilitates the execution of actions. However, not all actions are instrumental in the attainment of goals and purposes. S. Kaplan (1983) noted that people also have "inclinations," that is, things they like to do that are rewarding in themselves. Furthermore, some actions are required and necessary. Moments of reflection and thought are also part of everyday life. These different types of actions have different priorities over time. By choosing different places on the basis of judgments of their suitability, a person may achieve the desired congruence. The important role of perceptual-cognitive processes is to inform the person of how likely the environment is to support his or her actions, at a certain time, and over time. Studies of preference of places have provided some support for S. Kaplan's theory. For instance, Nasar (1983) had a heterogenous group of subjects rate their preferences for a number of residential scenes presented in wide-angle color slides. A panel of experts rated the same scenes on a number of descriptive adjectival scales. Multiple-regression analyses showed that rated unambiguousness was one factor that predicted the laypeople's preferences. In another study, Purcell (1986) found that preferences for places depended on their similarity to a schema or prototype. However, high preferences required some discrepancy from the schema, more so for architectural than for nonarchitectural students.

Research on place experiences has clearly advanced beyond the stage of establishing suitable response measures and correlations between the properties of places and these response measures. Theories that aim to explain such correlations have been proposed in recent years. In these theories, an affective response is conceptualized as being the end product of a sequence of information-processing stages. However, that an immediate affective response to places also may determine how the information is processed appears to have been overlooked. This point was made by Ulrich (1983) in his interpretation of some of the findings in research on aesthetic and affective responses to natural environments. He further noted that an immediate affective response of like or dislike to gross structural features of places (such as their complexity, defined as their number of varied elements, depth, and focal points attracting attention) may be affected by the emotional state of the person. Affective responses may also affect the emotional state of an individual by, for instance, reducing stress. Ulrich (1983) indicated how this view can be integrated with cognitive models of affective response like those reviewed above.

Images of Places. Methods similar to those used to investigate direct experiences of places have been employed in studies of images of places at various spatial scales. Such studies of images are typically carried out by means of interviews in settings other than those that are the subject of investigation. Thus, the subjects have, in effect, been interrogated about their cognitive-affective representations of places. The distinction between place experiences and images of places is operationally clear-cut. However, this does not mean that they are unrelated. As was noted above, cognitive-affective representations of places, based on previous direct (or indirect) experience, are assumed to play an important role in the direct experience of places. A direct comparison in a study by Lowenthal and Riel (1972) indicated that the representational systems for directly experienced places and for images of places were similar. However, even though the representational systems (i.e., the dimensions or categories that people use to represent the place) are the same, a particular place may not be perceived exactly as it has been remembered. Conditions may have changed from one occasion to another, and forgetting may occur. Forgetting may sometimes be schema-dependent. In such forgetting, the memory representation of a place is more schematized (Mainardi-Peron *et al.*, 1985); that is, it contains fewer details that distinguish it from the general class. Much more needs to be learned about how acquisition and forgetting influence images of places, both in adults and in children (cf. Axia, Baroni, & Mainardi-Peron, 1986).

Studies of images of places have significance in attempts to model people's choices of places at various distances (Halperin, Gale, & Gorenflo, 1988). These choices are assumed to be functions of the attractiveness or utility of the places and the distances to them. In image studies, a multiattribute utility conceptualization of decisions and choices is usually entertained; that is, the choice alternatives are conceived of as being decomposable into attractiveness values or utilities, associated with different attributes, which in some way are combined. Such research, therefore, has tried to find out about what these attributes are.

The most extensive research on images of places has been concerned with the retail environment. Downs (1970), in his pioneering study, postulated nine attributes or dimensions of the images of retail centers. In addition to shopping-related dimensions, such as price, service, and product quality, dimensions related to the perception of the centers, such as structure and design, ease of internal movement and parking, visual appearance, and atmosphere, were postulated. A factor analysis of adjectival scale ratings indicated that the factors could be grouped into retail establishment factors and factors related to structure and function. The former group of factors appeared to be more important than the latter.

Some subsequent studies have used the repertory grid (e.g., Hudson, 1974; Timmermans, van der Heijden, & Westerveld, 1982a) and MDS of similarity, dissimilarity, or preference ratings (e.g., Spencer, 1980). Results similar to those of Downs (1970) have been obtained with these techniques, which do not presuppose that attributes are specified in advance. For instance, using grid elements, presented in triads, Timmermans *et al.* (1982a) asked a sample of subjects to respond to a number of familiar shopping centers. For each triad, a subject indicated which important "constructs" made two of the elements similar to one another and dissimilar from the third element. The presentation was interrupted when no new constructs were elicited. The results, validated with another method called *factor listing,* indicated that retail factors like choice range, price, and quality of goods were among the more important. Accessibility and convenience were also mentioned. Cleanliness, cosiness, and atmosphere were factors mentioned as being related to the perception of the shopping centers.

At a general level, the results of studies of images appear to be similar to those obtained in studies of direct place experiences (e.g., Ward & Russell, 1981). A more important source of differences in results may be the type of places investigated and, in particular, the actions these different places are suitable for. Images of retail environments clearly entail aspects of shopping; in other environments, aspects of other actions are salient. Thus, the importance of an action-theoretical perspective in studies of environmental perception and cognition, as detailed in Canter (1983) and in Kaplan (1983), is highlighted. Image studies that also aim to predict and explain consumers' spatial choices are reviewed below. In such studies, an action-theoretical perspective is evident.

EMPIRICAL RESEARCH IN LARGE-SCALE ENVIRONMENTS

The cognitive maps acquired of large-scale environments like cities do not solely, or perhaps not even primarily, serve navigation and spatial orientation purposes. Where to shop, where to go for recreation, and where to buy a house are examples of intended actions that, before their execution, may sometimes require an extended decision and choice process. Cognitive maps may play an important role in this process as one of the sources of information that one draws upon. Once a choice of a location has been made, questions of how to get there are raised, and cognitive maps are likely to play a role in this process, too. As Lynch (1960) put it a long time ago, cognitive maps have an important function as organizers of many everyday actions.

Research demonstrating differences between the cognitive maps of cities and of environments at a smaller spatial scale are reviewed next, followed by a selective review of research on spatial preferences, decisions, and choices.

Cognitive Maps of Cities, Spatial Orientation, and Navigation

There are reasons to believe that cognitive maps of cities differ from cognitive maps of environments at a smaller spatial scale. Cognitive maps of cities are acquired under different circumstances. They are acquired when one is driving a car or is a passenger in a car, during underground transportation, and so on. Furthermore, verbal and printed communications may be as important as direct experience. For instance, Canter (1977) showed that the characteristics of the public subway maps strongly influenced Londoners' cognitive maps of cities. In another study, which compared seven different cities in countries all over the world, Canter and Tagg (1975) found evidence of the important role of road and rail systems. General topography and major geographical features were, however, also found to be important.

The work by Lynch (1960) suggested that cognitive maps of cities contain elements like landmarks, nodes, paths, districts, and edges. Several studies using statistically more sophisticated methods have replicated these original findings (e.g., Aragonés & Arredondo, 1985). Landmarks, nodes, and districts are all place identifiers; landmarks and nodes are place identifiers at the smaller scale, and districts are, at the larger, city scale. In contrast to landmarks and nodes, which may be represented in memory as imagelike structures, districts are a generalized, more abstract representation of a place. Thus, differences in representational formats may also be found between cognitive maps of environments at different spatial scales.

The large-scale environment can be regarded as consisting of hierarchically nested units. In the anchor point theory proposed by Golledge (e.g., Golledge, Gale, & Richardson, 1986a,b), it is assumed that individuals develop a hierarchical ordering of places, paths, and areas (called cues) in the environment. The position of a cue within this hierarchy depends on the amount of experience an individual has had with it as well as its significance to her or him. Primary nodes in the hierarchy, corresponding to smaller scale environments (e.g., home, one's own neighborhood, and one's workplace), may expand as direct and indirect experiences of the city accrue. There will, however, always be lower order nodes; the connections of such nodes are not spatial but networklike. Therefore, it is not possible to represent a cognitive map of a city geometrically (Golledge & Hubert, 1982). Golledge and Spector (1978) corroborated major predictions made from the anchor point theory, using methods such as structural equation modeling and ordinary regression analyses to discern what factors account for distortions and imprecisions in the spatial information represented in the cognitive maps of a particular city (Columbus, Ohio).

Navigation and spatial orientation in cities, in contrast to that in smaller scale environments, are probably not as intimately related to perceptions of the environment's physical-spatial properties. How long it will take to get from one place to another is often important to know, but knowledge of the direction from one place to another is seldom important. More salient questions are which route and which transportation mode should be followed. Signs, street maps, and communications with other people are likely to be involve when one travels in cities. Unfortunately, only a few studies have examined orientation and navigation in cities. Bronzaft, Dobrow, and O'Hanlon (1976) conducted an experimental study that demonstrated the difficulties people may have. The subjects participating in the study were asked to make trips through the New York City subway system. Not more than about 50% of the trips were carried out in optimal ways. In another study, Gärling, Böök, and Ergezen (1982) found that subjects to whom a city was more familiar had acquired accurate knowledge about the network of bus routes, whereas another group of subjects, less familiar with the city, had not. The latter subjects had an accurate knowledge of their habitual routes, but they were less successful in combining this knowledge to find efficient new routes. Giraudo and Péruch (1986) investigated taxi drivers in Paris (see also Pailhous, Lepecq, & Péruch, 1987). They found that less experienced drivers chose less optimal routes (as indicated by drawings on maps); differences in time estimates between less and more experienced drivers were also observed.

The question has also been raised whether orientation and navigation aids in cities need to and can be improved. For instance, Garland, Haynes, and Grubb (1979) found that transit maps could be improved by the combined addition of color coding and of street details, but that street details alone made trip planning less accurate. In another study with a similar aim, Bartram (1980) compared different modes of presenting information about bus routes. Maps—in particular, schematic maps—were found to be superior to a list format of presenting the information about interconnected bus stops. How information is acquired from cartographic maps was studied by Thorndyke and Stasz (1980). They first identified certain procedures used by subjects who showed superior acquisition; then, these procedures were successfully taught to other, less successful subjects. Although the authors did not do so, it may be possible to suggest ways of redesigning aids on the basis of such detailed observations of good performance.

Spatial Preferences, Decisions, and Choices

How objective attributes of the environment, such as spatial structure, affect spatial behavior at the city scale (for instance, residential mobility, travel, and consumer spatial choices) has been extensively investigated (Halperin *et al.*, 1988). Psychological approaches, in contrast, aim to predict spatial choices from a knowledge of individuals' preferences and decisions. The predictive power of models developed within such approaches may not be superior, but these models may be more valid as explanations (Golledge, 1981; Spencer & Blades, 1986). For instance, as Figure 4 illustrates, consumers' spatial choices have been conceptualized as being a process by which a person forms preferences for a set of shopping destinations on the basis of his or her cognitive representation of the environment, decides and chooses on the basis of these preferences and other constraining factors (e.g., budget con-

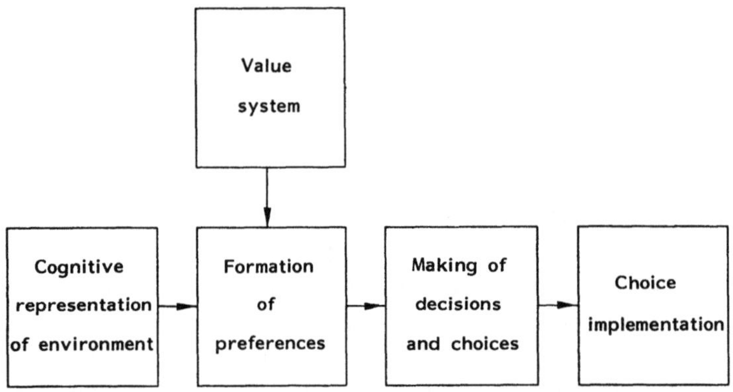

FIGURE 4. A conceptualization of consumers' spatial choices. (After Timmermans, 1982.)

straints), and, finally, implements the choice (Timmermans, 1982). A person's value system determines what he or she will prefer. Cognitive factors such as previous experience, representational ability, information-processing strategies, and decision rules affect the representation of the environment, as well as the processes of the formation of preferences, decision making, and choice implementation. Residential choices and choices of travel mode have been similarly conceptualized (Louviere & Henley, 1977; Louviere & Norman, 1977).

Much less research has focused on the link between cognitive representations, preferences, and actual spatial choices than on each of these factors separately (Lloyd, 1976). In two pioneering studies, Cadwallader (1975) and MacKay and Olshavsky (1975) demonstrated that the cognitive distances to a number of shopping locations were better predictors of consumers' choices of these locations than were the objective distances. Cadwallader (1975, 1981) proposed and verified a model in which preference was directly related to the psychological attractiveness of a shopping destination and was inversely related to how far away it was perceived to be.

Another related observation made by Cadwallader (1975) was that familiar shopping locations were more frequently patronized than less familiar ones. In other research on so-called consumer information fields (Timmermans, van der Heijden, & Westerveld, 1982b), it was assumed that consumers undergo a search process during which shopping locations are cognitively mapped with respect to both location and other attributes, and that the choice set of locations a consumer considers consists of the locations that he or she has learned about. Similar search processes with the same effects on choices have been assumed to be operative in connection with residential choices (Smith, Clark, Huff, & Shapiro, 1979).

The research on consumers' spatial choices has been confined to certain types of activities, whereas research directly focusing on environmental perception and cognition has not, in general, been so restricted. Consumers' spatial choices have also been viewed from a broader perspective, in which these choices are seen as being interrelated with other activities. Shopping choices may then be identified in daily travel patterns (Miller & O'Kelly, 1983), but it appears to be almost an insurmountable problem to find out from such observations what factors the choices are based on and what the constraints are. Nevertheless, some attempts have been made to infer the properties of the choice process from the objective characteristics of trips, such as length, the number and sequencing of stops, and complexity (Hanson, 1980). It has then been found, for instance, that people may travel farther than expected to purchase goods simply because they combine shopping with other errands. For the same reason, the attractiveness of shopping destinations may be influenced by the attractiveness of the surrounding facilities. Such results seem to confirm essentially what others have suggested (Gärling *et al.*, 1984; Passini, 1984; Russell & Ward, 1982), namely, that people carry out plans in their daily lives, possibly in many cases habitual plans. The conceptualization of choices described earlier needs to incorporate these features.

IMPLICATIONS FOR FUTURE RESEARCH AND APPLICATION

The review of environmental perception and cognition research in this chapter leaves the impression of a field of high productivity. The impression is also one of increasing scientific maturity. Theoretical perspectives have undergone changes in the direction of being more accurate conceptualizations of the real-life phenomena that are the major foci of the research. More important, many explanatory theories have been proposed in recent years, and purely descriptive studies of the relationships between environmental factors and measures of perceptual-cognitive-affective responses are increasingly being replaced by research aiming at investigating the validity of such explanatory theories. This replacement is evident in all subfields, but particularly salient examples are the theories of place evaluations proposed by Canter (1983) and Kaplan (1983), which constitute recent advances in the theoretical underpinnings of the assessment of places (Craik & Feimar, 1987; Holohan, 1986). A recent advance in research on the acquisition and use of cognitive maps is the implementation of computer simulations (Golledge *et al.*, 1985; Kuipers, 1978), which have contributed an increased precision and specificity to already existing theories (e.g., Gärling *et al.*, 1984; Golledge & Spector, 1978; Neisser, 1976; Siegel & White, 1975).

An increasing methodological sophistication is also contributing to the picture of a scientifically mature field. Very few, if any, cognitive mapping studies rely anymore on the sketch map technique introduced by Lynch (1960) in his pioneering study. A number of ingenious measurement procedures have been developed since then, not least in studies of children (e.g., Acredolo, 1985; Gibson & Spelke, 1983; Heft & Wohlwill, 1987). Moreover, cleverly designed field and laboratory experiments have replaced the case study design of the early days. In such efforts, techniques developed to simulate environments (reviewed by Craik & Feimar, 1987) have been instrumental (e.g., Evans *et al.*, 1984; Hunt, 1985b; Weisman, 1981). Advanced, multivariate data-processing methods are now routinely being used (e.g., Canter *et al.*, 1985).

What has been said so far should not obscure the fact that a unifying framework that integrates the diverse explanatory theories is lacking. There are also gaps in the knowledge accumulated so far. This review has tried to highlight commonalities between different subfields by consistently imposing the perspective of action theory (e.g., Frese & Sabini, 1985). Such a perspective, in which the overall functioning of the individual in the environment is stressed, seems desirable and appropriate. By far the most sophisticated theoretical treatment of the relationship between perception-cognition and action, in terms of decision-making conceptual models, can be found in research on spatial preferences and choices (Halperin *et al.*, 1988). An explicit recognition of the role of goals and purposes in theories of place evaluations (Canter, 1983; S. Kaplan, 1983) and the introduction into the cognitive-mapping research area of concepts such as action and travel plans (Gärling *et al.*, 1984; Russell & Ward, 1982), as well as decision-making frameworks (Passini, 1984), suggest the possibility that integrative frameworks will emerge in the coming years.

With respect to empirical studies, a few imbalances have led to some gaps in knowledge. Spatial preferences and choices have been studied at a city scale (Halperin *et al.*, 1988). Such studies should be equally valuable in smaller-scale environments. Knowledge has been obtained about spatial orientation and navigation in small-scale and medium-scale environments, but route selection, preferences for and choices of places, and related phenomena have scarcely been studied at this spatial scale. Studies of the acquisition of cognitive maps and place images, and of spatial orientation and navigation, are clearly underrepresented in large-scale environments. Much more needs to be known about place recognition in small-scale environments (Böök, 1986a), but it may also be of value to learn more about the perception of physical-spatial properties, which was of focal interest in pioneering studies of environmental perception (Holohan, 1982). To some extent, the information-processing theoretical perspective may have unduly diverted researchers' attention from basic perceptual processes, although this perspective certainly has, at the same time, been instrumental in providing a more encompassing and accurate conceptualization of environmental perception and cognition.

The application of research findings has undoubtedly been more complicated as the field of environmental perception and cognition has matured scientifically. It is seldom the case that any simple relationships are possible to discern between environmental factors, which are controlled by design, and perceptual-cognitive-affective responses by user populations. These relationships usually depend on a number of other factors. Such other factors are not only environmental but may be person characteristics, such as goals and purposes, affective states, previous experiences, and so on. Still, it is clearly a joint responsibility of researchers and practitioners to improve and to monitor the quality of human environments. Models based on theories that predict what environments may facilitate the functioning of perceptual-cognitive processes in connection with basic actions such as movements (e.g., Gärling *et al.*, 1986a), in combination with the skillful and knowledgeable use of such models by those directly responsible for the design of environments, are one, rather traditional solution. In other cases, the designer or some consultant may need to measure perceptual-cognitive responses in simulated environments before a design is implemented. As noted above, techniques suitable for this kind of measurement have been developed (e.g., Canter *et al.*, 1985; Weisman, 1981).

REFERENCES

Acredolo, L. (1985). Coordinating perspectives on infant spatial orientation. In R. Cohen (Ed.), *The development of spatial cognition.* Hillsdale, NJ: Erlbaum.

Acredolo, L. P., Adams, A., & Goodwyn, S. W. (1984). The role of self-produced movement and visual tracking in infant spatial orientation. *Journal of Experimental Child Psychology, 38,* 312–327.

Allen, G. L. (1981). A developmental perspective on the effects of "subdividing" macrospatial experience. *Journal of Experimental Psychology: Human Learning and Memory, 7,* 120–132.

Allen, G. L., & Kirasic, K. C. (1985). Effects of the cognitive organization of route knowledge on judgments of macrospatial distance. *Memory & Cognition, 13,* 218–227.

Appleyard, D. (1976). *Planning a pluralistic city.* Cambridge, MA: MIT Press.

Aragonés, J. I., & Arrendondo, J. M. (1985). Structure of urban cognitive maps. *Journal of Environmental Psychology, 5,* 197–212.

Axia, G., Baroni, M. R., & Mainard-Peron, E. (1986). *Children's and adults' memory for places.* Manuscript submitted for publication.

Baird, J. C., Wagner, M., and Noma, E. (1982). Impossible cognitive spaces. *Geographical Analysis, 14,* 204–216.

Bartram, D. J. (1980). Comprehending spatial information: The relative efficiency of different methods of presenting information about bus routes. *Journal of Applied Psychology, 65,* 103–110.

Biel, A. (1983). Children's spatial representation of their neighbourhood: A step towards a general spatial competence. *Journal of Environmental Psychology, 2,* 193–200.

Böök, A. (1986a). *Human recognition memory for nonverbal visual information.* Unpublished manuscript, University of Umeå, Department of Psychology.

Böök, A. (1986b). *Memory for scene perspectives and spatial orientation.* Unpublished manuscript, University of Umeå, Department of Psychology.

Böök, A., & Gärling, T. (1981). Maintenance of orientation during locomotion in unfamiliar

environments. *Journal of Experimental Psychology: Human Perception and Performance, 7,* 995–1006.
Bronzaft, A. L., Dobrow, S. B., & O'Hanlon, T. J. (1976). Spatial orientation in a subway system. *Environment and Behavior, 8,* 575–594.
Brunswik, E. (1956). *Perception and the representative design of psychological experiments.* Berkeley: University of California Press.
Cadwallader, M. (1975). A behavioral model of consumer spatial decision making. *Economic Geography, 51,* 339–349.
Cadwallader, M. (1981). Towards a cognitive gravity model: The case of consumer spatial behaviour. *Regional Studies, 15,* 275–284.
Canter, D. (1977). *The psychology of place.* London: Architectural Press.
Canter, D. (1983). The purposive evaluation of places: A facet approach. *Environment and Behavior, 15,* 659–698.
Canter, D., & Rees, K. (1982). A multivariate model of housing satisfaction. *International Review of Applied Psychology, 31,* 185–208.
Canter, D., & Tagg, S. K. (1975). Distance estimation in cities. *Environment and Behavior, 7,* 59–80.
Canter, D., Brown, J., & Groat, L. (1985). A multiple sorting procedure for studying conceptual systems. In M. Brenner, J. Brown, & D. Canter (Eds.), *The research interview.* London: Academic.
Chalmers, D. J., & Knight, R. J. (1985). The reliability of ratings of familiarity of environmental stimuli: A generalizability analysis. *Environment and Behavior, 17,* 223–238.
Conning, A. M., & Byrne (1984). Pointing to preschool children's spatial competence: A study in natural settings. *Journal of Environmental Psychology, 4,* 165–175.
Craik, K. H., & Appleyard, D. (1980). Streets of San Francisco: Brunswik's lens model applied to urban inference and assessment. *Journal of Social Issues, 36,* 72–85.
Craik, K. H., & Feimar, N. R. (1987). Environmental assessment. In D. Stokols & I. Altman (Eds.), *Handbook of environmental psychology.* New York: Wiley.
Doherty, S., & Pellegrino, J. W. (1985). Developmental changes in neighborhood scene recognition. *Children's Environments Quarterly, 2,* 38–43.
Downs, R. (1970). The cognitive structure of an urban shopping centre. *Environment and Behavior, 2,* 13–39.
Downs, R. M. (1981). Maps and mappings as metaphors for spatial representation. In L. S. Liben, A. H. Patterson, & N. Newcombe (Eds.), *Spatial representation and behavior across the life span.* New York: Academic.
Downs, R. M., & Stea, D. (1973). Cognitive maps and spatial behavior: Process and products. In R. M. Downs & D. Stea (Eds.), *Image and environment.* London: Arnold.
Evans, G. W. (1980). Environmental cognition. *Psychological Bulletin, 88,* 259–287.
Evans, G. W., & Pezdek, K. (1980). Cognitive mapping: Knowledge of real-world distance and location information. *Journal of Experimental Psychology: Human Learning and Memory, 6,* 13–24.
Evans, G. W., Marrero, D. G., & Butler, P. A. (1981). Environmental learning and cognitive mapping. *Environment and Behavior, 13,* 83–104.
Evans, G. W., Skorpanich, M. A., Gärling, T., Bryant, K. J., & Bresolin, B. (1984). The effects of pathway configuration, landmarks and stress on environmental cognition. *Journal of Environmental Psychology, 4,* 323–335.
Foley, J. E., & Cohen, A. J. (1984). Working mental representations of the environment. *Environment and Behavior, 16,* 713–729.
Frese, M., & Sabini, J. (1985). Action theory: An introduction. In M. Frese & J. Sabini (Eds.), *Goal directed behavior.* Hillsdale, NJ: Erlbaum.

Garland, H. C., Haynes, J. J., & Grubb, G. C. (1979). Transit map color coding and street detail: Effects on trip planning performance. *Environment and Behavior, 11,* 162–184.

Gärling, T. (1976). The structural analysis of environmental perception and cognition: A multidimensional scaling approach. *Environment and Behavior, 8,* 385–415.

Gärling, T., Böök, A., Lindberg, E., & Nilsson, T. (1981). Memory for the spatial layout of the everyday physical environment: Factors affecting rate of acquisition. *Journal of Environmental Psychology, 1,* 23–35.

Gärling, T., Böök, A., & Ergezen, N. (1982). Memory for the spatial layout of the everyday physical environment: Differential rates of acquisition of different types of information. *Scandinavian Journal of Psychology, 23,* 23–35.

Gärling, T., Lindberg, E., & Mäntylä, T. (1983). Orientation in buildings: Effects of familiarity, visual access, and orientation aids. *Journal of Applied Psychology, 68,* 177–186.

Gärling, T., Böök, A., & Lindberg, E. (1984). Cognitive mapping of large-scale environments: The interrelationship of action plans, acquisition, and orientation. *Environment and Behavior, 16,* 3–34.

Gärling, T., Böök, A., & Lindberg, E. (1986a). Spatial orientation and wayfinding in the designed environment: A conceptual analysis and some suggestions for postoccupancy evaluation. *Journal of Architectural and Planning Research, 3,* 55–64.

Gärling, T., Lindberg, E., Carreiras, M., & Böök, A. (1986b). Reference systems in cognitive maps. *Journal of Environmental Psychology, 6,* 1–18.

Genereux, R. L., Ward, L. M., & Russell, J. A. (1983). The behavioral component in the meaning of places. *Journal of Environmental Psychology, 3,* 43–55.

Gibson, E. J., & Spelke, E. S. (1983). The development of perception. In P. H. Mussen, Flavell, J. H., & Markman, E. M. (Eds.), *Handbook of child psychology* (4th ed.), Vol. 3. New York: Wiley.

Gibson, J. J. (1966). *The senses considered as perceptual systems.* Boston: Houghton & Mifflin.

Gibson, J. J. (1979). *The ecological approach to visual perception.* Boston: Houghton & Mifflin.

Giraudo, M.-D., & Péruch, P. (1986). *Spatiotemporal aspects of urban space representation.* Unpublished.

Golledge, R. G. (1976). Methods and methodological issues in environmental cognition research. In G. T. Moore & R. G. Golledge (Eds.), *Environmental knowing.* Stroudsburg, PA: Dowden, Huchinson & Ross.

Golledge, R. G. (1981). Misconceptions, misinterpretations, and misrepresentations of behavioral approaches in human geography. *Environment and Planning A, 13,* 1325–1344.

Golledge, R. G. (1987). Environmental cognition. In D. Stokols & I. Altman (Eds.), *Handbook of environmental psychology.* New York: Wiley.

Golledge, R. G., & Hubert, L. J. (1982). Some comments on non-Euclidean mental maps. *Environment and Planning A, 14,* 107–118.

Golledge, R. G., & Spector, A. N. (1978). Comprehending the urban environment: Theory and practice. *Geographical Analysis, 10,* 403–426.

Golledge, R. G., Smith, T. R., Pellegrino, J. W., Doherty, S. E., & Marshall, S. P. (1985). A conceptual model and empirical analysis of children's acquisition of spatial knowledge. *Journal of Environmental Psychology, 5,* 125–152.

Golledge, R. G., Gale, N., & Richardson, G. D. (1986a). *Cognitive maps of cities I: The nature and properties of cognitive maps.* Unpublished.

Golledge, R. G., Gale, N., & Richardson, G. D. (1986b). *Cognitive maps of cities II: Studies of selected populations.* Unpublished.

Groat, L. (1981). Meaning in architecture: New directions and sources. *Journal of Environmental Psychology, 1,* 73–85.

Groat, L. (1982). Meaning in post-modern architecture: An examination using the multiple sorting task. *Journal of Environmental Psychology, 2,* 3–22.

Hall, R., Purcell, A. T., Thorne, R., & Metcalfe, J. (1976). Multidimensional scaling analysis of interior, designed spaces. *Environment and Behavior, 8,* 595–645.

Halperin, W. C., Gale, N., & Gorenflo, L. J. (1988). Current issues in behavioral modeling of consumer choice. In R. G. Golledge & H. Timmermans (eds.), *Behavioral modeling approaches in geography and planning*. London: Croom Helm.

Hanson, S. (1980). Spatial diversification and multipurpose travel: Implications for choice theory. *Geographical Analysis, 12,* 245–257.

Hart, R. A., & Moore, G. T. (1973). The development of spatial cognition. In R. Downs & D. Stea (Eds.), *Image and environment*. London: Arnold.

Hazen, N. L. (1982). Spatial exploration and spatial knowledge: Individual and developmental differences in very young children. *Child Development, 49,* 623–636.

Hazen, N. L., Lockman, J. J., & Pick, H. L., Jr. (1978). The development of children's representations of large-scale environments. *Child Development, 49,* 623–636.

Heft, H. (1981). An examination of the constructivist approach and Gibsonian approaches to environmental psychology. *Population and Environment, 8,* 227–245.

Heft, H., & Wohlwill, J. F. (1987). Environmental cognition in children. In D. Stokols & I. Altman (Eds.), *Handbook of environmental psychology*. New York: Wiley.

Herman, J. F., Shiraki, J. H., & Miller, B. S. (1985). Young children's ability to infer spatial relationships: Evidence from a large, familiar environment. *Child Development, 56,* 1195–1203.

Herman, J. F., Norton, L. M., & Klein, C. A. (1986). Children's distance estimates in a large-scale environment: A search for the route angularity effect. *Environment and Behavior, 18,* 533–558.

Holohan, C. J. (1982). *Environmental psychology*. New York: Random House.

Holohan, C. J. (1986). Environmental psychology. *Annual Review of Psychology, 37,* 381–407.

Hudson, R. (1974). Images of the retailing environment: An example of the use of the repertory grid methodology. *Environment and Behavior, 6,* 470–494.

Hunt, M. E. (1985a). Enhancing a building's imageability. *Journal of Architectural and Planning Research, 2,* 151–168.

Hunt, M. E. (1985b). Environmental learning without being there. *Environment and Behavior, 16,* 307–334.

Ittelson, W. H. (1973). Environment perception and contemporary perceptual theory. In W. H. Ittelson (Ed.), *Environment and cognition*. New York: Seminar.

Kameron, J. (1973). Experimental studies of environment perception. In W. H. Ittelson (Ed.), *Environment and cognition*. New York: Seminar.

Kaplan, S. (1983). A model of person-environment compatibility. *Environment and Behavior, 15,* 311–332.

Kaplan, S., & Kaplan, R. (1982). *Cognition and environment*. New York: Praeger.

Kasmar, J. V. (1970). The development of a usable lexicon of environmental descriptors. *Environment and Behavior, 2,* 153–169.

Kirasic, K. C., Allen, G. L., & Siegel, A. W. (1984). Expression of configurational knowledge of large-scale environments. *Environment and Behavior, 16,* 687–712.

Kosslyn, S. M. (1980). *Image and mind*. Cambridge, MA: Harvard University Press.

Kuipers, B. (1978). Modeling spatial knowledge. *Cognitive Science, 2,* 129–153.

Kuipers, B. (1982). The "map in the head" metaphor. *Environment and Behavior, 14,* 202–220.

Lachman, R., Lachman, J. L., & Butterfield, E. C. (1979). *Cognitive psychology and information processing*. Hillsdale, NJ: Erlbaum.

Landwehr, K. (1988). Environmental perception: An ecological perspective. In D. Canter, M. Krampen, & D. Stea (Eds.), *Environmental perspectives*. Guildford: Gower.
Leff, H. L. (1974). *Experience, environment, and human potentials*. New York: Oxford University Press.
Levine, M., Jankovic, I. N., & Palij, M. (1982). Principles of spatial problem solving. *Journal of Experimental Psychology: General, 111*, 157–175.
Levine, M., Marchon, I., & Hanley, G. (1984). The placement and misplacement of you-are-here maps. *Environment and Behavior, 16*, 139–157.
Lindberg, E., & Gärling, T. (1982). Acquisition of locational information about reference points during locomotion: The role of central information processing. *Scandinavian Journal of Psychology, 23*, 207–218.
Lindberg, E., & Gärling, T. (1983). Acquisition of different types of locational information in cognitive maps: Automatic or effortful processing? *Psychological Research, 45*, 19–38.
Lloyd, R. E. (1976). Cognition, preference, and behavior in space: An examination of the structural linkages. *Economic Geography, 52*, 241–253.
Louviere, J. J., & Henley, D. A. (1977). Information integration theory applied to student apartment selection decisions. *Geographical Analysis, 9*, 130–141.
Louviere, J. J., & Norman, K. (1977). Applications of information processing theory to the analysis of urban travel demand. *Environment and Behavior, 9*, 91–106.
Lowenthal, D., & Riel, M. (1972). The nature of perceived and imagined environments. *Environment and Behavior, 4*, 189–207.
Lynch, K. (1960). *The image of the city*. Cambridge, MA: MIT Press.
MacKay, D. B., & Olshavsky, R. W. (1975). Cognitive maps and spatial behavior of consumers. *Geographical Analysis, 7*, 19–34.
Magnusson, D. (Ed.), (1981). *Towards a psychology of situations*. Hillsdale, NJ: Erlbaum.
Mainardi-Peron, E., Baroni, M. R., Job, R., & Salmaso, P. (1985). Cognitive factors and communicative strategies in recalling unfamiliar places. *Journal of Environmental Psychology, 5*, 325–333.
Marans, R. W., & Ahrentzen, S. (1987). Developments in research design, data collection, and analysis: Quantitative methods. In E. H. Zube & G. T. Moore (Eds.), *Advances in environment, behavior, and design*, Vol. 1. New York: Plenum Press.
Mervis, C. B., & Rosch, E. (1981). Categorization of natural objects. *Annual Review of Psychology, 32*, 89–115.
Miller, E. J., & O'Kelly, M. (1983). Estimating shopping choice models from travel diary data. *Professional Geographer, 35*, 440–449.
Moar, I., & Bowers, G. H. (1983). Inconsistency in spatial knowledge. *Memory & Cognition, 11*, 107–113.
Moore, G. T. (1979). Knowing about environmental knowing: The current state of theory and research on environment cognition. *Environment and Behavior, 11*, 33–70.
Moore, G. T. (1987). Environment and behavior research in North America: History, developments, and unresolved issues. In D. Stokols & I. Altman (Eds.), *Handbook of environmental psychology*. New York: Wiley.
Nasar, J. L. (1983). Adult viewers' preferences in residential scenes: A study of the relationship of environmental attributes to preference. *Environment and Behavior, 15*, 589–614.
Neisser, U. (1976). *Cognition and reality*. San Francisco: Freeman.
Ohno, R. (1986). Notion of duality in visual systems and its implications for environmental design. In W. H. Ittelson, M. Asaki, & M. Ker (Eds.), *Crosscultural research in environment and behavior*. Tucson, AZ: University of Arizona Press.

Pailhous, J., Lepecq, & Péruch, P. (1987). Human spatial reference systems. In P. Ellen & C. Thinus-Blanc (Eds.), *Cognitive processes and spatial orientation in animal and men* (Vol. 2). Dordretch, The Netherlands: Nijhoff.

Passini, R. (1984). *Wayfinding in architecture.* New York: Van Nostrand Reinhold.

Péruch, P., Pailhous, J., & Deutsch, C. (1985). How do we locate ourselves on a map: A method for analyzing self-location processes. *Acta Psychologica, 60,* 71–88.

Presson, C. C., & Hazelrigg, M. D. (1984). Building spatial representations through primary and secondary learning. *Journal of Experimental Psychology: Learning, Memory and Cognition, 10,* 716–722.

Purcell, A. T. (1986). Environmental perception and affect: A schema discrepancy model. *Environment and Behavior, 18,* 3–30.

Rieser, J. J., Lockman, J. J., & Pick, H. L. Jr. (1980). The role of visual experience in knowledge of spatial layout. *Perception & Psychophysics, 28,* 185–190.

Rieser, J. J., Guth, D. A., & Hill, E. W. (1988). Sensitivity to perspective structure while walking without vision. *Perception* (in press).

Russell, J. A., & Ward, L. M. (1982). Environmental psychology. *Annual Review of Psychology, 33,* 651–688.

Russell, J. A., Ward, L. M., & Pratt, G. (1981). The affective quality attributed to environments. *Environment and Behavior, 13,* 259–288.

Sadalla, E. K., & Staplin, L. J. (1980). An information storage model for distance cognition. *Environment and Behavior, 12,* 183–193.

Schank, R. C. (1982). *Dynamic memory.* Cambridge: Cambridge University Press.

Siegel, A. W., & White, S. H. (1975). The development of spatial representations of large-scale environments. In H. W. Reese (Ed.), *Advances in child development and behavior* (Vol. 10). New York: Academic.

Sime, J. D. (1986). Creating places or designing spaces? *Journal of Environmental Psychology, 6,* 49–63.

Smith, T. R., Clark, W. A. V., Huff, D., & Shapiro, P. (1979). A decision making and search model for intraurban migration. *Geographical Analysis, 11,* 1–22.

Smith, T. R., Pellegrino, J. W., & Golledge, R. G. (1982). Computational process modeling of spatial cognition and behavior. *Geographical Analysis, 14,* 305–325.

Spencer, A. H. (1980). Cognition and shopping choice: A multidimensional scaling approach. *Environment and Planning A, 12,* 1235–1251.

Spencer, C., Blades, M. (1986). Pattern and process: A review essay on the relationship between behavioural geography and environmental psychology. *Progress in Human Geography, 10,* 230–248.

Spencer, C., & Darvizeh, Z. (1981). The case for developing a cognitive environmental psychology that does not underestimate the abilities of young children. *Journal of Environmental Psychology, 1,* 21–31.

Spencer, C., & Weetman, M. (1981). The microgenesis of cognitive maps: A longitudinal study of new residents of an urban area. *Transactions of the Institute of British Geographers, 6,* 375–384.

Stokols, D. (1978). Environmental psychology. *Annual Review of Psychology, 29,* 253–295.

Thorndyke, P. W., & Hayes-Roth, B. (1982). Differences in spatial knowledge acquired from maps and navigation. *Cognitive Psychology, 14,* 560–589.

Thorndyke, P. W., & Stasz, C. (1980). Individual differences in procedures for knowledge acquisition from maps. *Cognitive Psychology, 12,* 137–175.

Timmermans, H. (1982). Consumer choice of shopping center: An information integration approach. *Regional Studies, 16,* 171–182.

Timmermans, H., van der Heijden, R., & Westerveld, H. (1982a). The identification of factors influencing destination choice: An application of the repertory grid methodology. *Transportation, 11,* 189–203.

Timmermans, H., van der Heijden, R., & Westerveld, H. (1982b). Perception of urban retailing environments: An empirical analysis of consumer information and usage fields. *Geoforum, 13,* 27–37.

Tolman, E. C. (1932). *Purposive behavior in animals and men.* New York: Appleton-Century-Crofts.

Tversky, B., & Hemenway, K. (1983). Categories of environmental scenes. *Cognitive Psychology, 15,* 121–149.

Ulrich, R. S. (1983). Aesthetic and affective response to the natural environment. In I. Altman & J. F. Wohlwill (Eds.), *Human behavior and environment* (Vol. 6). New York: Plenum.

Ward, L. M., & Russell, J. A. (1981). The psychological representation of molar physical environments. *Journal of Experimental Psychology: General, 110,* 121–152.

Weisman, J. (1981). Evaluating architectural legibility: Wayfinding in the built environment. *Environment and Behavior, 13,* 189–203.

Weiss, L., & Baum, A. (1987). Physiological aspects of environment-behavior relationships. In E. H. Zube & G. T. Moore (Eds.), *Advances in environment, behavior, and design,* Vol. 1. New York: Plenum.

Wicker, A. W. (1987). Behavior settings reconsidered: Temporal stages, internal dynamics, and context. In D. Stokols & I. Altman (Eds.), *Handbook of environmental psychology.* New York: Wiley.

V

ADVANCES IN RESEARCH AND DESIGN METHODS

8

Facility Programming

HENRY SANOFF

Facility programming is about information. The idea of a program is not new to the world of architecture. Its earliest application can be found in 1865 in a set of instructions for architects competing to design a new Court of Justice in London. Before that, the courts that had been built had been found inadequate from the start. The competing architects were advised to inspect the existing facilities and were warned that the arrangement of the courts and offices was extremely important; on it mainly depended the success or failure of the legal process: "Light and quiet were major considerations and were to be indicated for each room on a three-point scale"; the primary concern was "to facilitate the efficient dispatch and accurate transaction of the law business of the country" (Port, 1968, p. 81).

Programming is generally viewed as an information-processing system setting out design directions that will accommodate the needs of the user, the client, the designer, or the developer (Brever, 1986). The nature and scope of design decisions and design information have changed rapidly, and the role of programming has thus changed as well. The uses of programming have been extended from primarily acquiring and organizing information (Heimsath, 1977) to investigating and developing information, analyzing client and user needs, and evaluating projects after construction and occupancy (Freidman, Zimring, & Zube, 1978; Wener, 1988).

Facility programming has been seen as a valuable resource for a

Henry Sanoff • School of Design, North Carolina State University, Raleigh, North Carolina 27695-7701.

systematized process that provides a structured framework for accumulating and classifying data. As an analytical process, it encourages decision making through objective procedures rather than on individual assumptions or personal prejudices. A report by the Building Research Board of the National Research Council (1986) on programming practices states that "Programming services may not always result in new construction or changes to the physical building, but in organizational or managerial changes that achieve the same objectives" (p. 1).

DIVERSITY OF VIEWPOINTS

There is considerable diversity in the use of the terms *space programming, facility programming,* and *functional programming* and in their meaning within the design professions. In addition to the differing terms that identify programming, there are also philosophical differences regarding the meaning of *programming*. Expressions such as "Programming is design," "Programming is not design," and "Programming is getting ready for design," underlie the diversity of purposes and places of programming in the design process.

Everyone, however, would accept the view that a *program* is the organized collection of specific information that involves developing, managing, and communicating. Most will also agree that *programming* is the process of identifying and defining the needs of a facility. Although general features have emerged, programming should be recognized as a dynamic and interactive process.

The basic references for facility programming lie in the published volumes of Palmer (1981), Pena (1977), Preiser (1978, 1985), Sanoff (1978), and White (1972). The principal objective of Preiser's edited collections *Facility Programming* (1978) and *Programming the Built Environment* (1985) is to provide an authoritative overview of the user-oriented programming approaches that are currently to be found at work in architectural and environmental design. Both books describe the professional programming activity conducted by architectural and programming firms. The topics covered range from problem definition, cross-cultural programming, and postoccupancy evaluation to adaptive reuse and other more specific examples of facility programming. Each chapter is largely self-contained and represents various attitudes about programming and about the breadth, scope, and prospects of the field.

As a complement to the Preiser volumes, Sanoff's *Methods of Architectural Programming* (1978) is about the "nuts and bolts," or the technical aspects of programming. Here, the material moves from data-gathering techniques, through methods of synthesizing and organizing data, to a field application of programming techniques that makes use of user expertise. The volume stresses a general flexibility of approach, in which the techniques may be combined and merged, depending on the situation at hand.

Palmer's *Architect's Guide to Facility Programming* (1981) covers much of the same ground as the Preiser and Sanoff volumes, through attempts to integrate information-gathering techniques with case studies. Although it was written 3 years later than Sanoff's book, it fails to show the development of facility evaluation studies and their relationship to facility programming.

By contrast, Pena's *Problem Seeking* (1977) is a presentation of one approach, the CRS (Caudill, Rowlett, and Scott, Architect and Planners) method of programming. The five-step process is presented in lock-step fashion with partially worked examples. Rather than adopting a flexible attitude toward the organization of information, the book views the role of the programmer as being conformity to a predetermined format.

In an effort to bridge the theory–practice gap, White (1982) conducted interviews with architects to assess practitioner attitudes toward programming practice. In an open-ended telephone survey of 73 architects in the United States, programming strengths were described by the respondents as including "thorough rigorous analytic process, strong client/user participation, programming tailored to each project, strong integration with design and successful projects, and happy clients." The recurring programming problems were reported to be "finding the true needs of the client, getting clients to make decisions, and clients don't appreciate programming at program-design connection" (p. 37).

According to White (1983), clients are frequently unaware of their needs and sometimes are impatient or do not understand why the information is needed. Some clients will not permit the staff to participate in the programming process. There are also occasions when the client says one thing and the programmer hears another; the client often faces difficult choices and is not prepared to work as quickly as the process requires. These difficulties cited by practitioners clearly suggest the need for considerable improvement in communication with the client, as well as collaborative involvement in the programming process.

When asked about the reasons for offering programming services, beyond providing a better building, the respondents in the White study said that such a service "facilitates the design process, marketing, project management tool, client confidence in project and firm, and saves client and firm time and money" (p. 52). The majority of architects agreed that programming provided the client a way to participate in the project's planning process. They also indicated that programming leads to improvements in the client's operation; therefore, programming firms sometimes serve also as management consultants. Hellmuth, Obata & Kassabaum (Silver & Klein, 1985) described the programming process as being able to "promote confidence, teamwork and consensus; convey a sense of care and attention to users; minimize space-allocation abuses; and leave a clear trail of decisions and approvals for superiors to review" (p. 3).

COMPARING PROGRAMMING MODELS

Architectural programming is not a rigidly defined process. Each programmer has his or her own style and emphasis, and each project requires a certain amount of custom fitting of any model that a programmer may have. Because a number of programmers have described their approach, the purpose of this section is to examine seven of these approaches, to compare their similarities and differences, and to combine them into a composite model.

THE SEVEN MODELS

The models reviewed are by Davis & Szigetti (1979), Farbstein (in Palmer, 1981), Kurtz (in Preiser, 1978, p. 41), McLaughlin (1976, p. 42), Moleski (in Palmer, 1981), Pena (1977, p. 43), and White (1972, p. 44). These authors were selected because their programming work is documented in the literature and they follow a well-delineated process.

Gerald Davis. Davis's programming model (Davis & Szigetti, 1979) is a 21-step process (Figure 1) that begins with preprogramming and moves through evaluating the facility in use. It is directed toward the planning of corporate facilities. The first part of the process, as with both Moleski and Kurtz (below), is to become familiar with the client's business organization, operation, activities, and needs, both present and long-range. The programming steps include gathering data on the operating facilities; on physiological needs, such as those affecting health, safety, and performance; and on behavioral requirements, such as those affecting motivation, learning, and attitudes. The impact of the project is evaluated in terms of its effect on the functioning of the organization as well as its effect on the community and the environment. Cost estimates are made, and essential criteria for design are established. Throughout the design process, the programmer provides feedback to the designer. After the facility is built, the programmer assists management in moving

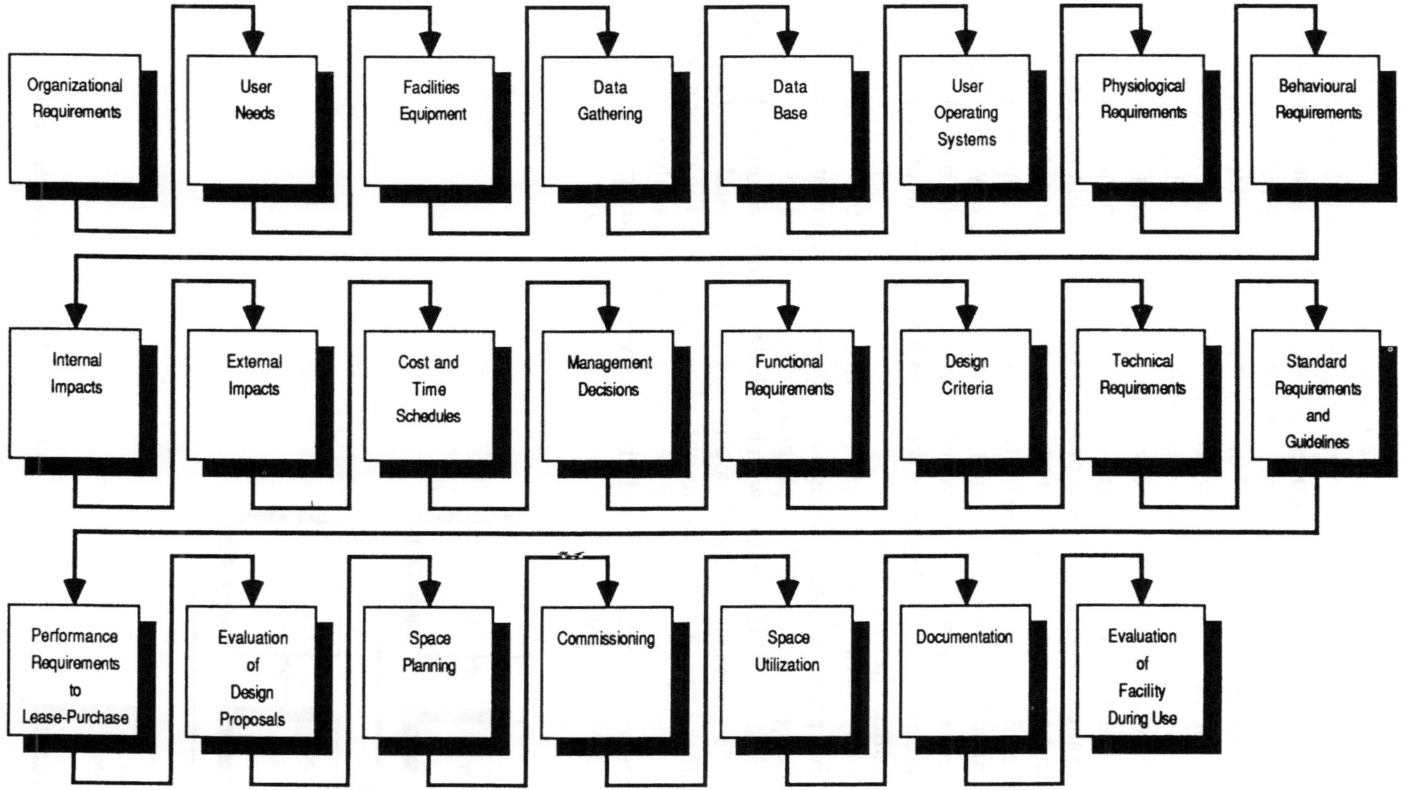

FIGURE 1. Davis's programming process.

into the facilities and in their fine tuning. The facilities are evaluated after they have been in use for a period of time, usually within 6 months.

Jay Farbstein. Farbstein's process (Palmer, 1981) consists of five steps (Figure 2) that identify client needs. His first step is to search the existing literature for information on the building type. The next step is to identify the users of the facility and their activities, attitudes, and characteristics. Then performance criteria are established for area needs, circulation, ambient environment, safety and security, surfaces, furnishings, flexibility, and site design. After the client reviews the performance criteria, the design issues are identified, the program options are identified for each issue, and each option is measured in terms of costs, benefits, and tradeoffs. The client is consulted again to assess the options. Space specifications and adjacency relationships are developed. Finally, the client is consulted once more for approval of the program and the budget. Farbstein's recent work (1985) stresses the periodic assessment of the facility's performance after occupancy.

Kaplan, McLaughlin, Diaz (KMD). McLaughlin (1976) of KMD believes that programming is design and that evaluation is an integral part of the design process. KMD's programming procedure consists of three phases (Figure 3), the first of which includes identification of the user, the user's organizational philosophy and objectives, and the financial feasibility of the project. The second phase considers the client's openness to new ideas, the physical context, a functional analysis, and aesthetic demands; this phase includes a survey of building types and all factors that may influence the form and content of the building. The

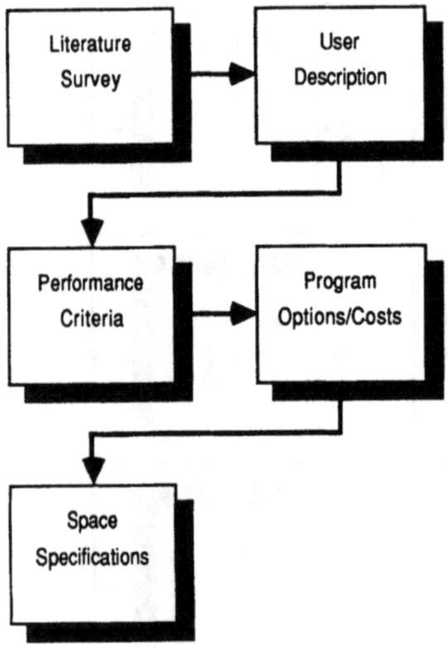

FIGURE 2. Farbstein's programming process.

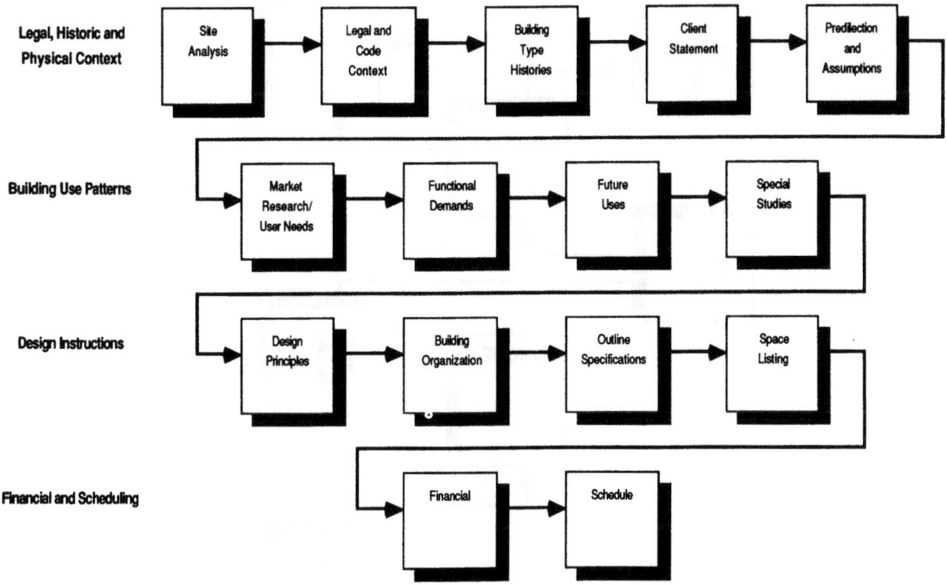

FIGURE 3. McLaughlin's programming process.

final phase corresponds more closely to what is typically referred to as *project development,* in which studies of building organization, schedules, outline specifications, and budgets are conducted. McLaughlin and KMD also follow a three-phase process of evaluating their buildings that begins in the programming stage of the project, with an exploration of the future occupants' expectations of the new facility. After the building has been occupied for several months, another survey of user expectations is conducted. The third survey of the users occurs after they have adjusted their work processes to the new facility.

John M. Kurtz. The programming model presented by Kurtz (Preiser, 1978) emphasizes iterative programming, which continues into the design phase (Figure 4). It is Kurtz's belief that generalized long-range programmatic decisions should be made at the beginning of a building project. The first step in this model, as in that of Moleski (below), is for the programmer to be familiar with the client's operation, philosophy, and objectives. The basic program includes a literature search on the building type, a determination of the client's operating requirements, and a preliminary program of the building's organization, space sizes, and relationships. The basic program is presented to the client for feedback, and the program is revised and presented again until general

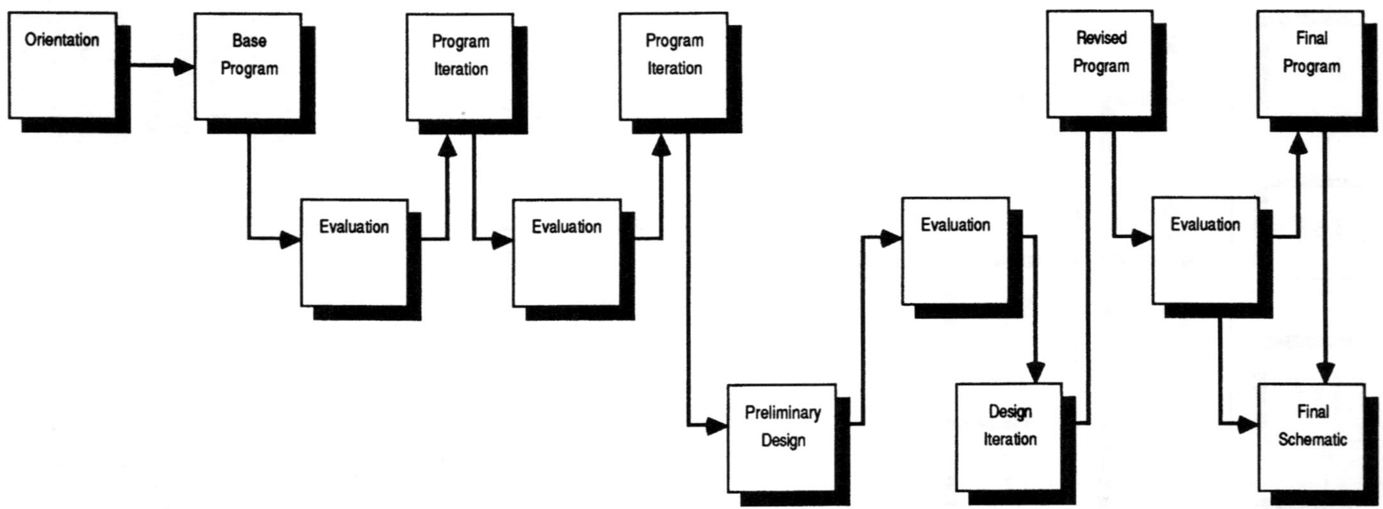

FIGURE 4. Kurtz's programming process.

agreement is reached. The design phase is considered part of programming because the client continues to provide feedback in reviews and revisions of preliminary designs.

Walter H. Moleski. Moleski's approach (Palmer, 1981) consists of four stages and two intermediate reviews with the client, the architect, and the programmer (Figure 5). The activities are (1) awareness; (2) diagnosis; (3) first review; (4) strategy; (5) second review; and (6) action. Initial familiarity with the client's operation occurs in Step 1, through reviews with the client and through interviewing the client to determine the nature of the organization, its functions, and its satisfactions and dissatisfactions with the present facility. Problem areas are identified, about which information is to be gathered. Information is then gathered in Step 2 through the use of such techniques as interviews, observations, and questionnaires. The information gathered is analyzed to show activities, relationships, problems, and needs. The data are organized, and concepts are formulated in a preliminary program that constitutes the awareness and diagnosis stages of the process. The first review, which occurs at Step 3 in the process, is a meeting with the client to discuss the preliminary problem statement and concepts and to select concepts for further development. The fourth step is to develop a strategy or to establish performance criteria. A more specific set of design needs is identified, and recommendations are made to the designer concerning spatial characteristics, physical conditions, symbolic attributes, and spatial arrangements. The second review (Step 5) is a client meeting to discuss the recommended program, to establish the architectural intent of the facility, and to obtain final approval of the program. The programmer's responsibilities extend into the action or follow-up stage, in which he or she consults with the designer on the intent of the program and evaluates the design solutions.

William T. Pena of CRS, Inc. Pena's programming procedure (1977), developed for CRS, is a comprehensive process of five steps: (1) establishing goals; (2) collecting facts; (3) uncovering concepts; (4) determin-

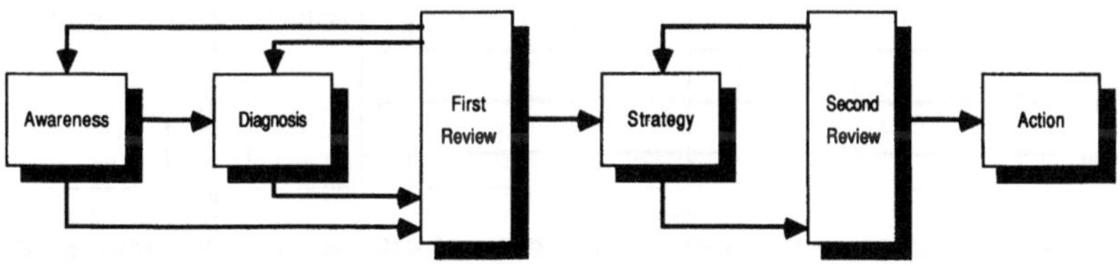

FIGURE 5. Moleski's programming process.

ing needs; and (5) defining the problem through a consideration of the five determinants of design: function, form, economy, time, and energy (Figure 6). A total of more than 132 considerations covers many aspects of a project, such as the reason for the project, space requirements, and site analysis. According to Pena, programming reduces the guesswork of designing for the user's needs and seeks rather than solves design problems. The process incorporates work sessions that bring together all parties involved in the project for a specific length of time. These sessions have been referred to as the "squatters' technique" because they occur in the client's backyard. The programming squatters use graphic analysis cards and "brown" sheets as working tools to determine the space program. The brown sheets indicate the space needs, which have been derived from the project goals, from facts, and from concepts. They represent area requirements and serve as worksheets during work sessions. The analysis card technique is a method of graphically recording the information that must be discussed, decided upon, and sometimes discarded during the programming phase of the project. A formal presentation of the program is made to the client, with a wall display of the analysis cards and the brown sheets.

Edward T. White. White's programming method (1972) consists of a series of tasks divided into three phases: preprogramming, programming, and postprogramming (Figure 7). In preprogramming, the client and the programmer agree on the process, the rules, the responsibilities, and the content of the program. Any previous work on the project is collected and organized. The information to be gathered is identified, the gathering sequence is decided on, and a team is put together, each member being assigned certain tasks. The programming stage consists of information gathering, analyzing, evaluating, correlating, organizing, and documenting. The information is then submitted to the client for

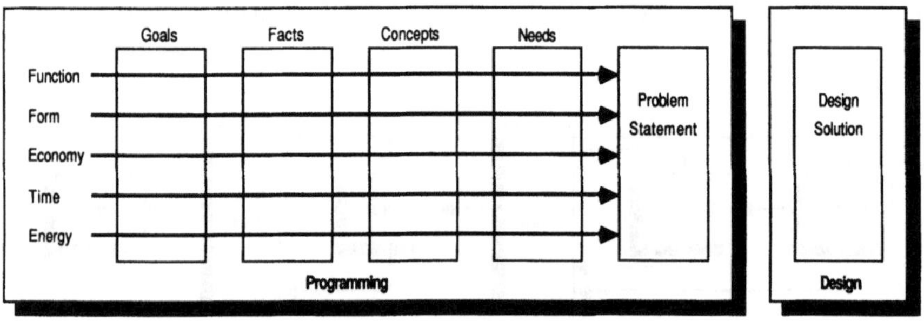

FIGURE 6. Pena's programming process.

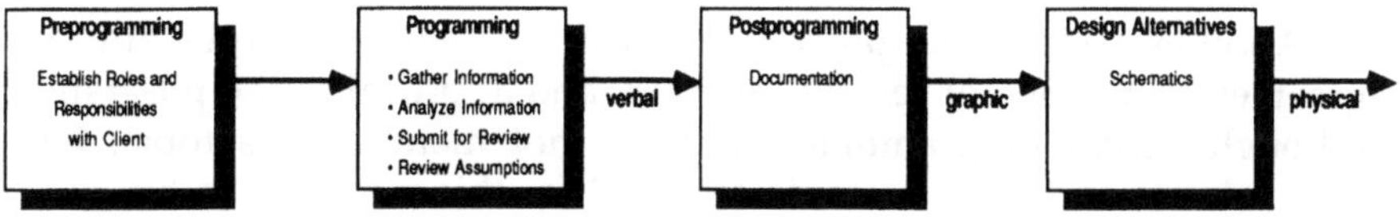

FIGURE 7. White's programming process.

review and approval. Space needs are tested against the budget, and planning and design implications (which are reviewed with the client) are drawn. Directives for the designer and alternative organizational concepts for the building are formulated and are again reviewed with the owner. Postprogramming is concerned with producing and distributing the program document and with giving presentations of the program, if appropriate.

COMPARISON OF THE SEVEN MODELS

All seven models make explicit the steps of collecting, analyzing, and documenting facts in a written program for the owner's approval and the designer's use. Pena described this process as collecting and analyzing facts and stating the problem. McLaughlin referred to actual and assumed user needs and their characteristics. White referred to programming, and Moleski considered the steps of diagnosis and collecting, analyzing, and organizing data. Kurtz made less of a point of data collection; yet, it is understood as being part of orientation and the base program. Davis referred to selective data-gathering and data bases for facilities. Although Farbstein's process begins with a literature survey, his other steps assume data collection.

Five of the programmers made a point of identifying needs relevant to the budget. Pena located economic considerations under the "determine needs" step, and McLaughlin's economic consideration is described as financial feasibility and occurs during various stages in the process. In his programming section, White (in Palmer, 1981) stated, "Test space needs against budget" (p. 40). Davis included three cost statements, under costs, elapsed time, and schedules, and he referred to cost again in discussing the functional requirements at each stage. Farbstein (in Palmer, 1981, p. 34) outlined, "program options and cost" to "measure costs/benefits/tradeoffs of options" and "review cost effectiveness of options," and his final step is for the client to approve the budget. Neither Moleski nor Kurtz made any references to budget considerations.

Another stage in the process in which there appears to be author agreement is the owner review, revision, and approval of the program. Although Pena's programming steps do not mention this topic, his brown sheets and analysis cards are used for this purpose. In his programming phase, White (in Palmer, 1981) mentioned that one should "submit information to sources for review and approval," "recycle allowable space information back to sources," "review planning assumptions with client," and finally, "review total program with client" (pp. 40–41). One of Moleski's major characteristics is review, during which the client discusses the preliminary problem statement and the solution concepts, reviews the recommended program, and finally, may revise the program during the design phase. Kurtz (in Preiser, 1978, p. 140) also emphasized review in his category "iterative programming," where the client provides feedback and "revision of program until general agreement is reached." "Design-as-feedback" is devoted to changing the program during the design phase. Davis was less concerned with review; however, in his step "evaluation of design proposals," he did provide feedback to the design team. Farbstein had the client reviewing after performance criteria, and after program options and costs, and finally approving the program and the budget.

Although the main objective of any building program is to establish goals, only five of the authors made this an explicit step. Pena's first 37 steps deal with establishing such goals as the accommodation of privacy or group interaction, the desired use of the existing facilities on the site, return on investment, and historic preservation. Moleski (in Palmer, 1981) was explicit only after his second review, when he said, "establish architectural intent of the project" (p. 37). Kurtz (in Preiser, 1978) started out his first step, orientation, by saying, "determine project objectives" (p. 139). Farbstein identified user objectives for the facility in his user description step, which appears to be the most comprehensive analysis of the users' social and behavioral characteristics, although McLaughlin and Davis also sought a thorough understanding of user characteristics.

Another important programming activity is a review of the literature on similar building types and operations. White reviews client data and conducts literature searches on similar facilities. Moleski, Kurtz, and McLaughlin review literature on similar buildings, and Farbstein conducts a literature survey of the pertinent research findings on specific behavior environments and professional practices related to the facility type.

There is much less consensus on what other steps are important. Two of the programmers (McLaughlin and White) believe that it is important to start the programming process by planning the program in which the objectives of the program are defined; the programming team is assembled, roles are defined, and responsibilities are assigned; information sources are identified; and a schedule is worked out.

Moleski, Kurtz, McLaughlin, and Davis have directed their methods toward the programming of organizations. They made a point of the programmer's understanding the nature, philosophy, and image of the client, its communication processes, its business plans, and its satisfactions and dissatisfactions with the present facility (Kaplan & McLaughlin, 1981).

Another important function of the program is to develop design concepts, but not all of the programmers made a specific point of this. Pena devotes 34 steps to uncovering and testing concepts dealing with function, form, economy, time, and energy, and McLaughlin devotes his whole third programming phase to project development. In his programming phase, White extracts planning and design implications, sets up directives for the designer, and designs alternative organizational concepts for the building. As a part of diagnosis, Moleski organizes data to discover and develop concepts for the solution of problems and the satisfaction of client needs; then, after the first review, the programmer selects concepts for further development. Farbstein identifies design issues and develops program options for each issue.

Only McLaughlin and Davis are concerned with assessing the impact of the project. Davis examines the internal impact on the functioning of the organization and its operating systems and procedures, as well as the external impact on the public that the building will serve, the surrounding community, and the natural ecology. This factor clearly advances facilities programming toward a new level of responsibility for the designer and the client.

Finally, very few of the programs studied regarded anything beyond the design of the building as being part of the programming process except Davis, who considered it the responsibility of the programmer to assist management and the users to move smoothly into the facility, to fine-tune at move-in, and to follow up with the users or the occupants as required. Davis planted the seeds of facility management, which is a concern about the operation of a facility beyond initial occupancy. He also regarded an evaluation of the facility during its use as being part of the programming process. He evaluates the building in terms of its original context, purpose, and requirements and in terms of the context and purposes currently applicable to its use. Evaluation is also an integral aspect of the McLaughlin process; a good part of this early evaluation work was conducted by the Environmental Analysis Group, a firm headed by Davis. Both Farbstein's and Moleski's work lend additional support to facility evaluation.

COMPOSITE PROGRAMMING MODEL

Clearly, the form, description, and emphasis of each of these programming models differ. Some programmers view the programming process as being distinct from the design process, and others view it as being part of a continuing and iterative process, of which design is a part. Regardless of whether programming is considered separate from or integral to the design, the intent of each programming model is to outline the categories of information to be collected, analyzed, and organized to assist in the design of a facility. If all seven of these models are combined into one, the result is a series of steps ranging from preprogramming to an evaluation of the facility in use. The following is a composite of these models, with a listing of each step referred to by each author (Figure 8).

The composite programming model is an attempt to distill and integrate the contributions of each approach. Although the model is organized sequentially in nine steps, the process is iterative and is not linear. The combined views do represent significant advances beyond each model, and they also categorically identify many key issues.

No doubt, there are programming approaches other than those identified (Davis & Szigetti, 1986). Of particular importance is the work of Gould (1986), Stone, Marraccini, and Patterson (1985), and Preiser (1985). However, their focus and contributions lie in the techniques of information organization, management, and designer–client interactions.

Robinson and Weeks (1984), whose work is more theoretical than that of the previous practitioners, argued that programming and designing should be linked in an interactive process, and that any separation may result in the premature definition of a problem. By incorporating a "hypothesis" approach, in which images are generated to be "proved" or disproved, these authors believed, formal ideas can be included in the early stages of the programming process. The combination of verbal and graphic exploration (Figure 9) is intended to allow the exploration of physical form and its implications. The process requires that the design ideas be challenged and explored at their inception and throughout the stages of the design process although their work has not yet been tested in practice. The architectural firm of Hellmuth, Obata & Kassalbaum (HOK) advocates a similar integrated approach (Figure 10), in which program and design development occur simultaneously. Although programming strategies are apt to vary with different project requirements, HOK programmers prepare clearly articulate work plans for each project.

COMPOSITE LISTING OF PROGRAMMING MODELS

	Davis	Farbstein	KMD	Kurtz	Moleski	Pena	White
Plan The Program							
Identify participants & organize programming team	●	●	●		●	●	●
Identify programming objectives	●	●	●	●	●	●	●
Identify program context	●	●	●		●		●
Identify information needed	●	●	●	●	●	●	●
Define with client - process, sequence, tasks, schedules, rules, responsibilities	●	●	●		●		●
Identify primary information sources	●	●	●			●	●
Understand Client's Organization & Philosophy							
Nature of organization, image & philosophy	●	●		●	●		
Organization function & communication process	●	●		●	●		
Satisfaction & dissatisfaction with present facility	●	●	●		●		
Identify user objectives	●	●	●		●		
Establish Project Goals							
Functional goals	●	●		●		●	
Form related goals						●	
Economic goals	●	●				●	
Time related goals						●	
Organizing Information Search							
Collect & organize project related facts	●	●	●		●		●
Questionnaires & interviews	●	●	●	●	●	●	●
Literature search	●	●	●	●	●	●	●
Observation of existing operations & facilities	●	●	●		●		
Background information from client	●	●	●		●		●
Review similar building types & operations	●	●	●	●	●	●	
Collect facts related to building function	●	●	●	●	●	●	●
Collect facts related to building form	●	●	●	●	●	●	
Collect facts related to building economy	●	●	●			●	●
Collect facts related to time (historical, present, future)	●	●					
Analyzing Information							
Analyze collected facts	●	●	●		●		●
Functional space standards	●	●					
Tabulation of space requirements	●	●	●	●			
Spatial diagrams		●	●		●		
Interaction patterns among activities		●	●	●	●		
Written description of functional units	●	●	●		●		
Conceptual Development							
Uncover, test & develop conceptual alternatives	●	●	●		●		●
Develop functional concepts			●			●	●
Develop form, economy & time related concepts						●	
Identify Budget Related Problems & Needs							
Functional needs	●	●			●	●	●
Form needs		●			●	●	
Economic needs	●	●				●	●
Project Impact							
Impact on client's organization & operation	●						
Community & ecological impact	●						
Program review and revision	●	●	●	●	●	●	●

FIGURE 8. Composite programming model: A listing of the procedural steps used by seven programmers.

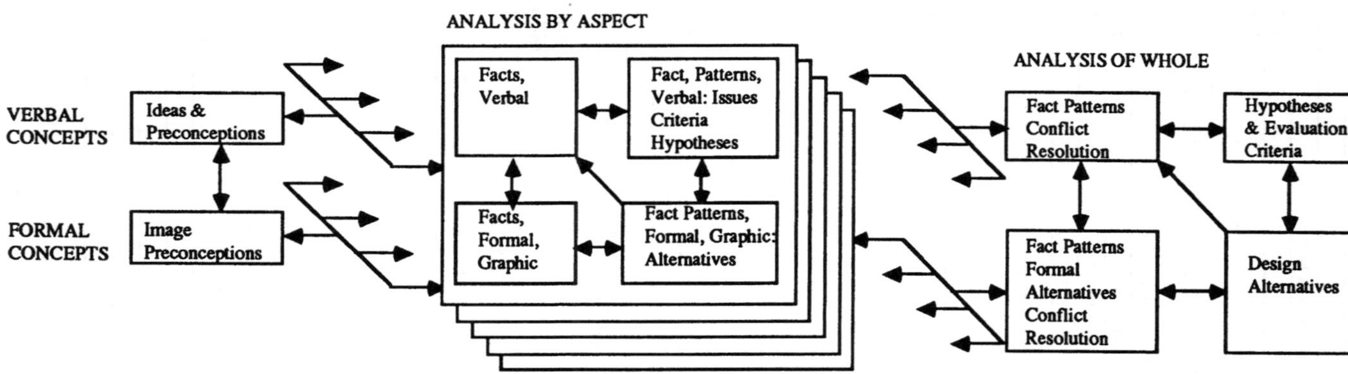

FIGURE 9. Robinson and Week's process: Interaction between programming and designing.

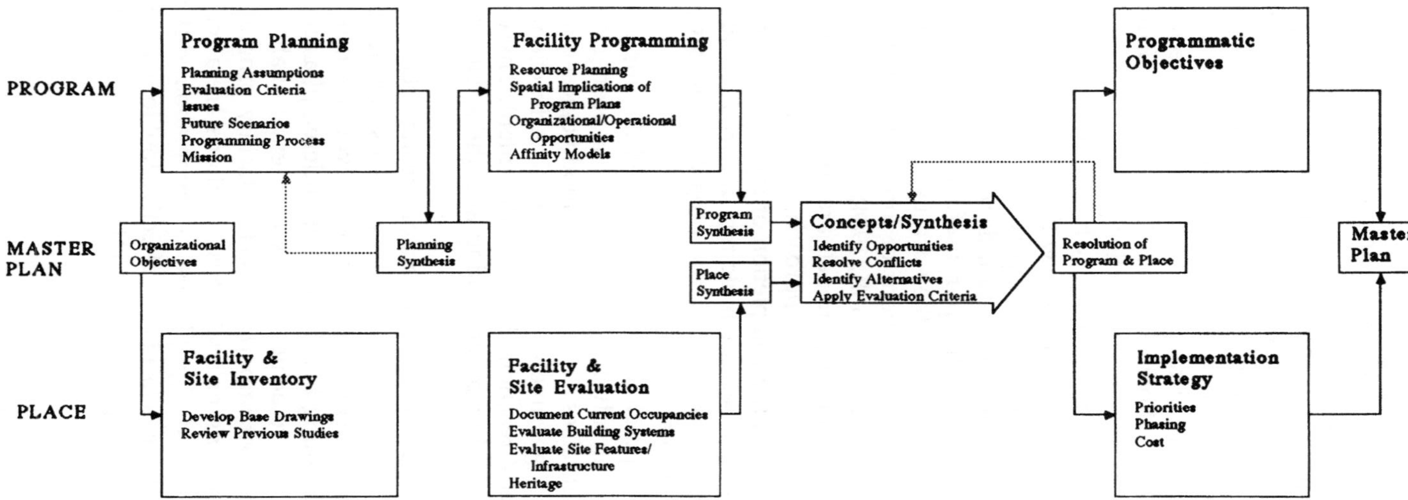

FIGURE 10. Master-plan integrated process. (From Hellmuth, Obata, & Kassabaum, 1985.)

The hypothesis approach is not unique to the field of architecture. Noted economist Downs (1976) advanced the idea of using arbitrary, but bold and imaginative judgments and assumptions in analyzing behavior ior patterns and in formulating social policies. He asserted that extremely arbitrary—even ridiculous—simplifying judgments are often useful in advancing perception, analysis, or prescription. This approach clearly has its parallels in facility programming, especially in the approaches advanced by Robinson and Weeks (1984). Examples of this approach in practice have been effectively introduced by this author in the planning of numerous facilities, where ideas about building form have been developed simultaneously with building function (Figure 11). This approach clearly represents a departure from conventional practice, which makes information gathering independent of formal judgments. Not only are these opportunities purposeful for the designer in challenging conventional thinking, but they are useful for the client, who becomes sensitized to the implications of form and meaning in the early stages of the design process (Sanoff, 1984).

Many of the programming models are characterized by a predominantly client-centered turnkey approach, in which the architect's responsibility ceases with the completion of the building. Because clients are not often users, and because buildings undergo considerable growth and change during their lifetime, the programming process needs to take these factors into account.

Although most practitioners have stressed the importance of client participation, direct involvement of the user has been given insufficient attention because it raises a number of methodological questions. Issues such as who should get involved are as crucial as the client's willingness to act on employees' reactions to the environment. These concerns address the core of facility programming and the architect's role in the process.

Not only questions of user satisfaction are important; so are satisfaction and management policies affecting the use of the environment. Therefore, the remainder of this chapter is devoted to clarifying aspects of facilities programming that have been underemphasized in practice, yet that hold the key to achieving environmental quality and satisfaction. Perhaps the most important facet of facility programming is the knowledge that existing or analogous situations are a major source of input into subsequent design phases. Learning about different ways in which the environment is used is an effective tool for gathering information and developing helpful insights into a building's performance.

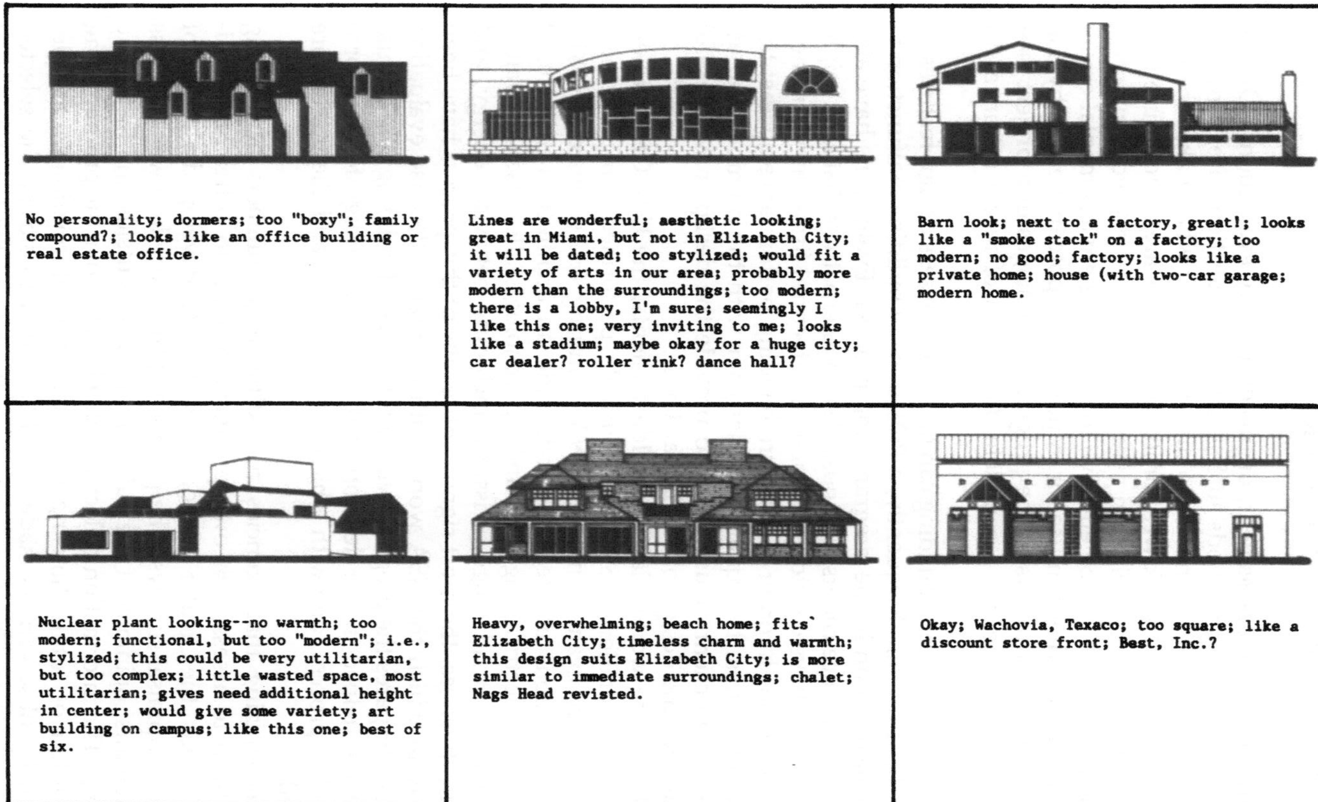

FIGURE 11. Building image ideas: Form studies are generated early in the programming process to sensitize clients to the interrelatedness of form and function.

LINKING EVALUATION TO PROGRAMMING

Environmental assessment, or postoccupancy evaluation (POE), is the practice of using methods such as surveys, questionnaires, and observations of people's behavior to discover exactly what makes the designed environment work well for its users. POEs are a procedure that involves the users in their own assessment of their everyday physical environment. POEs can be effective in correcting environmental errors by examining buildings in use, or in preventing potential errors by use of the information results in the programming stages of a project (see Wener, this volume, for a complete discussion of recent advances in POE).

An evaluation of a building can be used to assess design features as well as the program. Frequently, a POE traces a building malfunction to a false assumption in the programming process. Evaluations have also helped to convince clients to choose design alternatives that they might not otherwise have considered. Some architectural firms carry out their own evaluations in order to measure building performance against the original programs, to acquaint the designer with the opinions and attitudes of the client or user, and to provide the designer with useful feedback for the design of similar facilities.

One of the world's largest building clients, the U.S. General Services Administration (GSA), has routinely performed comprehensive postoccupancy evaluations since 1976 (Zimring & Reizenstein, 1981). These evaluations use teams that include an architect, a mechanical engineer, an electrical engineer, and a psychologist. Because the evaluations cover a broad range of issues, including behavior, energy, and cost-effectiveness, the most significant results of this process have been the way in which the GSA works with architects. Because evaluation process compares the program with the completed building, the evaluations have helped in the development of a more straightforward programming process, one with which both architects and client are more pleased.

Recognizing that agencies providing large numbers of housing units need to assess their own products in order to better fulfill residents' needs, the Greater London Council (GLC) developed a system of housing appraisal surveys. The purpose of the procedure was to assess GLC tenants' opinions of the built environment outside and inside the building. The information collected is used to improve the programming and design of future housing projects. This information is complemented by observations made by GLC staff who survey the selected housing projects. The social survey method used for the Program of

Housing Surveys is the Housing Appraisal Kit, which was developed jointly by the GLC and the Department of the Environment (Ravetz, 1980).

New Zealand's Ministry of Works and Development (MWD) and Victoria University at Wellington's School of Architecture (VUW) undertook a multiyear project to develop methods for the postoccupancy evaluation of government buildings (Daish, Gray, Kernohan, & Salmond, 1982). The MWD believed that an effective POE process could improve the quality of information, communication, and decision making it its building delivery activities and could thus enhance the quality of the program and the building. Central to the success of their POEs was the development of an appropriate process for gathering information. One method of evaluation used is described as a "walk-through" (see Gray, Watson, Daish, & Kernohan, 1985). This is a common method, often found in practice, of evaluating a building's performance by walking through and noting its salient features. The process involves two sets of people with distinct roles: a number of participant groups and one Task Group. The role of the members of the participant group is to evaluate the building from this group's point of view. Its members are drawn from among those with an interest in the occupied building. The role of the task group, which consists of the VUW research team, is to help the participant group in making evaluations. The task group plans and manages the evaluation tasks, records and evaluates the data, and writes the subsequent review. In planning for evaluation, a number of walk-throughs are scheduled, each with a small group of participants representing a particular interest in the building or project. Each walk-through involves the participant group and the task group in three connected activities: introductory meeting, a touring interview, and a review meeting. Following the walk-through, the participants and the task group meet to review the comments made during the walk-through and to form recommendations concerning the building and its functioning. When all walk-throughs are completed, the task group transcribes and classifies the participant groups' recommendations for reference and action.

LINKING DATA COLLECTION TO PROGRAMMING

Effective programming depends on knowing what types of data are needed and on selecting the appropriate means of obtaining and documenting them. There are a variety of techniques for obtaining information directly from the source (Michelson, 1975; Sanoff, 1978; Zeisel, 1981;

see especially Marans & Ahrentzen, 1987, and Low, 1987), of which walk-throughs are one. User reports through surveys and questionnaires are used for collecting opinions and attitudes, and observation methods, data logs, and standardized forms seek descriptive information. Both descriptive and evaluative data can be applied at various stages in the programming process. Background information from other than user reports is equally necessary and can be derived from analyzing existing data, literature, and records (Jones, 1970), as well as from the programmers' observations and insights.

A unique approach to health-care-facility programming is used by Stone, Marrachini, and Patterson (Tusler, Schraishuhn, & Meyer, 1985), who described two different user groups: patients (the primary users of health-care facilities) and the service providers, including nurses, physicians, administrators, and technicians. Each group has unique needs that relate to its particular role. Matching needs to resources, illustrated by a pattern or a framework for organizing information, analysis, and documentation (Figure 12), is the programmer's task. The four basic programming stages—utilization, function, systems, and space, combined with the areas of study—form a matrix that defines the programming pattern.

Data Collection

To illustrate techniques for collecting, organizing, and displaying programmatic information, a process has been developed in a project conducted for the Division of Health Services of the North Carolina Department of Human Resources (Kammeraad & Sanoff, 1979). The gathering and translation of information was integrated with a system of involvement in which staff and users were grouped into different levels corresponding to their areas of expertise. Patients and community members provided practical information on how the facility should be programmed by describing what actually works best. The major activities of the health department were described by those who were actually engaged in the work, and program directors or department supervisors were responsible for the information-gathering phase of the programming process, which included staff, patients, and information flow.

The planning process was developed and tested as a result of a grant from the Coastal Plains Regional Commission for a new Craven County Public Health Department facility. Program directors of the various health department services were selected to collect and summarize the data gathered from the programming process. A smaller decision-making group was formed to be responsible for the final decision-mak-

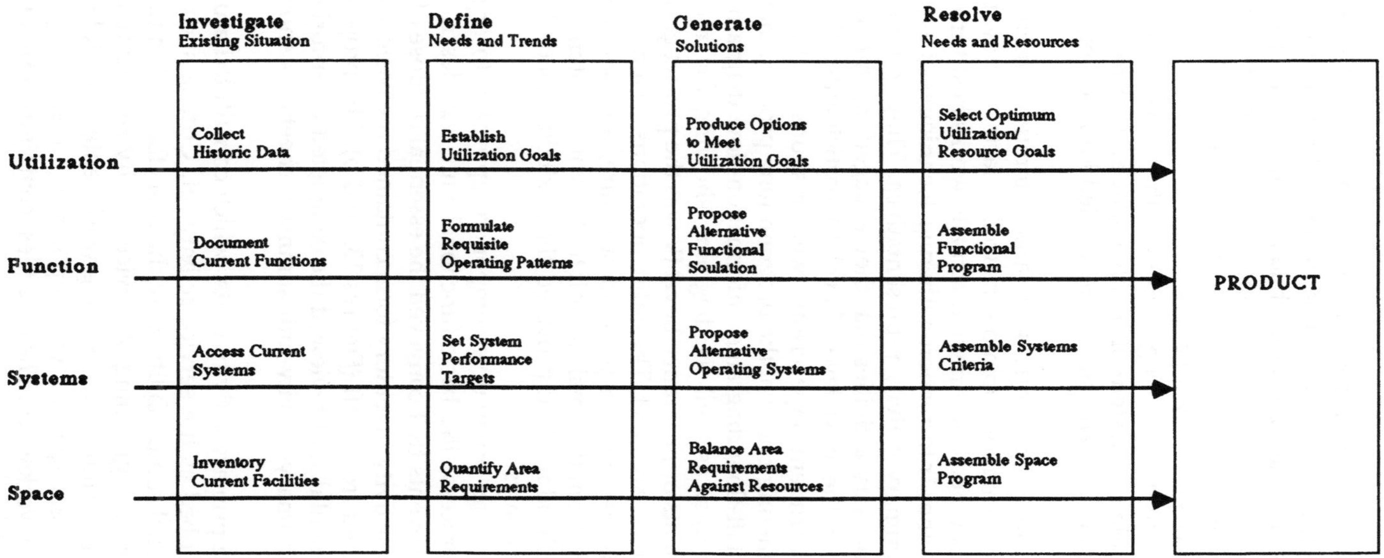

FIGURE 12. Programming-pattern overview. (W. H. Tusler, Jr., R. Schraishuhn, & E. R. Meyer, 1985.)

ing process. The director of the Craven County Health Department assumed the role of planning coordinator and administrative advocate. The supervisors of the adult health, child health, and the environmental health programs represented their respective departments. A set of public health goals and operational concepts was established following the first contacts between the two parties.

Participation in the planning process of the Craven County Health Department required a system of involvement and a procedure for reporting by which clients and staff were divided into four organizational levels.

Level I members (patients, staff, and community members) provided practical information (opinions as well as facts) on how the facility should be programmed and on what actually works best through questionnaires and other information-gathering techniques. Level II members were the primary activity representatives. They completed forms describing the primary activities that they engaged in, thus providing a basis for work at the next level. Level III participants, the program directors or department supervisors, were responsible for the whole information-gathering phase of the programming process, including the completion of staff-flow diagrams, information-flow diagrams, and patient-flow diagrams, as well as diagrams showing the relationships among departments or programs. Level IV was the project group, which included the project architect. The entire operations system of the health department was viewed as a pyramid structure, a series of functional modules ranging from the smallest unit to the whole-system conglomerate. This hierarchical method provided the flexibility needed for subsequent stages of analysis.

By the use of an integrative-component approach to organize the programming process, the basic functional unit was described as the primary activity events that conveyed the essential purpose of the facility. The primary-activity inventory form contained the basic activity data that were collected from staff (Figures 13a and b). The project architect and the departmental staff reviewed the completed inventories in order to prepare the primary-activity-unit summary sheets, as shown in Figure 14.

Interrelated primary activity units, which constituted a unit of service, were described as the service activity set. Criteria such as spatial and environmental compatibility were the basis for combining activities in a specific space. A grouping of service activity sets made up a functional module. For example a "nursing work-center" module was composed of a treatment space, a work-up space, a disposal space, and a nurses' stand-up workspace. The data files, containing primary-activity-

PROGRAM

FAMILY PLANNING

GENERAL ACTIVITY DESCRIPTION

A brief statement naming and defining the objectives of the activity

Examination

PARTICIPANTS

People involved in the activities at any one time

Family Planning Nurse
Patient
Doctor

Notes

MOLECULAR ACTIVITIES

Those subactivities necessary to accomplish the general action

Interview
dress/undress
Administer tests
give pills

Notes

SPATIAL REQUIREMENTS

Statement and projections of space needs to fulfill activities

Provide examination rooms
Provide for easy clean up
Provide a space for dressing & undressing
Provide a separate storage area for medicine
Provide a space nearby examining rooms for a doctor

Notes

MISCELLANEOUS NOTES

Any comment about the activity which is important, but has not been previously mentioned

Notes

FIGURE 13a. Primary-activity-inventory form (Side 1): Program representatives record information about participants, activities, and spatial requirements.

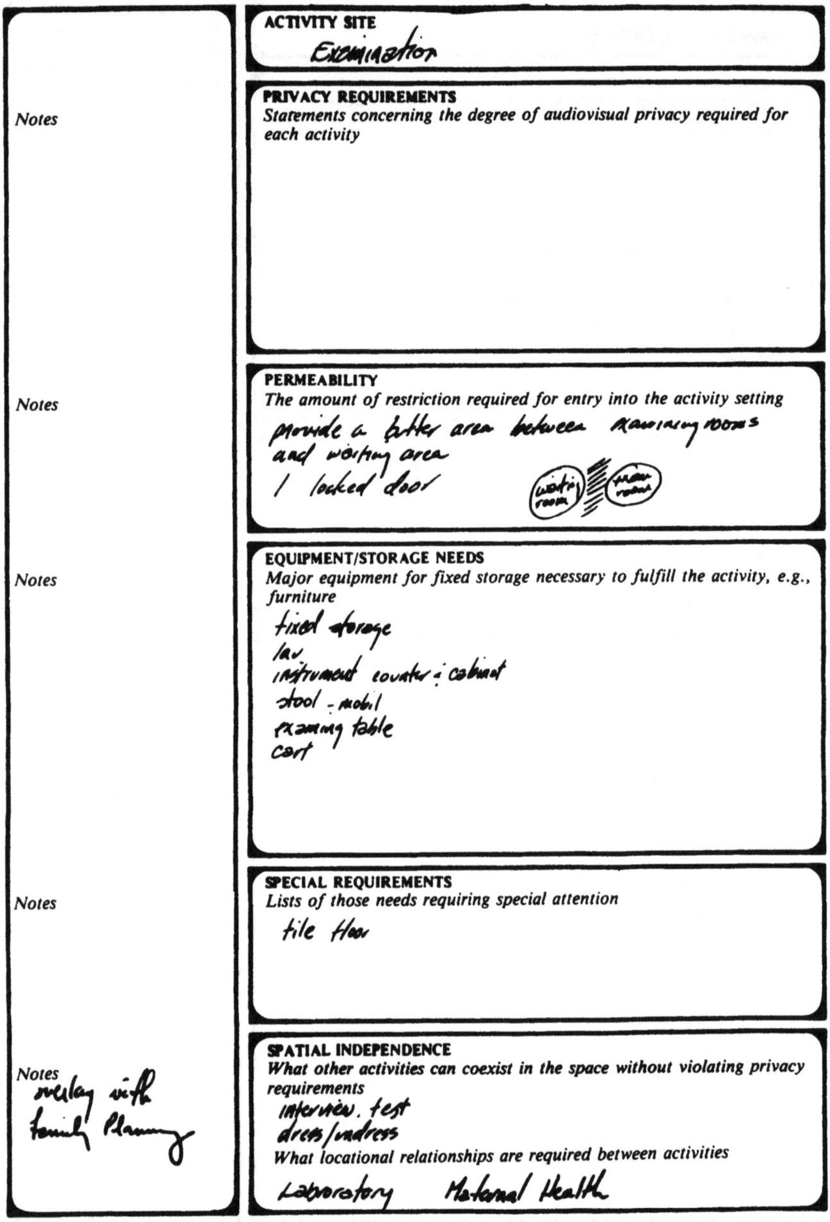

ACTIVITY SITE
Examination

Notes

PRIVACY REQUIREMENTS
Statements concerning the degree of audiovisual privacy required for each activity

Notes

PERMEABILITY
The amount of restriction required for entry into the activity setting
provide a better area between examining rooms
and waiting area
/ locked door

Notes

EQUIPMENT/STORAGE NEEDS
Major equipment for fixed storage necessary to fulfill the activity, e.g., furniture
fixed storage
lav
instrument counter & cabinet
stool - mobil
examing table
cart

Notes

SPECIAL REQUIREMENTS
Lists of those needs requiring special attention
tile floor

Notes
overlay with family Planning

SPATIAL INDEPENDENCE
What other activities can coexist in the space without violating privacy requirements
interview, test
dress/undress
What locational relationships are required between activities
Laboratory Maternal Health

FIGURE 13b. Primary-activity-inventory form (Side 2): Program representatives record information about privacy requirements, permeability, and spatial independence.

PRIMARY ACTIVITY UNIT SUMMARY SHEET

ACTIVITY UNIT Gynecological Examination

Molecular Activities

Prepare exam room
Fill speculum bucket with soap solution
Change roll paper on exam table before new patient
Have patient undress and put on gown
Have patient get on exam table
Refold sheet to cover patient
Read x-rays
Conduct exam or procedure
Take specimens
Record information on patient chart
Finish exam and have patient get dressed
Throw away soiled or used supplies
Wash and sterilize instruments and specula

Furniture & Equipment

Examining table; adjustable with stirrups
Examining light; cool cone 100 v.
Stool
Table for supplies
Otoscope
Writing or dictation station
Casework with sink
Soiled linen and specula recepticles
Waste basket
Foot stool
X-ray viewer
Sterilizer
Bulletin board
O_2 tank and equipment
Cabinets

Environmental Requirements

Temperature: 72°-$78^{o}F \pm 1.5^{o}$
Air changes: 8-12/hour
Relative humidity: 30-55% ± 10%
Relative pressure: Negative or zero
Filtration requirement: Medium
Lighting level: Minimum foot-candles
General - 50
Exam table - 100

Floorspace Requirements

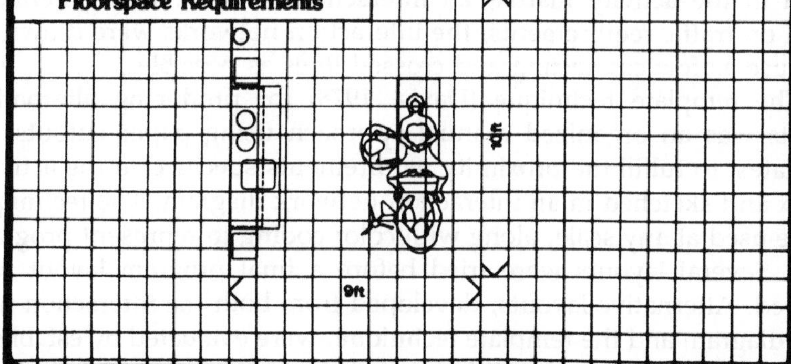

FIGURE 14. Primary-activity-unit summary sheet: Gynecological examination.

unit data, service activity sets, and functional module schemes, were organized by the staff, the architect, or a combination of both.

The next procedure was the completion of the routing or flow of people and information. The flow diagrams illustrated the most desired procedures, even when they differed from existing procedures. An intraprogram flow diagram (Figure 15) showed all the molecular activities that a patient or staff member normally encountered. Each procedure was written chronologically, from top to bottom, as shown on the patient–staff-flow data sheet (Figure 16). The interprogram-flow diagram showed the frequency of interaction among different programs to be housed in the same facility (Figure 17) and was completed by the program directors. The diagrams suggest the probability of patient–patient, patient–staff, and staff–staff interconnections.

Data Transformation

The planning for space needs required a summarization of program flow, which was a listing of the activities that a patient must encounter in each primary activity unit for each program (Figure 18). All primary activity units were reviewed for location compatibility, and any privacy requirements (noted in the spatial-independence section of the primary-activity-inventory form shown in Figure 13b) were taken into account (Figure 19). An activity matrix was used to tabulate the activity adjacencies necessary to fulfill the compatibility requirements recorded on the various activity forms (Figure 20). From the adjacency relationships coded in the activity matrix, an interaction network was generated. Based on traffic requirements, the interaction networks were drawn so as to use as few connecting and crossed lines as possible.

The template technique (Davis, 1978) for producing alternative layouts was an organized manual approach using paper cutouts, or templates, to fulfill the proximity requirements specified in the activity matrix and sketched in an interaction network diagram. The technique can be used at any scale, along with color coding to represent program areas. Several layouts were tried before a final program layout was adopted. Alternative layouts, developed from both the interaction network diagram and the template technique, were evaluated by establishing a set of weighted criteria that stated the objectives to be achieved, along with other influential factors (Figure 21).

A significant modification of the template technique incorporates graphic symbols corresponding to the activity and its relative area (Sanoff, 1984). With the use of graphic symbols, spatial planning can effectively simulate circulation, activity adjacencies, and conceptual variations in spatial layouts (Figure 22).

PROGRAM

INTRAPROGRAM FLOW DIAGRAM

For each program show all possible procedures or modecular activities that a patient and/or 2 staff member normally encounters. Procedures or molecular activities are represented by circles

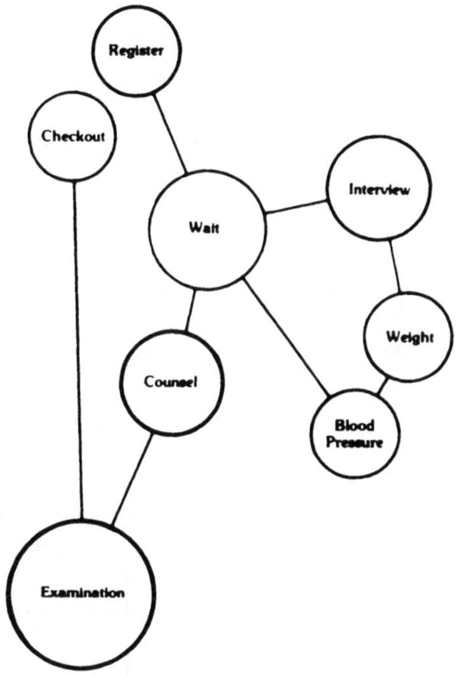

FIGURE 15. Intraprogram flow diagram.

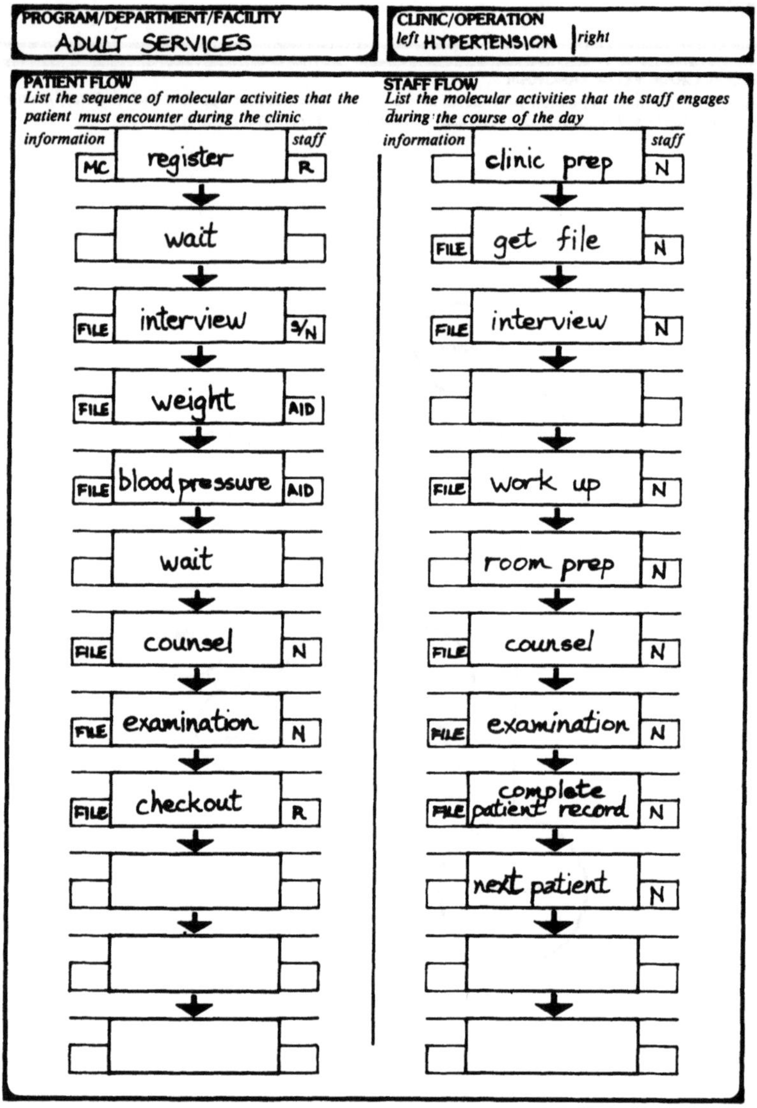

FIGURE 16. Patient–staff-flow data sheet: A chronological listing of the procedures and corresponding information and material exchanged by staff.

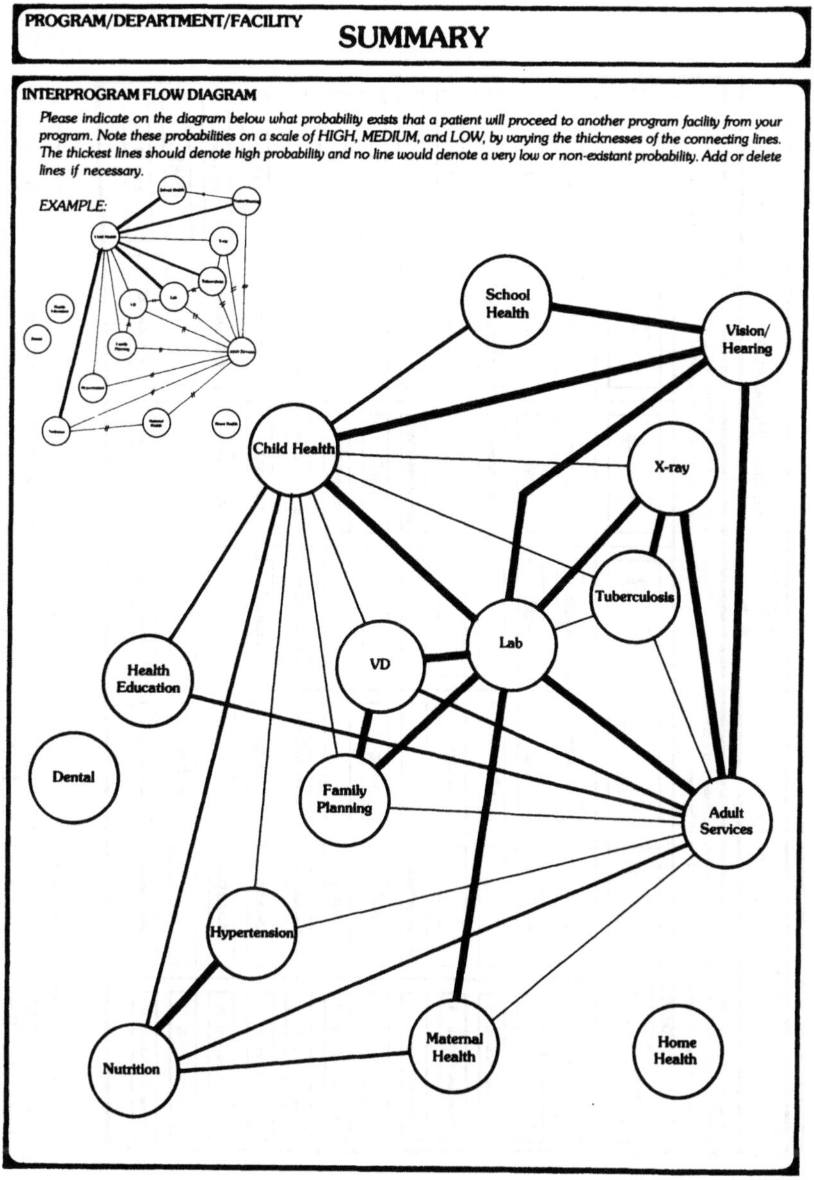

FIGURE 17. Interprogram-flow diagram.

PROGRAM FLOW SUMMARY FORM

PROGRAM/DEPARTMENT ADULT HEALTH

CLINIC FAMILY PLANNING

PROGRAM FLOW *List the sequence of molecular activities that a patient must encounter in each PRIMARY ACTIVITY UNIT for each program.*

Record in the CUMULATIVE PROGRAM FLOW SUMMARY column the cumulative list of molecular activities (MA's) encountered at least once in the program.

PRIMARY ACTIVITY UNIT / MOLECULAR ACTIVITIES	Primary Activity Unit 1	Primary Activity Unit 2	Primary Activity Unit 3	Primary Activity Unit 4	Primary Activity Unit 5	CUMULATIVE PROGRAM FLOW SUMMARY
Molecular Activity 1	register	register	register	register	register	MA1
Molecular Activity 2	wait	wait	wait	wait	wait	MA2
Molecular Activity 3		interview		interview	interview	MA3
Molecular Activity 4		weight			weight	MA4
Molecular Activity 5	blood pressure			blood pressure		MA5
Molecular Activity 6	wait	wait		wait	wait	MA6
Molecular Activity 7	counsel			counsel		MA7
Molecular Activity 8	examination	examination	examination	examination	examination	MA8
Molecular Activity 9	counsel	counsel	counsel	counsel	counsel	MA9
Molecular Activity 10	education			education	education	MA10
Molecular Activity 11		billing		billing	billing	MA11
Molecular Activity 12	checkout	checkout	checkout	checkout	checkout	MA12

FIGURE 18. Program-flow summary form: The procedure for completing this form, for each health-department program, began with a listing of the molecular activities for each primary activity unit as recorded on the patient–staff-flow data sheet in Figure 16. Next, in the primary activity unit columns, all molecular activities that took place at least once in any of the clinics were recorded in sequence.

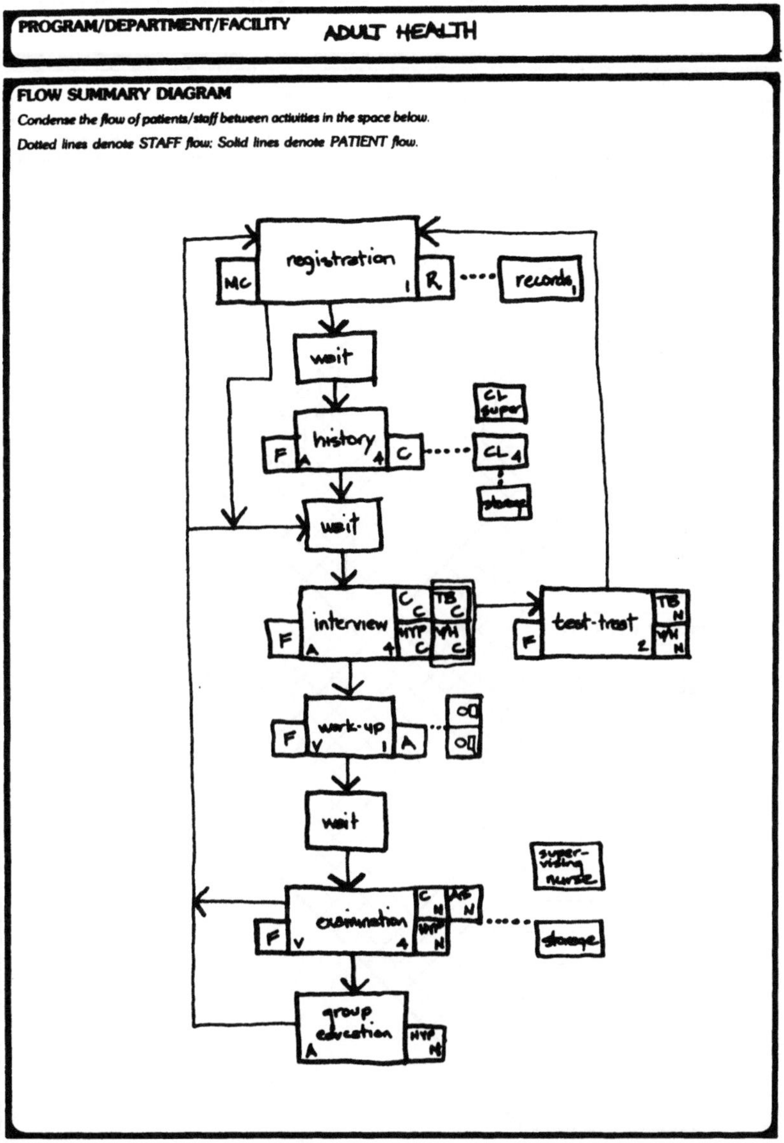

FIGURE 19. Flow summary diagram: The sequence of molecular activities generated from Figure 18 was transferred to this graphic format. Where personnel were utilized was noted in small boxes on the right side of the activity box, as on the patient–staff-flow data sheet in Figure 16. From the primary activity inventory forms (Figures 13a and b), the number of primary activity unit spaces was transferred to the lower right corner of the activity box. The type of privacy required by each molecular activity was identified as visual (V) or auditory (A) in the lower left corner of the activity box. Finally, the support activity cells were connected to the appropriate molecular-activity boxes by dotted lines.

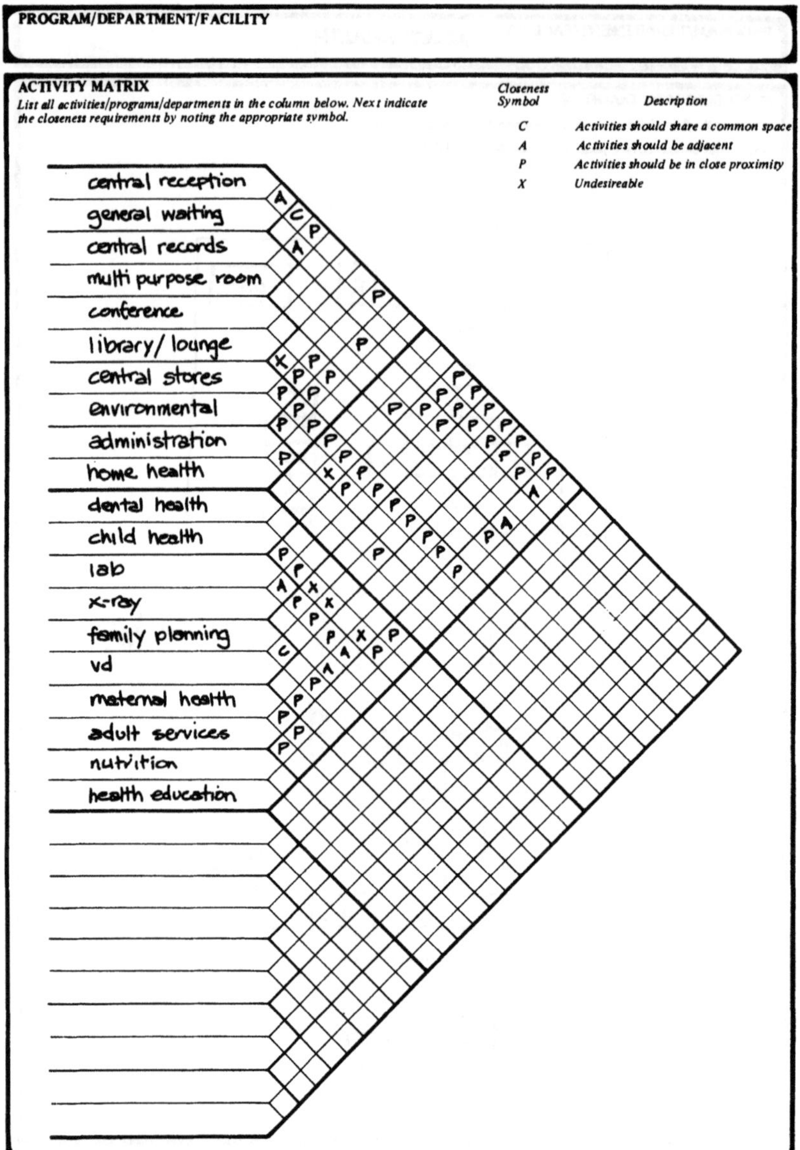

FIGURE 20. Activity matrix: The notations on the matrix were defined as follows: *C*—Coexistence of activities in space. Activities take place within the same boundary. A boundary is a vertical division between spaces; it may be a wall, a partition, or a change in floor or ceiling level, depending on the privacy and permeability requirements. *A*—Adjacency of activities. Activities take place separately, but with a common boundary. *P*—Proximity of activities. Activities take place separately but are connected by a link or pathway. *X*—Adjacency undesirable. Activities have characteristics that are in conflict with each other's environmental and spatial needs.

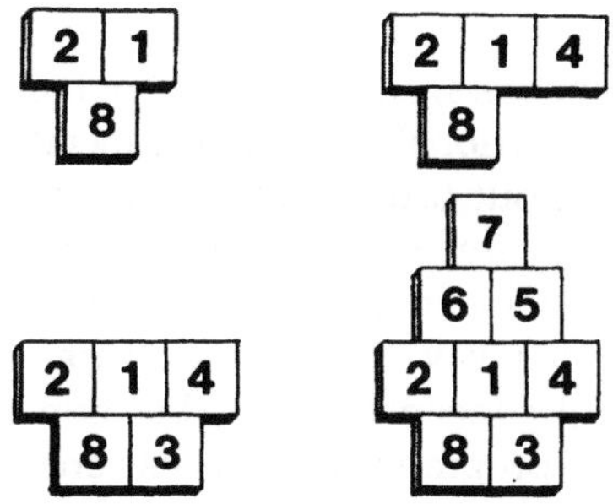

FIGURE 21. Examples of the template technique: The determination of the above location programs began by selecting the template having the greatest number of adjacencies, or "A" relationships. The template having an "A" relationship with the initial template and the greatest number of other "A" relationships was selected and placed next to the first template. The next template selected had the highest combined relationship with the two already selected. All templates were placed as near as possible to the templates with which they had the closest relationships. The next template selected was the one whose combined relationship with those already placed was the highest. The procedure continued in this way until all program templates were placed in the layout.

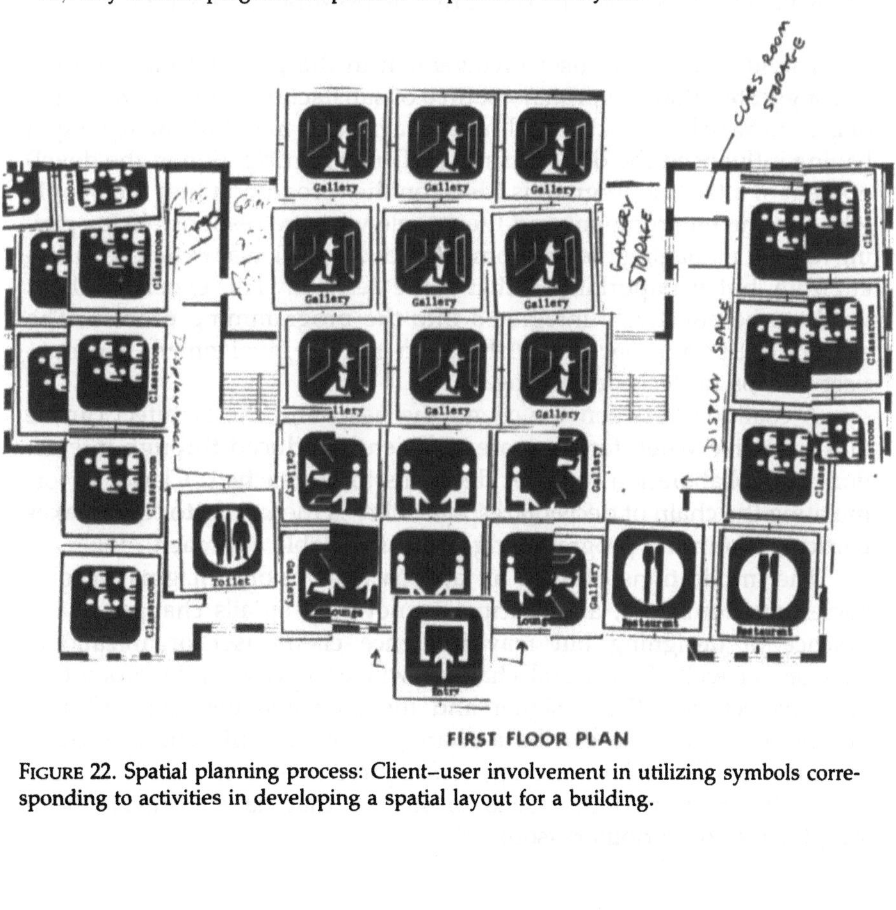

FIRST FLOOR PLAN

FIGURE 22. Spatial planning process: Client–user involvement in utilizing symbols corresponding to activities in developing a spatial layout for a building.

Project group members, department personnel, and architects helped to identify the functional problems associated with a new public-health facility by acting out future patient and staff roles. In repeated plan-layout sessions, squares of paper, color-coded to represent functional areas and the flow of people, were used to simulate the life of a health department. In this way, the two- and three-dimensional implications of different layout concepts quickly became apparent.

Participation in the programming process ensured that there would be no major surprises or disappointments for the interested parties, who were kept informed about why decisions and compromises had to be made. Health department personnel involved in the process understood the short- and long-term goals and appreciated the consistent or changeable elements of the facility program, which represented the centers of possible growth in the future.

USER PARTICIPATION IN THE PROGRAMMING PROCESS

The ideal of client–user involvement in the programming process has suggested that the primary source of satisfaction is not so much the degree to which the individual needs have been met, but the feeling of having influenced the decisions made. Direct participation in the development of a facility program is based on the recognition that the people who use the environment and the people who are responsible for paying for and managing the environment have a particular expertise different from, but as important as, that of the designer. This expertise, then, needs to be effectively integrated into the programming process. The essence of this approach is collaboration and the involvement of client–user groups in decision making.

A coalition between the client, the user, and the architect can be most effective when tensions are sufficiently reduced through interaction and collaboration to allow decisions to be reached. Clarity in formulating the chain of decision responsibility is the key factor that makes participation in the programming process a viable approach.

The implications of such an approach for more imaginative outcomes are numerous, as participation not only entails changes in the practice of designing but may influence client–user organizational changes as well. Procedural changes will be necessary to promote a dialogue between the designer and the user and between different groups, so that there can be a sharing of value differences through openness of discussion. The opportunity for people to openly discuss their differences permits issues to surface that have normally been swept aside for various reasons.

ROADBLOCKS TO PARTICIPATION

Because the ideas associated with participation are new to the design field, there are detractors, skeptics, and a more general body of professionals who regard participation as a passing fad. As a result, there are numerous arguments against participation. Some of them are valid, and others are simply based on false assumptions. The most persistent argument against participation is the proposition that teams cannot design. The camel is facetiously cited as a horse designed by a team or a committee. However, the emphasis on individual genius has been overromanticized by designers who report the creative act in terms of inexplicable insight, neglecting the daily routine that has given the insight an underpinning. Overemphasis on the group as an ultimate creative context can be equally detrimental, whether the group is called a team, a task force, or a committee.

Quite often, groups without a disciplined approach degenerate to the level of choosing the safest or the most obvious solution. The thinking that underlies most naive group activity is that the capabilities of the participants are additive, that is, that the moderately creative individuals are equal to one highly creative individual. Similarly, another mistaken belief is that, if one member does not think of a solution, another may. If none arrive at a solution, then the blame is equally spread throughout the group.

Many of these problems can be resolved practically if every member of the group is personally committed to solving the problem in the best possible way, so that the participants can shift their emphasis from personal capabilities to collective capabilities. A willingness to openly recognize better solutions is the key ingredient of group success. Similarly, to be effective, groups need to shape the collective expertise of each of the members. Psychologist William J. J. Gordon, author of *Synectics* (1961), described a process aimed at the development of creative capacity and substantiated the idea that the individual process in the creative act has a direct analogy in the group process.

Because groups attempt to expand the expertness of the individual, it is necessary to consider the limitations of the experts, whether they represent the user population, the client, the government, or other designers. Quite frequently, complaints result from the experiences that the wants of user groups or committees are rooted in their experience, that many groups' aspirations are commonplace and mundane, and that the designer's responsibility is to expand the horizons of the group. As these observations are certainly valid, it may be useful to clarify these beliefs.

The quality of expertness consists of access to special knowledge of a subject, which supposedly raises the individual's awareness above the obvious. In reality, the expert is the least able to create a new idea, as the problem is often described in the technical terms of the expert's language, and it is impossible to view the problem in a new way. Yet, the knowledge of the expert is necessary in stating the obvious or the commonplace in order to expand the narrowness of vision often found in highly trained people. Sometimes, an attack on convention is viewed as an attack on the people themselves and results in experts who do not function as individuals but as images of themselves. Communication is necessary at the personal as well as at the professional level if group activity is to achieve more than the obvious.

Another roadblock is to people's residential and job mobility. Because the time span for the renovation or construction of a new building through occupancy may be several years, opponents of participation have argued that the original participants may have left or will soon leave, so that their involvement in the design process is invalidated. For many years, this recurring myth has plagued designers: The design tailored to the departing individual—whether resident or employee—will have to be redone. That there have been similar proclamations about residential recreation projects, as well as student housing, suggests that, because there are individual differences among people, the best way to cope with this situation is to ignore individual differences and to design for no one in particular. Student housing is frequently cited as having a transient population. Yet, there are features of students' rooms that continue to create dissatisfaction because decisions were made without a clear understanding of students' needs and desires. Numerous studies (Moos, 1978; Null, 1979; Van der Ryn & Silverstein, 1967) have shown that students prefer to study alone, and that a sizable number of students use the bed extensively for studying and the desk for storage. This research suggests that student needs can be identified and generalized, although the population is mobile.

Quite often, building programs do not produce an accurate assessment of existing facilities in terms of people's use, preferences, or complaints. Lacking such information, the program usually relies on an idealized stereotype of the building's occupants. Institutional clients rely on building committees to advocate the user's point of view. Such committees are often far removed from the needs of those who actively use the building.

Because people have such different needs, attempts to create a single, standardized, "ideal" environment works to everyone's disadvantage. However, simply creating physical variety does not solve this problem. The user must participate in this assessment process as an agent to ensure variety.

In a study reported in *Ideas,* Sommer (1979) noted that allowing employees to select their own furniture from sample items assembled in a vacant warehouse resulted in a layout that was decentralized, modest, and personal, with the individual station at its core. The office had an unplanned quality, as the total environmental arrangement evolved from the sum of individual decisions. Different employees had different equipment and furnishings and were more satisfied with their work setting than were those in a comparable sample of employees who worked in a setting furnished from a single furniture system prescribed by expert space planners. Of particular interest here is that the first building has been denied design awards, whereas the latter has received several. One juror described his denial of the award on the basis of the plan's "residential quality" and "lack of discipline and control of the interiors." In this case, visual order and social control become the goal, not productivity or user satisfaction. The argument that employees want everyone to be treated in a visibly identical fashion does not hold up when employees are allowed a genuine opportunity to make informed choices. Similarly, the appearance of order is based on the premise that the designed environment is created for users who are more-or-less identical. Yet, we have seen from the results of this and other research that there are many differences among individuals, and that these differences should be reflected in the complexities and variations of the environmental support applied to them.

Consequently, designers can create places that encourage individual choice in spite of situations that involve high occupancy turnover. People's desires for personal territory, privacy, flexibility, and appropriateness are all design concerns that are often difficult to measure directly in relation to the institutional balance sheet. Often, it is the user and the community that pay the psychic and social consequences of ill-fitting environments.

Examples of User Participation

A participatory process developed by Halprin and Burns (1974) was the focus of the design of a Catholic secondary school in Berkeley, California. This process entails involving the people who will use the project in a series of participatory workshops at the very outset of conceptual planning. Halprin and Burns's approach to collective creativity is based on synectics (Gordon, 1961); Halprin and Burns develop techniques that aid communication and sharing and that unleash energies that are otherwise underutilized. Their work emphasizes the various accomplishments of both the individual and the group and then engages users and clients in a take-part process.

For 10 years, the students, faculty, parents, administrators, and neighbors of St. Mary's College High School worked together to determine the changes to be made in the environment of the 12-acre campus. The results include a brothers' residence, a student center, a central plaza, and site planning. At the initial workshop in 1974, students, teachers, administrators, parents, and community people began their planning process by discussing the physical and nonphysical qualities of St. Mary's; as a basis for future decisions, they were asked to prepare "real-life" plans for the campus on scaled maps. With the help of a professional team, these people prepared plans to demonstrate their recommendations for the campus. The plans were then presented for discussion and were commented on, and the workshop conductors offered a summary of the agreements and differences that emerged. The workshop took one full Saturday, with more than 80 people participating. When the campus plan was developed, workshop representatives reviewed it to ensure that it would reflect the agreements that had previously been reached. Soon after, actual work began on several projects. The new brothers' residence was itself the subject of a smaller workshop for the residents.

In 1981, Marquis Associates and Burns (1983) returned to St. Mary's to involve people in updating and changing the campus plan according to newly stated needs. The professional team reinvolved people in the wide range of campus-planning issues, changing previous decisions, and making new decisions with the aid of a take-apart scale model of the campus.

Commenting on user participation as an aid to professional programming, Marquis's associate partner, Caldwell, stated that the "workshops quickly gave us information for which we otherwise would have worked for weeks, and some of it we would have never discovered, buried as it was in people's personal feelings" (Burns, 1983, p. 63).

In a recent article describing Moore and Centerbrook's Hood Museum at Dartmouth, Campbell (1986) stated that "an unusual design process produced the Hood, greatly influencing its final form" (p. 39). This process occurred when Centerbrook set up an office in the adjacent building next to a snack bar. Here, the architects could observe college life while developing more intimate contact with the museum's building committee. Even as basic a decision as where to site the museum came out of this collaborative client–architect process. Unfortunately, there is no discussion of how the process was organized; this lack is symptomatic of many articles appearing in architectural journals.

Architects Lewis and Gindroz (1974) described their participatory experience in the design of a human resource center in Pontiac, Michigan, as one that deeply influenced their subsequent mode of practice. In the course of a complex and open process involving a great many people, sometimes in bitter confrontation, sometimes in quiet working sessions, a building form emerged that included learning areas for children and adults, a health center, theaters and a library, a community fieldhouse, a food coop, workshops and studios, a food service for the elderly, and a public restaurant and lounge.

What Lewis and Gondroz learned there—and have since seen repeated in a number of situations—is that, when consumers and citizens are openly enfranchised early in the design process and are invited to assume creative and responsible roles, planners and architects suddenly find available to them undreamed-of resources in the form of local perceptions and wisdom and the reinforcement of the community.

Purpose of Participation

Simple questions, such as who should participate, how they should be involved, where participation should lead, and when it should occur, are basic to the effective use of people's knowledge and experience. The purpose of participation, then, becomes an important concern because the objectives will vary from time to time. Participation can be used to generate ideas, to identify attitudes, to provide information, to review a proposal, or to resolve some conflict. Once the purpose is clarified, specific techniques can be used to achieve the desired objectives. An effective process for involving people must be carefully designed. The random involvement of people without a clear sequence of events and without clearly understood roles can result in chaos.

A few essential ingredients are necessary for success in any type of participatory effort. Initially, there needs to be a shared view of what the goals of the project are and what the participants want to achieve. As the process moves ahead, the goals may change; yet, the structure should be adhered to, as open-ended processes that permit people to join and drop out usually end in frustration. To ensure the continuity of the process, a steering committee, an executive board, a citizens' council, or a building committee should be formed at the outset. Its role should also include maintaining open communication between all participants at all times. Open dialogues often provide protection against any hidden agendas that may emerge. It is also quite important for the process to have a clear beginning and end, so that the participants understand their responsibilities and how they are interconnected with each other and with the designer. In this process, the designer is not only the process facilitator but also the technical specialist, who makes recommendations and develops the necessary design documents. Because the process is open to laypeople, it must use communications systems that everyone can clearly read.

Thus, the designer's role is to facilitate the user group's ability to reach decisions about the environment through a communicable procedure.

IMPLICATIONS FOR FUTURE RESEARCH AND APPLICATION

It is evident that our changing world has challenged the design community to effectively control our everyday environment. The view that new knowledge about the effects of the physical environment on people can aid the designer in achieving optimal human environments has serious limitations. The fact remains that we cannot easily predict much about future users, their behavior patterns, and their value systems (Hershberger, 1985). We have very little generalized knowledge about the future users or about the environments necessary to facilitate their needs. Although there is an ongoing effort to explore new techniques for making decisions, it is also necessary to explore the new roles that designers can assume in a changing society, such as helping organizations to effectively utilize their existing environments.

This concept is called *environmental management* and is based on the recognition that individuals other than designers effect behavioral and experiential changes in the physical environment. Management decisions can substantially alter human environments, even though the physical environment and its parts have been designed. A minor change, such as the moving of lights in an exhibition space, can alter the meaning of the entire environment for people. In fact, many occupations and professions are involved in making similar environmental decisions on a daily basis. People in occupations that perform a direct service to people, such as restaurant managers, hotel managers, and space or office managers, make environmental decisions. Many of these decisions are made by a broad group of professionals other than the users. Although these decisions influence human environments, the people who are affected have no control over decisions affecting their welfare. It is evident that the current ideal of environmental management should be replaced by management by participation, and designers need to realize that the management of physical environments is not only possible but desirable. Until they do, designers will continue to see their work mismanaged.

Clients, too, will continue to be disappointed in the failure of designers to meet unforeseen circumstances until they see the essential need for environmental management in their own operations. The concept of *design management* has been advanced because a building has a life of its own after occupancy. Organizing and altering space has a continuing affect on people's behavior.

Generally, design professionals have little or no direct control over or responsibility for a facility once it is completed. Ongoing space management, as a form of environmental design involving decisions about the use of facilities, is recognized, but it has not been institutionalized in a meaningful way (Becker, 1981). Environmentally related decisions are routinely made in organizations without any attention to their implications for the behavior of the people in the organization. Management is usually concerned about motivation, alienation, and communication, whereas designers and planners usually deal with environment and space. Concerns about environment have been structurally separated from concerns about program, about personnel, and about management. The role of the workplace as a medium of communication, as well as its structuring of interaction and communication patterns, remains largely overlooked. Understanding the physical environment as enabling and facilitating human actions requires thinking about the workplace as an interdependent part of an encompassing physical and social setting, which it influences and which influences it. Decisions or design changes in one part of a system can influence behavior in other parts.

Quite often, designers who attempt to create an effective environment for users restrict the easy manipulation of the environment in favor of built-ins and master plans that prevent the users from creating their own environment. The sense of order sought by many designers limits the possibility of individual diversity and consequent productivity. The appearance of order is based on an assumption that all users are alike and that the designed workplace will be perceived by the user as the designer intends.

The role that the environment plays in people's performance, however, is different for everyone. The level of an individual's performance is a function of individual capabilities and environmental support. Those who need more privacy, for example, can be located in areas away from major traffic and circulating paths. This uniqueness is made apparent by taking into account individual differences in aggressiveness, sociability, and privacy needs, as these have been assessed through various types of observations, surveys, and interviews. Organizing the environment to meet these needs is a means of accommodating individual differences within job categories. It shows that variable spatial provisions can be made within the same setting, so that individuals will obtain environmental supports that closely match their own needs, although these needs may be different from those of the person sitting at the next desk.

For the environment to facilitate the variations in human behavior and in individual work styles, certain attributes need to be identified and organized to support individual or group action. Becker (1981) identified four major environmental categories that interact with the social and organizational system: spatial form, activity patterns, communications, and ambience. The form of the workplace—its size, shape, and boundary conditions—is perhaps the most obvious environmental factor, although circulation paths and support facilities all contribute to spatial form. People's behavior—described by the intensity of traffic flow, occupancy, space use, and time scheduling—represents the activity patterns in workplaces. Communications are the signs and symbols that convey messages about roles, regulations, and the status of various people in the organization. The ambience of the workplace is reflected in the physical features that support human activities, such as microclimate, illumination, color, and sound transmission.

Creating effective programs for environments that are responsive to individual control depends on an ability to view the total operation of a facility. The holistic view includes the ability to evaluate support systems, to assess performance, and to create the support systems that will make the physical environment a viable entity.

As organizations and operations change, the original fit between the environment and its users can deteriorate; therefore, continuous assessment is necessary to identify this deterioration. It is, however, equally important to state what spaces and characteristics are needed. It is evident that, in order for this process to work effectively, the users need to be trained in stating their requirements and in reviewing design documents so that they can avoid any misunderstandings or misinterpreted comments.

Facility programming can be more than an inventory of behavioral activities and spatial requirements. It can lead to organizational self-assessment, so that the congruence between goals, objectives, performance, and satisfaction is continuously monitored and altered by the building's occupants. Creating new and improved environments could be linked to the changing requirements of the people who use the facility. The work environment would then act as a mediator between the fulfillment of tasks and the social well-being of the users. Continuous space management can effectively project space needs, reallocate space, consolidate activities, and resolve shortages.

Because the primary source of improved methods of facility programming is information, it becomes increasingly evident that more

The Participants: In order to facilitate initial contacts this heading is to give the most basic information on the people and organizations likely to be involved.

In addition to the participants or "project team," there will be related groups concerned with some aspects of the project and it is important that the participants should have information on the roles and organizations of such bodies.

Project participants:

- Client
- Occupiers/users
- General manager/administrator (if not the client)
- Briefing consultant
- Other consultants
- Builder

Related groups

- Central government
- National/international agencies
- Local government
- Planning and building authorities
- Financiers
- Groups/persons with special interest
- Site owners and tenants
- Neighbours and their consultants
- Press
- Insurers

The project organization and methods: The organizational framework is given some priority in the checklist because certain aspects of it most be clearly established before later parts of the list can be effectively considered.

The overall management of the project is particularly important at its beginning as it is by the proper exercise of this role that comprehensive identification of, and information on, authorative, informed and interested participants is achieved. The task is essential for good communication, motivation, coordination, and effective control of time, cost, and quality. Overall management is a client responsibility that should only be delegated for exceptional reasons and then with great care.

Participants/management:

- Authority, responsibilities, roles, skills
- Organizational diagram/relationships
- Client responsibilities and consultant involvement
- Procedures of selection of participants
- Methods of communication, coordination, — control and recording (including time and costs)
- Contracts/legal agreements
- Design evaluation and other technical procedures

The project background: A knowledge of past events and decisions will ensure that the participants are made aware of matters which may have an influence on the intended project.

History:

- Relevant information about the clients/user enterprise, past events, studies investigations, decisions, and attitudes

Current situation:

- Client/user activities and situation
- Existing sites, facilities, buildings
- Ongoing investigations

Reason for intended change:

- Perceived need, demand, market
- Legislation
- Pressures

Existing commitments: (client/user)

- Organizational arrangements
- Contractural arrangements

FIGURE 23. Checklist excerpts from International Organization for Standardization.

documentation of the facility-programming process is needed. Only through well-recorded decision processes can we learn about and improve the present approaches used by architects. Although design professionals are not accustomed to describing the programming process, it seems that self-reflective descriptions can be more informative to educators and practitioners then concealing decision processes and are an important reservoir for growth. Organizations that normally use various participatory methods in the programming process need to convey their experiences so that learning can occur across organizations and in the design profession.

A recognition that many of the problems associated with the satisfactory completion of a building arise from an inadequate communication of the clients' and users' needs prompted the International Organization for Standardization (1984) to propose a checklist for programming. The checklist is offered as an international standard or agenda, in which the interests of all participants in the project are expressed and conflicts are resolved. (See Figure 23 for an example of the contents of the program.)

The acceptance of such guidelines further suggests the need for the international community to communicate its experiences with the programming process.

REFERENCES

Bechtel, R., Marcus, R., & Michelson, W. (1987). *Methods in environmental and behavioral research.* New York: Van Nostrand Reinhold.

Becker, F. (1981). *Workspace: Creating environments in organizations.* New York: Praeger.

Brauer, R. (1986). *Facility planning: User requirements method.* New York: American Management Association.

Building Research Board. (1986). *Programming practices in the building process: Opportunities for improvement.* Washington, DC: National Academy Press.

Burns, J. (1983). Designing for (and with) the Christian Brothers. *Architecture, 72*(2), 60–63.

Campbell, R. (1986). An architecture of verve and invention. *Architecture, 75*(1), 33–39.

Daish, J. R., Gray, J., Kernohan, D., & Salmond, A. (1982). Post-occupancy evaluation in New Zealand. *Design Studies, 3,* 2.

Davis, G. (1978). A process for adapting existing buildings for new offices. In W. F. E. Preiser (Ed.), *Facility programming.* Stroudsburg, PA: Dowden, Hutchinson & Ross.

Davis, G., & Szigeti, F. (1979). Functional and technical programming, when the owner/sponsor is a large or complex organization. In J. G. Simon (Ed.), *Conflicting experiences of space: proceedings of the Fourth International Architectural Psychology Conference.* Louvain: Catholic University of Louvain.

Davis, G., & Szigeti, F. (1986). Planning and programming offices: Determining user requirements. In J. D. Wineman (Ed.), *Behavioral issues in office design*. New York: Van Nostrand Reinhold.

Downs, A. (1976). *Urban problems and prospects*. Chicago: Markham.

Farbstein, J. D. (1985). Using the program: Applications for design, occupancy, and evaluation. In W. F. E. Preiser (Ed.), *Programming the built environment*. New York: Van Nostrand Reinhold.

Freidman, A., Zimring, C., & Zube, E. (1978). *Environmental design evaluation*. New York: Plenum Press.

Gordon, W. J. J. (1961). *Synectics: The development of creative capacity*. New York: Collier.

Gould, B. P. (1986). Facilities programming. In S. Braybrooke, (Ed.), *Design for research*. New York: Wiley.

Gray, J., Watson, C. G., Daish, J. R., & Kernohan, D. (1985). Putting a POE to work: A case study in which a POE is combined with participatory programming. In S. Klein, R. Wener, E. S. Lehman (Eds.), *Environmental change/social change*. Washington, DC: Environmental Design Research Association.

Halprin, L., & Burns, J. (1974). *Taking part: A workshop approach to collective creativity*. Cambridge, MA: MIT Press.

Heimsath, C. (1977). *Behavioral architecture*. New York: McGraw-Hill.

Hellmuth, Olsata, & Kassabaum. (1985). *Master-plan integrated process*. Unpublished manuscript.

Herschberger, R. G. (1985). Values: A theoretical framework for architectural programming. In W. F. E. Preiser (Ed.), *Programming the built environment*. New York: Van Nostrand Reinhold.

International Organization for Standardization. (1984). *Functional/user requirements and performance in building construction*. London: British Standards Institution, ISO/TC 59/SC3.

Jones, J. C. (1970). *Design methods: Seeds of human futures*. London: Wiley.

Kammeraad, D., & Sanoff, H. (1979). Facility planning. In R. J. Wallace, D. Kammeraad & D. R. Polston (Eds.), *Design/proposal guide for capital project planning by county health departments*. Raleigh, NC: North Carolina Division of Health Services.

Kaplan/McLaughlin (1981). Feasibility study and program for a health clinic. In M. Palmer (Ed.), *The architect's guide to facility programming*. New York: McGraw-Hill.

Kurtz, J. (1978). Habitable schools: Programming for a changing environment. In W. F. E. Preiser (Ed.), *Facility programming*. New York: Van Nostrand Reinhold.

Lewis, D., & Gindroz, R. (1974). Toward a design process that re-enfranchises citizens and consumers. *American Institute of Architects Journal, 62*(5), 28–31.

Low, S. M. (1987). Developments in research, design, and analysis: Qualitative methods. In E. H. Zube & G. T. Moore (Eds.), *Advances in environment, behavior and design*, Vol. 1. New York: Plenum.

Marans, R., & Ahrentzen, S. (1987). Developments in research, design, and analysis: Quantitative methods. In E. H. Zube & G. T. Moore, (Eds.), *Advances in environment, behavior, and design*, Vol. 1. New York: Plenum.

McLaughlin, H. (1976). Programming. In R. A. Class & R. E. Koehler (Eds.), *Current techniques in architectural practice*. Washington, DC, and New York: American Institute of Architects and Architectural Record.

Moss, R. H. (1978). Social environments of university student living groups: Architectural and organizational correlates. *Environment and Behavior, 1*, 109–126.

Null, R. L. (1979). Determinants of student perceptions of the social and academic climate of suite living arrangements in university residence halls. In A. D. Seidel & S. Danford

(Eds.), *Environmental design: Research, theory and practice*. Washington, DC: Environmental Design Research Association.
Palmer, M. (1981). *The architect's guide to facility programming*. Washington, DC, and New York: American Institute of Architects and Architectural Record.
Pena, W. (1977). *Problem seeking: An architectural programming primer*. Boston: Cahners Books International.
Port, M. H. (1968). The New Law Courts Competition 1866/67. *Architectural History, 2*, 81.
Preiser, W. F. E. (Ed.). (1978). *Facility programming*. New York: Van Nostrand Reinhold.
Preiser, W. F. E. (Ed.). (1985). *Programming the built environment*. New York: Van Nostrand Reinhold.
Ravetz, A. (1980). PSSHAK 18 months on. *The Architects' Journal, 171*(9), 425–439.
Robinson, J., & Weeks, S. (1984). *Programming as design*. Minneapolis: University of Minnesota, School of Architecture.
Sanoff, H. (1978). *Methods of architectural programming*. New York: Van Nostrand Reinhold.
Sanoff, H. (1984). *Arts center workbook*. Raleigh, NC: North Carolina Arts Council and North Carolina State University, School of Design.
Silver, G. H., & Klein, R. A. (1985). Get your space act together. *Office Systems '85*. Springhouse PA: Springhouse.
Sommer, R. (1979). Participatory design. *Ideas* (Herman Miller, Ann Arbor, MI), *3*, 4.
Tusler, W. H., Schraishuhn, R. T., & Meyer, E. R. (1985). Stone, Marraccini and Patterson's health care facility programming. In W. F. E. Preiser (Ed.), *Programming the built environment*. New York: Van Nostrand Reinhold.
Van der Ryn, S., & Silverstein, M. (1967). *Dorms at Berkeley: An environmental analysis*. Berkeley: University of California, Berkeley, Center for Planning and Development Research.
Wener, R. (1988). Environmental evaluation. In E. H. Zube and G. T. Moore (Eds.), *Advances in environment, behavior, and design*, Vol. 2. New York: Plenum Press.
White, E. T. (1972). *Introduction to architectural programming*. Tucson, AZ: Architectural Media.
White, E. T. (1982). *Interviews with architects about facility programming*. Talahasee: Florida A&M University, School of Architecture.
White, E. T. (1983). *Project programming: A growing architectural service*. Talahasee: Florida A&M University, School of Architecture.
Zeisel, J. (1981). *Inquiry by design*. Belmont, CA: Brooks/Cole.
Zimring, C. M., & Reizenstein, J. E. (1981). A primer on post occupancy evaluation. *American Institute of Architects Journal, 70*(13), 52–58.

9

Advances in Evaluation of the Built Environment

RICHARD WENER

The goal of this chapter is to examine the current evaluation of built settings, also known as the postoccupancy evaluation (POE). This chapter deals with post occupancy evaluation, rather than postconstruction evaluation, accepting the distinction made by Anderson and Butterfield (1980) that the latter deals with the technical and structural aspects of facilities, and the former attends to social and behavioral issues.

This chapter uses past reviews (Friedman, Zimring, & Zube, 1978; Marans, 1984; Moore, 1982; Zube, 1980) as benchmarks, especially in terms of the conceptual frameworks they have provided. It focuses on advances in the *process* of completing evaluations rather than the *content* of those evaluations. That is, it does not take as its primary task reporting substantive findings. Instead, it spotlights progress in the way in which evaluations are conducted and used.

This chapter first discusses the increasing sophistication in assessments of the setting (social, organizational, and historical) within which evaluations are performed. Next, it looks at advances in research design, including extensions beyond single-site or single-time case studies, and at advances in evaluation techniques and instruments, data analysis, and theoretical models.

Richard Wener • Environment-Behavior Studies Program, Social Sciences Department, Polytechnic University, Brooklyn, New York 11201.

DEFINING EVALUATION

Evaluation is an attempt to pause and reflect on the impact of an intervention. It differs from architectural criticism, as Zimring and Reizenstein (1980) pointed out, by being data-based, by being more than the simple reflection of one person, and by addressing more than aesthetic issues. The goal is identical to that ascribed to program evaluation by Rossi and Freeman (1985): to assess concept, implications, and utility, and to judge and improve planning, effectiveness, and efficiency of programs.

Evaluation can provide feedback to clients and designers on the impact of settings on behavior. Feedback is critical to positive change and improvement in any system—environmental, social, physiological, or mechanical. The more immediate and discrete the feedback, the easier it is to change and improve the system. Without feedback, change is difficult or impossible; with it, it is inevitable.

Lack of feedback has placed designers at a disadvantage in discerning the impact of their creations on user groups. Human satisfaction or comfort is not easy to measure. The design process does not assign the responsibility for environment–behavior fit to any professional group. This lack of feedback on human responses makes progress difficult. Shibley (1985) noted that design without evaluation becomes action without the chance for reflection. Without reflection, design cannot evolve and grow.

It is this gap in the feedback loop that environmental evaluation tries to fill. Several recent efforts have addressed this problem directly and provide a prototype for future work, by instituting the use of evaluation in the planning processes of the institutions responsible for many facilities (see Sanoff, this volume).

One broad way of classifying evaluations is to describe their purpose as either comparative or generative. In *comparative* evaluations, the main task is to assess the status of a setting, comparing it with other sites (e.g., Francescato, Weidemann, Anderson, & Chenowith, 1979), other users (e.g., Margulis, 1981), other times (Bakos, Bozic, Chapin, & Neuman, 1980), or some independent set of criteria (Preiser, 1981). In *generative* evaluations, the main tasks are to generate new ideas to aid in a planning or programming process (e.g., Gray, Watson, Daish, & Kernohan, 1985; see other examples in Sanoff, this volume), and to provide data for an open-ended search to increase knowledge about a setting and to generate ideas, models, and hypotheses (see Low, 1987, for examples; Table 1).

These functions are often overlapping but not identical. Most

TABLE 1. Types of Evaluations and Related Research Designs and Techniques

Type of evaluation	Functions	Research designs	Research techniques	Examples
Comparative	Assess status, benchmark	Case studies, multisite or across time	Quantitative, with standardized instruments	Moos & Lemke (1984)
	Test hypotheses	Experimental, quasi-experimental	Quantitative, with standardized instruments	Margulis (1981)
Generative	Planning and programming	Case studies	Qualitative—open-ended	Gray, Watson, Daish, & Kernshaw (1985)
	Discovery—basic, descriptive research	Case studies	Qualitative—open-ended	Low (1987)

important, there are essential and nontrivial differences in the kinds of methodologies and research instruments that are best used for these different evaluation functions. For example, in serving to compare (i.e., to audit status or to provide benchmarks), evaluation responds best to closed-ended, standardized questions, with established reliability and validity. An experimental or quasi-experimental research design is more appropriate to testing hypotheses and to looking for significant differences among situations.

In evaluating to generate new ideas (i.e., to develop an initial understanding or model of a space, or to program a space), open-ended, qualitative techniques are most appropriate (of the kind described by Low, 1987), in the context of descriptive case studies. In programming, for example, qualitative methods may allow one to ask subjects to project beyond their current situation to future requirements. These ideas can provide the fuel for the inductive leap to a new formulation, which is a part of the creative design process.

Making these distinctions may help to reduce misunderstanding among the consumers of design research information. For example, planners may find the data from a standardized evaluation questionnaire too narrow and fixed to be of use for programming, compared to responses from open-ended interviews. Social scientists, on the other hand, may be dismayed by the difficulty of comparing responses across times or sites when qualitative research methods are used. By making these distinctions clear, and fitting them to the goal of a particular project, it is possible to provide appropriate as well as useful information.

Of course, the preceeding is not to suggest that any evaluation should restrict itself to one type of data-gathering technique. Projects can and have made fruitful use of both types, usually when the projects have both comparative and generative goals (e.g., Weidemann & Anderson, 1980). Rather, the point here is to identify how some research methods and techniques can best fit evaluation goals.

ADVANCES IN ASSESSING SETTING CONTEXT

Shibley and Schneekloth (1988) noted:

> No decision can be evaluated in terms of its appropriateness or technical competence without first understanding the processes by which it was developed and the socially constructed reality within which it is applied. (p. 306)

Since the late 1970s, there has been important progress in understanding the need for attention to the environmental context, and in applying this approach to evaluations.

Friedman et al, (1978) provided a framework for conducting evaluations that places a premium on describing the historical and social context of the setting being evaluated, including the proximate environmental context, the users, and the process of the design activity. The proximate environmental context, for example, includes nearby environmental factors such as air quality, noise, or traffic routes, as in the study of urban plazas by Whyte (1980), where the location of plazas near heavy pedestrian traffic was found to be important to plaza usage.

Keys and Wener (1980) noted the importance of understanding organizational structures in order to keep research from being ignored and unused, for example, an understanding of the agenda, roles, and power relationships of the various players, or a knowledge of their previous experience in dealing with researchers.

Schneekloth (1987) selected five case studies of exemplary design research practice, based in part on their recognition and appreciation of contextual issues. These examples demonstrate the need to understand the complex nature of formal and informal social and environmental relationships (i.e., who the decision makers are and what their rewards are) in order to successfully act with and for others.

The implications of the role of context are several, and they deeply affect the evaluation process. At one level, it suggests another domain of data to be collected, relating to a background understanding of the situation. At a deeper level, however, it suggests a change in the basic ways of doing evaluation, "risking collaboration" (Shibley & Schneekloth, in press) by getting those involved in the setting to be a part of and to "own" the research. These authors argued that a precondition to commitment to a setting or to the results of a study is collaboration in setting the goals, values, and methods of the study. Ownership comes from the "ability to influence or exercise free and informed choice over the results of the collaboration." Kantrowitz and Nordhaus (1980) described a specific case in which user involvement in the research process improved the implementation of the results in an urban housing project.

This ownership of the research and design process is part of the development of what Sommer (1984) called the "questioning community." It can help demystify the research process and provide a means by which the community can continually challenge, question, and change its own setting.

ADVANCES IN RESEARCH DESIGNS

There are a considerable number of research methods available to the evaluator that can be applied to an environmental assessment (see Low, 1987; Marans & Ahrentzen, 1987). Identifying the best possible research design for a given situation involves creativity and, occasionally, serendipity. Particularly in field settings, ingenuity is critical if one is to make use of the resources at hand and develop a design that can maximize understanding. A prerequisite is an understanding of the philosophical and statistical assumptions that underlie the testing of hypotheses and the drawing of conclusions, so that the right conditions can be met.

Experimental and Quasi-Experimental Designs

The discussion of applications of experimental and quasi-experimental designs to field settings is not new (Campbell & Stanley, 1966; Cook & Campbell, 1979). Their application to design evaluations, however, is an encouraging trend. Marans and Ahrentzen (1987) described in some detail experimental and quasi-experimental research designs that can be applied to environment–behavior research. Margulis (1981) provided an extensive discussion of the advantages of using nonequivalent-control-group designs to enhance the quality of POE data, along with a "how to" description for evaluators, and Marans and Ahrentzen (1987) cited examples of evaluations using nonequivalent-control-group design.

One rare example of a true experimental design in environmental evaluation, which shows both the difficulties and the potentials of controlled research, is a study of a geriatric center by Bakos *et al.* (1980). They used control groups and random assignment, in a repeated-measures study, while still adhering to the principles of full user participation in the planning process. Residents and staff from two floors were randomly selected for design workshops, which proceeded to plan spatial modifications. The resultant plan was implemented first on one floor and then, after a delay, on the other (Table 2).

Data on behavior and user satisfaction were collected at a number of points: before the workshops, after the workshops but before the changes, after the changes on the first floor, after the changes on the second floor, and again as a follow-up after several months. Because there were people on each floor who had and had not been in the workshops, the researchers could independently assess the effects of participation and environmental change. They found that participation alone had a positive effect on behavior, but that this effect seemed to fade if not reinforced by the implementation of the design changes (Figures 1, 2 and 3).

One of the strongest features of this example is the ability to take advantage of naturally occurring conditions and to manipulate them in ways that tremendously amplify the quality of the data. The fluidity of events in field settings makes it especially important for researchers to be ready to take advantage of the opportunities that the setting offers. For example, when a judge ordered a federal jail to change double-bed rooms to single beds, researchers who had completed a jail evaluation before that ruling returned to reevaluate the space (Wener & Keys, 1988; Wener & Olsen, 1980). Although the primary focus of the original study

TABLE 2. Research Design Used by Bakos, Bozic, Chapin, and Neuman (1980)[a]

	Collect data, Time 1	Planning workshops	Collect data, Time 2	Implement design, Floor A	Collect data, Time 3	Implement design, Floor B	Collect data, Time 4
Floor A							
Group 1	O_1	X_1	O_2	X_2	O_3		O_4
Group 2	O_1		O_2	X_2	O_3		O_4
Floor B							
Group 3	O_1	X_1	O_2		O_3	X_3	O_4
Group 4	O_1		O_2		O_3	X_3	O_4

[a]Notation follows Campbell and Stanley (1966), where *O* stands for observation (or evaluation measurement), and *X* stands for treatment (or a design change to be evaluated).

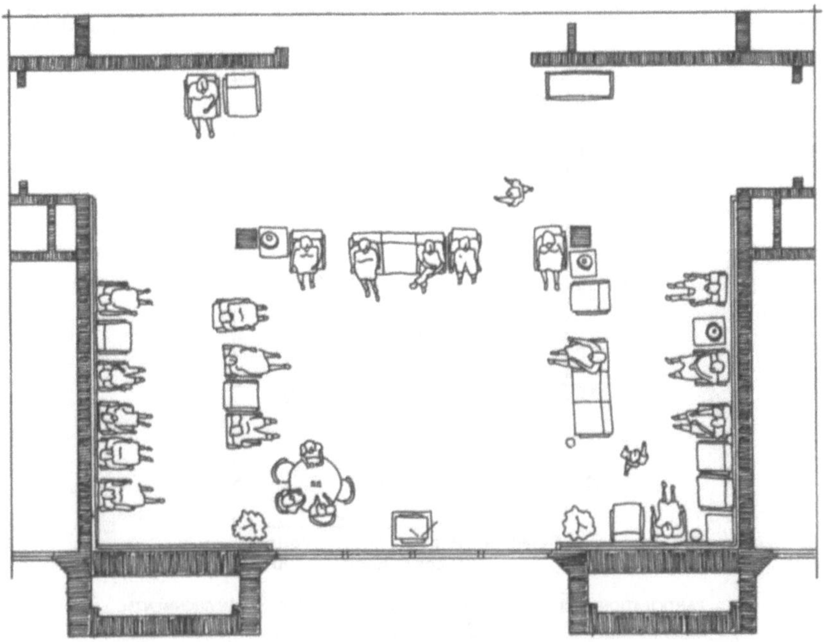

FIGURE 1. Geriatric center dayroom, before design modifications. (Printed with permission of *Hospital and Community Psychiatry*.)

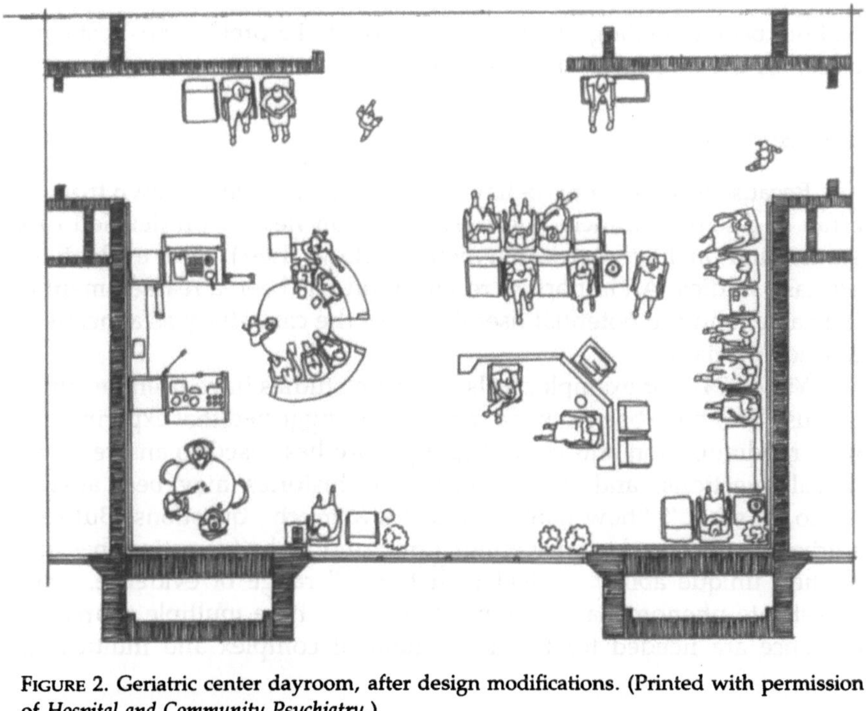

FIGURE 2. Geriatric center dayroom, after design modifications. (Printed with permission of *Hospital and Community Psychiatry*.)

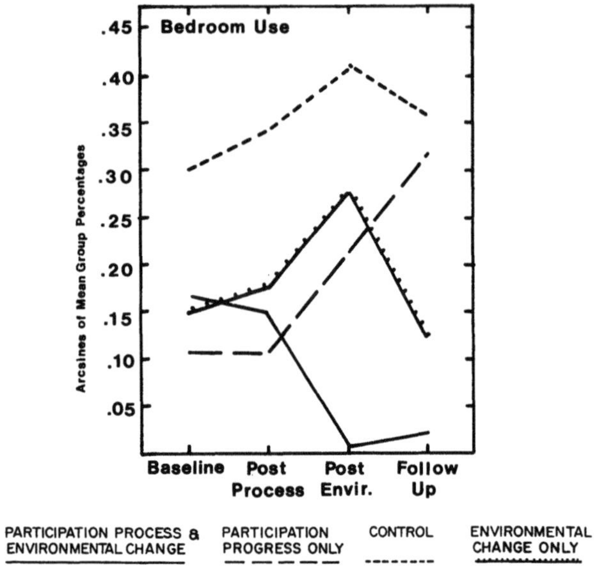

FIGURE 3. Use of bedrooms in geriatric facility. (Adapted from Zimring and Wener, 1985.)

had not been crowding, the follow-up provided a pretest–posttest evaluation of the effects of changes in density.

Case Studies

Because environment–behavior research has always drawn from an array of disciplines, including those that lean heavily on detailed case studies and qualitative research methods (Low, 1987), most evaluations are case studies. An important recent trend has been a reassessment of the value, and the potential usefulness, of the case study as a means of developing data.

Yin (1984), for example, feels that case studies have been devalued because they have been misunderstood. He suggested that experiments, with randomization and control groups, are best used to answer basic causal questions, and that surveys and histories may best answer "who," "what," "how many," and "how much" questions. But case studies are best used to answering "how" and "why" questions because of their unique ability to deal with the full range of evidence. They investigate phenomena in real-life contexts, where multiple sources of evidence are needed for the assessment of complex and multicausal events. Criticisms of case studies as lacking rigor or generalizability show a misunderstanding of the method. Generalizability, for example, can depend on how sites are chosen for the case study:

> The essence of a case study . . . is that it tries to illuminate a decision or set of decisions; why they were taken, how they were implemented, and what was the result. (Schramm, 1971, cited in Yin, 1984, p. 22)

In evaluation research, Yin wrote, case studies have four different applications: (1) they are used to "explain the causal links in real-life interventions that are too complex for survey or experimental strategies" (p. 25); (2) they can describe the context in which an intervention has occurred; (3) they can illustrate the intervention through description; and (4) they can explore situations "in which the intervention being evaluated has no clear, single set of outcomes" (p. 25).

In Yin's categorization, case studies can be of single-case or multicase design. Single-case designs can choose critical situations in which predictions are made from theory to rare, unusual, or extreme settings. They can be holistic (looking at global issues only) or "embedded" (looking at specific issues within the global context). Multiple sites don't have to be chosen randomly. They can be selected by replication logic to represent key questions, just as multiple experiments are conducted. In a study of National Park Service visitor centers, Zube, Crystal, and Palmer (1976) chose 12 centers based on criteria including (1) geographic distribution; (2) representation of natural, historic, and recreational themes; (3) examples of several construction periods; and (4) variability in quality according to nominations by the client.

Winkel (1983) also defended the appropriateness of case studies, answering in the affirmative the question: Can the diversity of a setting be investigated in a way that allows the study of environment–behavior relationships? Case studies, he wrote, can address causal questions by addressing the uniqueness of the setting. What is needed is appropriate theory from which to derive testable hypotheses. Variations and irregularities within an environment allow comparisons that add to an understanding of how that setting affects behavior. In studying the qualities of urban plazas and parks, Whyte (1980) chose a range of parks, varying in a number of design features hypothesized to be important in the level of use (including sun, seating, and proximity to the street).

Experimental and quasi-experimental research designs are available to environmental researchers who can take advantage of serendipitous situations and apply greater control to data collection. It is clear, however, that case studies should not be considered a research design of default. For an open exploration of complex settings, they may be the design of choice, whereas experimental and quasi-experimental designs are more appropriate to the confirmation of hypotheses.

Extension of Studies across Time and Place Barriers

Zimring and Reizenstein (1980) presented a 2 × 2 table as a way of classifying how studies have addressed sampling issues. One dimension represents sampling in single or multiple settings, and the other represents sampling in longitudinal or cross-sectional methodologies.

Here, this framwork is expanded into a 2 × 4 matrix (Table 3). The number of settings is again represented by single- versus multiple-site studies. The time dimension has been expanded to bring focus to the purpose and use of time as a dimension in research.

Many evaluations (most, in fact) remain single-site posttest-only case studies. Others have collected data over time, but rarely has time itself become a variable for study. Wicker (1987) pointed out that behavior settings are dynamic and that development takes place over time.

Table 3. Matrix of Time and Sites in Environmental Evaluations[a]

	Single site	Multiple sites
Posttest only	$X–O$	$X_1–O$ $X_2–O$ . . $X_n–O$
Pretest and posttest	$O_1–X–O_2$	$O_1–X_1–O_2$ $O_1–X_2–O_2$ . . $O_1–X_n–O_2$
Longitudinal	$O–O–O–X–O–O–O$	$O–O–O–X_1–O–O–O$ $O–O–O–X_2–O–O–O$ . . $O–O–O–X_n–O–O–O$
Iterative	$O–X_1–O–X_2–O–X_3–O$	$O–X_1–O$ $O–X_2–O$ . . $O–X_n–O$

[a]Notation and terminology follows Campbell and Stanley (1966) where *O* stands for observation (or evaluation measurement), and *X* stands for treatment (or a design change to be evaluated).

This development is rarely seen, however, because of our lack of repeated time measurements. Most of the cross-time evaluations available are "pretest–posttest" studies. Although these sample at two different times, the design change is the variable of interest, rather than time itself. Passage of time is also incidental and is not of direct interest in "iterative projects" (Wener, in press). These projects address the research-based modification of a design concept or prototype as it is used and changed over succeeding generations of facilities. There are few studies that look at time itself as a variable of interest.

In this taxonomy, the term *longitudinal* has a use similar to that in Cook and Campbell's time-series design (1979) and is reserved for studies that collect data at various times, so that dynamic changes within a setting can be assessed.

Single-Site Pretest–Posttest Assessments. There is an increasing array of studies that have assessed specific sites before and after environmental interventions, probably because an increasing acceptance of environment–behavior research in design and planning has provided increased opportunities to see plans realized. Even though simple pretest–posttest studies are usually subject to biases of statistical regression, history, and maturation (Cook & Campbell, 1979), they offer an extension of the single-time case study and an opportunity to verify the impacts of change. Two kinds of pretest–posttest studies appear in the literature: (1) design projects in which the pretest evaluation data were used in the plans for a setting change, and the posttest study was used as a follow-up assessment of outcome, and (2) projects that were not involved in design, but that opportunistically conducted evaluations before and after a planned environmental change.

One good example of a project in which the pretest evaluation was part of the change process is the renovation of the Exxon Mini-Park by Project for Public Spaces (1978). PPS was called in to redesign an urban plaza that had become overrun with drug dealing and was little used by local office workers and midtown pedestrians. PPS completed a detailed behavioral mapping of the site. The observations confirmed that drug dealing was common and that local workers avoided the space.

The design attempted to make the space less desirable for drug dealing by making it more attractive to local office workers. More varied and flexible space was provided, including an entertainment area (and entertainment programming) and food service. Behavioral observations made after the redesign found that drug dealing had been virtually eliminated, and that overall use had risen dramatically across a broad range of age groups (Figure 4).

An evaluation by Anderson, Lewis, Barker, Lund, and Roth (1984)

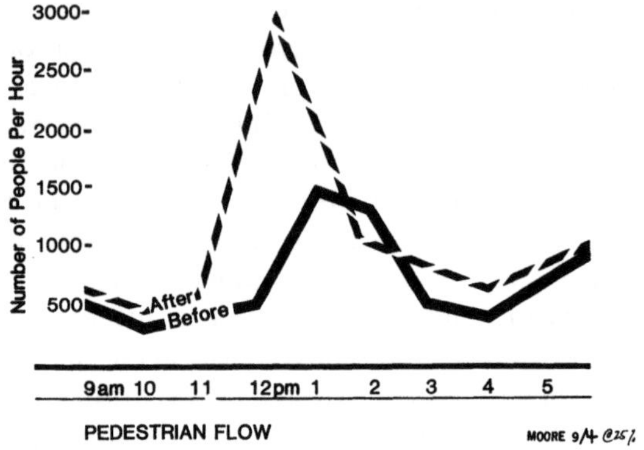

FIGURE 4. Pedestrian flow at Exxon MiniPark, before and after design modification. (Figure courtesy of Project for Public Spaces.)

of a new correctional setting for youths used pretest data that were not at all related to the environmental change. The opportunity arose at the last moment to evaluate the staff and residents of the youth center before a planned move. The researchers used standardized instruments to quickly conduct an evaluation of overall satisfaction, crowding, privacy, and safety. The evaluations uncovered existing problems in privacy and crowding, and significant improvements were made in both areas in the new institution.

Single-Site Longitudinal Assessments. A few studies have looked at change over time, measuring behavior at a number of points. These studies allow a view of settings as dynamic situations that may vary over time.

Several single-site pretest–posttest studies provide longitudinal information because they took several measurements at several data points before and after the environmental change. Picasso (1983) surveyed workers in an office setting over five different time periods. Data were collected before a design change, immediately after the change, and three more times, up to 2 years after the change. Because the respondents were identifiable, Picasso was able to assess not only overall changes in time, but differences among groups based on when they had entered the system.

The immediate postchange data showed a positive change after the design modifications, with some significant exceptions. For example, problems in acoustical privacy in the new setting were identified, and new design modifications were implemented in response. The follow-up surveys found that the users continued to have positive perceptions of the new design over a 2-year span. Those who had been involved in the original planning were more positive than newer employees.

Knight, Weitzer, and Zimring (1978) and Bakos *et al.* (1980) also generated longitudinal data within a pretest–posttest change-framework, but for different purposes. In the Knight *et al.* study, the collection of data over time provided an extended baseline against which the effects measured after the design changes became more clear. In the Bakos *et al.* study, measurements over time were used both to create an extended prechange baseline, and to see how well positive behavioral changes endured.

For Fitzhugh and Anderson (1979), the assessment of the impact of time on resident attitudes in a housing project was a prime research goal. They used two time measurements, 3 months apart, to assess change over time in order to make causal statements about the relations among the variables.

Single-Site Iterative Assessments. Within one site, an investigator can track behavior across a series of planned, executed, and evaluated changes in a series of linked pretest–posttest assessments. One example of such a study evaluated the visitor use of an aviary and found problems in the orientation and layout of the aviary, which limited visitor enjoyment of the exhibit (O'Reilly, Shettel-Neuber, & Vining, 1978). From these data, the researchers made recommendations for design modification and saw some of their recommended changes implemented. They then reevaluated the setting and made further recommendations, some of which were also implemented, and there were plans for continued evaluation of the space.

Multiple-Site Posttest-Only Assessments. By comparing two or more sites at one point in time, research can provide a context from which the data for each site are seen in better perspective. The influences of specific site differences can be compared and contrasted, so that a better overall understanding of that setting type is possible.

Margulis's evaluation (1981) of industrial (factory-built) housing may be the largest and best controlled multiple-site study available. The goal was to assess consumer satisfaction with the Operation Breakthrough housing program of the U.S. Department of Housing and Urban Development. Four separate studies were conducted: an initial survey of visitors to the models; an assessment of dwelling repairs; exit interviews with persons leaving Operation Breakthrough housing; and, finally, a study of user satisfaction with these houses.

In the final survey, 698 Operation Breakthrough residents and 551 conventional-housing residents were interviewed. In contrast to previous studies, which Margulis wrote indicated that consumers would not accept "industrialized housing," the Operation Breakthrough evaluation showed considerable favorable response when compared to the matched sample of conventional housing. Operation Breakthrough owners were at least as satisfied as, and in many areas were more satisfied with their housing than, conventional owners.

Other major multiple-site studies include the evaluation of 37 public-assisted projects in 10 different states, as well as of 1,900 subjects by Francescato *et al.* (1979); assessments of a broad national range of child care facilities and play environments by Moore, Cohen, and McGinty (1979); and evaluations of National Park Service visitor centers by Zube *et al.* (1976).

Multiple-Site Pretest–Posttest Studies. Concurrence of results over several points in time and/or across several different settings increases confidence in the validity of conclusions, even without the use of randomization and control groups. The most logical extension of an evaluation study should be to follow a group of related sites over time, collecting data at several points. Arranging such a project, however, is costly and time-consuming, and there are no available examples of studies that have monitored multiple sites over extended periods of time.

Several projects have looked at multiple-site pretest–posttest changes. A large, multidisciplinary project (White, 1981) evaluated interior office environments in the U.S. Senate Office Building before and after changes aimed at improving morale, efficiency, and productivity while projecting a positive image to the public.

The project design included data from interviews of almost all staff in these offices, as well as detailed behavioral observations. The comparative analysis allowed trends across offices and across time periods to become noticeable in terms of general improvements in satisfaction. For example, the report focused on problems relating to trade-offs between communication and privacy that became apparent in the post-occupancy evaluation of the offices. By comparing the several offices, the researchers were also able to assess the impact of various levels of participation on satisfaction with the final product.

Brill, Margulis, and Konar (1984) administered a 500-item questionnaire to over 5,000 office workers at more than 70 sites at two times: 2–4 months before a move to new offices and 8 months after the move. The results related a number of environmental factors to such response categories as job satisfaction, environmental satisfaction, ease of communication, and job performance. Job satisfaction, for example, was found to be related to furniture, noise, participation, comfort, communication, lighting, floor area, and heating, ventilation, and air-conditioning systems. The posttest study concluded that 15%–17% increases in productivity could be attributed to environmental change.

Multiple-Site Iterative Assessments. Several researchers have reported projects in which a design type was followed and evaluated across several successive sites, and in which the environmental design research was used to inform each later iteration of the building type. Although each iteration of the building type might involve considerable passage of time, time was not a variable in itself. Rather, each succeeding setting served as a replication and extension of the previous one. Such research can assess the reliability of the initial findings and can search for construct validity in testing the outcome of design changes that had been based on the conclusions of previous evaluations. As Winkel (1983) noted, the use of theory can provide a link that makes each succeeding case potentially valuable in generating generalizable results.

A series of evaluations of jail environments provided tests of a new model of correctional environment design and management known as *direct supervision* (Wener, Farbstein, & Frazier, 1985). This model changed the role of the officers—placing them in direct contact with the inmates—and provided physical settings that allow greater privacy and self-control for the inmates. An explicit attempt was made to change the expectations conveyed by the physical setting to one of prosocial behavior. The assessments of half a dozen jails supported the assumptions behind the direct supervision model. The incidence of vandalism and graffiti in these settings was very low, and the number of assaults was greatly reduced (by as much as 90%) in comparison with previous, non-direct-supervision settings.

Lawton, Fulcomer, and Kleban (1984) conducted a series of projects in which design and behavior change were looked at in several phases at a geriatric institution. The overall design was based on a broad environment–behavior model, and narrower conceptualizations were also used in assessing behavioral responses. Initially, small-scale changes were made and assessed, which fed information into the design of a new facility. The assessment of the new setting confirmed and extended the initial results. A number of positive changes in resident, staff, and visitor behavior were described, and some unexpected changes were also noted and explained. For example, in some cases, staff–resident contact was decreased because the new setting provided much greater visibility, so that staff could monitor residents without immediate contact.

ADVANCES IN RESEARCH TECHNIQUES AND INSTRUMENTS

An important area of growth in environmental evaluation has been in the development of techniques to focus the scale of the evaluation on the needs of the client, and to develop generic, standardized instruments for data collection.

Tailoring the Evaluation to the Needs of the Client

Environmental evaluation has been adopted by several large organizations, such as government agencies, that have a purview over many facilities. One emerging trend in these organizations has been to modify evaluation techniques in timing, scope, and scale to meet their own special needs. Often, these needs include the rapid and inexpensive assessment of a large number of settings. These modified techniques provide a significant increase in the flexibility of the assessment procedures available.

One approach to solving the problem of generating information quickly has been the "walk-through" assessment procedure, used as part of a multilevel evaluation. To quickly obtain an overview of function and a consensus on critical issues, Gray *et al.* (1985) combined a thorough tour of the setting with on-tour interviews of relevant users and posttour group meetings. This approach has been successfully integrated into the planning process used by the New Zealand Ministry of Public Works and Development.

Goldman, Zimring, and Dunne's three-phased approach (1987) for evaluating prisons was closely modeled after the New Zealand system. Their procedures include (1) profile assessments that provide a brief audit of performance and identify strengths and problems in the setting; (2) POEs and design evaluations that provide a more detailed assessment of facilities, looking for causes and solutions of problems; and (3) issue studies within or across settings that study specific issues, such as double-celling or energy use.

Another similar, institutionally based, phased evaluation system was developed by Harvey and Zeisel (Canadian Hospital Evaluation Program, 1986) for Health and Hospital Canada to use to evaluate health-care facilities.

Standardized Instruments and Instrument Packages

Attempts at quick, institutionally based evaluation procedures have dovetailed with a recognition of the need for more reliable, valid, and generalizable research tools. It is often impossible to collapse, pool, and/or compare data among many evaluations because of the uniqueness of the data collection instruments used. Indeed, the procedures mentioned above are impossible without standardized research procedures and instruments.

The Multiphasic Environmental Assessment Procedure (MEAP) of Moos and Lemke (1984) contains nine instruments, of tested and published reliability, for assessing sheltered-care homes. The MEAP addresses organizational type, policy, physical environment, programs, and user attitudes. Moos and Lemke have used the MEAP in a number of facilities to develop a normative group that forms the basis for the standardized scores. With this "kit," an evaluator can quickly compare one setting to many others on a range of important issues.

Other examples of packages of standardized instruments are a system of instruments for evaluating correctional environments (Farbstein & Wener, 1982), which has been adopted whole or in part by other correctional researchers (Anderson *et al.*, 1984), and Canter and Kenny's User Survey Evaluation Package (USEP; 1977), which assesses staff and patient responses to hospital environments.

Single instruments have been developed to measure responses to a specific kind of setting and have been used with many subjects across a wide variety of environments, for example, the Revised Occupant Satisfaction and Attitude Procedure (ROSAPS) for multiple-family housing (Francescato *et al.*, 1979; Weidemann & Anderson, 1980) and the Building Performance Assessment Survey for offices (Picasso, 1987).

A curious anomaly of environment–behavior research is that the nature of the physical environment is often dealt with in only the grossest of terms. For example, settings may be identified as "high-rise housing" or "open offices." It is not surprising that results from what would seem to be similar settings are often widely divergent, given that these broad terms can encompass settings of wide variety. High-rise housing can range from 4 to 90 stories and can have a corresponding range in services, amenities, and layouts. *Open offices* can mean a broad expanse of unseparated desks or an attractively landscaped setting with acoustical partitions.

One approach to measuring the environment is to develop lengthy checklists of setting features, as has been done in the MEAP and by Farbstein and Wener. These checklists yield scores on conceptually important physical dimensions like controllability (e.g., the ability to turn lights off and on) and privacy (e.g., access to private spaces).

Checklists are limited, however, by the need for subjective observer ratings, which may reflect social-environment as much as physical-environment characteristics. Another, more objective approach to rating the physical setting has been taken by Archea (1984) and Benedict and Burnham (1980). They have developed related, although procedurally different methods for objectively coding an environment according to dimensions of accessibility. Archea's system identifies the visual access

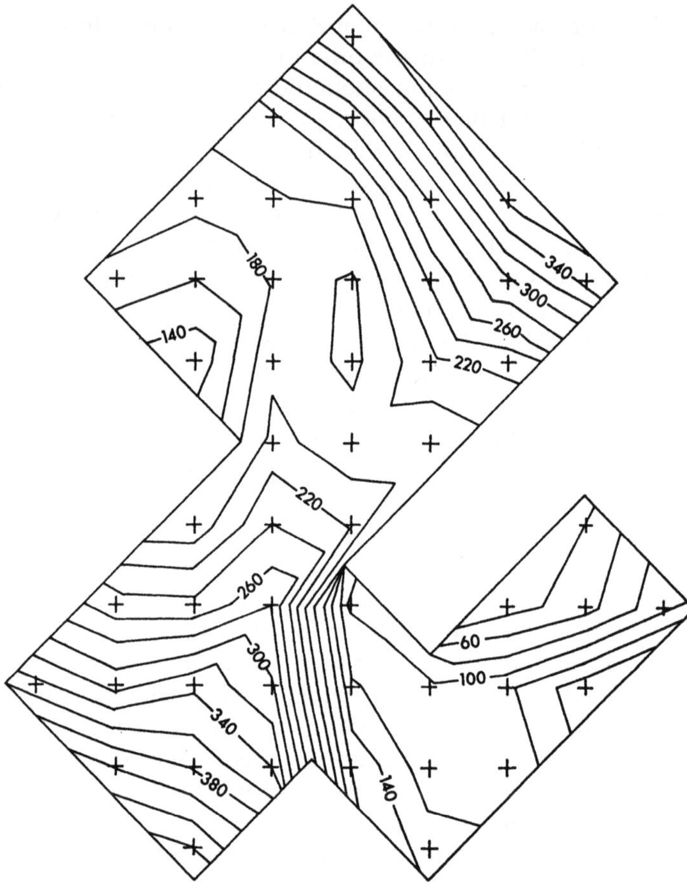

FIGURE 5. First-order visual access of prototypical space. (Figure courtesy of John Archea.)

and exposure ratio for each area in a setting, representing the degree to which any person can be seen (exposure) or can see others (access). When connected to specific behavioral hypotheses (e.g., that people seek high-access, low-exposure settings), this system can predict behavior and assess the behavioral implications of settings from floor plans.

A study of residents in the Philadelphia Geriatric Center validated the ability of Archea's technique to predict the actual numbers of people that the subjects could see or by whom they could be seen. The residents' behavior corresponded closely to the predictions of seeking high visual access and low exposure (Figures 5 and 6).

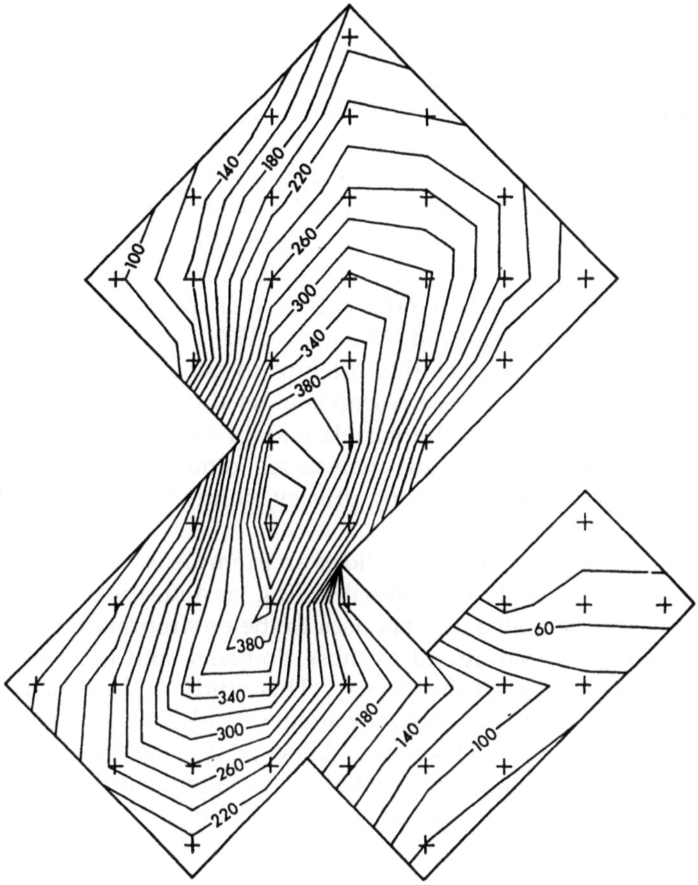

FIGURE 6. First-order visual exposure of prototypical space. (Figure courtesy of John Archea.)

ADVANCES IN DATA ANALYSIS

Typically, environmental evaluations have relied on descriptive data analysis, as befits the descriptive goals of most studies. The development of more reliable instruments makes feasible the use of more sophisticated data-analysis tools to extract the most information from a data base.

The work by Weidemann and Anderson (1980), using the Revised Occupant Satisfaction and Perception Survey (ROSAPS), serves as a model for the use of multivariate analysis techniques in design evaluation. None of the statistical procedures of these techniques are new to social science research, but rarely have they been used with as much good effect in environmental evaluation.

In one study (Weidemann & Anderson, 1980), 57 residents of a family housing complex were interviewed with the ROSAPS. Five questions were used for a "satisfaction index." The remaining 200 items were subjected to a principal-components analysis, which yielded 25 factors. These factors were used as predictor variables in a multiple-regression equation, and the satisfaction index was used as the dependent variable.

A study by Fitzhugh and Anderson (1979) made use of two advanced statistical techniques that attempt to generate causal analyses from correlational data. These researchers used cross-lagged panel correlation (which compares the correlations of several variables across several points in time) and path analysis techniques to develop a causal model of the relationships among privacy, control, security, and satisfaction.

Kenny and Canter (1981) provided an example of using facet analysis, a derivative of multidimensional scaling, as a means of understanding the possible domains of responses to the environment, of developing questionnaires that attend to those domains, and of providing an analysis of responses to the survey. This study addressed issues of environmental design in nurses' evaluations of hospital wards by surveying 1,921 nurses on 144 wards in 23 hospitals. The system, too complex to describe in detail here, provides a unique quantitative rationale for conceptualizing and assessing physical design impacts on user perceptions.

USE OF THEORETICAL MODELS

Theory development is critical to the advancement of any scientific endeavor. Theory allows disparate data to be organized and interpreted and makes gaps in knowledge clear. With it, a field can go from describing to understanding a situation. Theories also make analysis portable, allowing results to be generalized across settings.

Moore (1987) noted that the multidimensionality of environment–behavior problems requires theoretical formulations that encompass physical, cultural, and social dimensions, at scales of settings ranging from the micro- to the macrolevel, including time.

Moore (1987) also provided a distinction among levels of theory, ranging from broad conceptual approaches (theoretical orientations), arrays of relationships among findings (frameworks), and predictive simulations of the real world (conceptual models), to explicit systems that provide hypotheses and link behavior to environment (explanatory theories). He suggested that there is a movement in environment–behavior studies toward the development of theories that will help to explain the accumulating base of environment–behavior data.

Both Stokols (1984) and Winkel (1987) have written on the need for better and more complex theoretical formulations to account for the contextual nature of environmental design issues. For Stokols, contextual research means identifying additional variables drawn from the situational context that can help to explain the primary-target predictor variable. Theory becomes crucial when one is selecting related supplementary variables from the almost infinite array available.

Winkel (1987) suggested that we see these complexities as opportunities, rather than as threats to validity. The variations in settings should be viewed as challenges for theoretical development. He noted that, in trying to evaluate the "transferability" of constructs from one setting to another, the investigator is obliged to develop a theoretical rationale that can justify linking the instruments and findings across settings. It is in interpreting the meaning of naturally occurring variations within and across settings that we can begin to understand the relationships of people and settings.

Environmental evaluations can take a variety of approaches to the use of theoretical models. First, they can purport to be purely empirical—collecting and reporting data independent of frameworks or conceptual models (although an implied model must always be in effect, if only to allow a set of questions to be chosen for inquiry and reporting from the almost infinite array of possible items). The early dormitory assessments of Miller and Wheeler (Wheeler, 1985) did an excellent job of describing student interests, desires, and habits, but without much recourse to a conceptual organization of the information, so that the generalizability of these descriptions to other settings and situations was limited.

Second, evaluations may make explicit use of frameworks or conceptual models in organizing the data that have been collected so as to make sense of the results. Most evaluations operate at this level, using the available conceptual models to organize information, in part because of the dearth of broader explanatory theories. For example, in the Weidemann and Anderson (1980) housing study, concern about the lack of fences between yards was interpreted in terms of privacy regulation and social control.

Third, an evaluation may begin with an overall guiding framework that leads directly to the questions asked and the interpretations given, and that may allow for direct testing of theoretically derived hypotheses. Although some classic setting-specific studies, such as the study of large and small schools by Barker and Gump (1964), have tested predictions developed directly from theory, few environmental evaluations have such clear theoretical underpinnings.

One study that was able to make use of a conceptual model specific to a site type was the evaluation by Sinha (1985) of children's playgrounds. She developed a set of hypotheses based on a conceptual model of continuity and branching, developed by Moore *et al.* (1979) on how differing playground designs may affect children's play behavior. The study evaluated the behavior at several playgrounds, which varied along the dimensions deemed important by Moore *et al.* and provided important support for the validity of that model.

IMPLICATIONS FOR FUTURE RESEARCH AND APPLICATION

Many of the advances noted above represent the cutting edge of environmental evaluation. For this field to progress, there needs to be a continued efforts to advance by developing more sophisticated instruments, more general and applicable explanatory theories, and so on. In addition, there must be an effort to disseminate these innovations more widely, so that they will become part of the mainstream of evaluations.

The framework developed by Friedman, Zimring, and Zube (1978) is useful in this regard because it provides an outline that evaluators can follow to be sure they have addressed the key aspects of a situational context. The context issues dealt with by Shibley and Schneekloth (in press) are more difficult to turn into checklists or guidelines, as they deal with process issues requiring considerable organizational sophistication and sensitivity.

For the most practical of reasons, it is vital that evaluations become more attuned to the use of theory. As Winkel (1983) and Yin (1984) have pointed out, much useful practical and theoretical work can be done by means of case studies. Evaluations, however, need to use theoretical frameworks in choosing sites and research instruments that allow divergent situations to be linked and that provide a context for defining extremes or special situations in design, users, or organization. For these kinds of studies to become more common, there must be (1) a greater development and dissemination of theoretical frameworks useful for environmental evaluation and (2) a greater application of standardized instruments to the description of the setting and to the collection of satisfaction and perceptual data, so that settings and responses can be more easily compared across studies. If the use of such instruments can generate large data bases, single-site case studies can more easily relate their data to the larger context. Each individual case can then be seen as representing one cell in the larger matrix of setting types × user types × organization types.

Calling for an increased use of standardized instruments is not contradictory to suggestions for greater attention to qualitative methods (Low, 1987), as evaluations are often required to meet diverging goals that call for these varied data-gathering techniques.

By making the distinction between the comparative and generative purposes of evaluations, it is easier to develop instruments and procedures that can be applied to the specific needs of a study. It may be necessary to advance in two opposite directions at once, for instance, developing broad and shallow but highly quantitative data bases useful for "benchmarking" and, at the same time, becoming more sophisticated in providing fast information of use to designers, such as that elicited by "walk-through" assessments.

Zimring and Reizenstein (1980) noted that there is clearly a dearth of studies that view changes in settings over time, particularly across several sites. The number of studies looking at pretest–posttest changes is increasing, but there are still very few that use time as a variable, looking at growth or change as a function of passage of time. There is also a need for studies that more closely compare multiple sites, especially across time.

Such studies are enormously expensive and time-consuming and are not likely to be conducted as part of any regular design practice. This is especially true if we expect methodological standards similar to those used by Margulis (1981). It may be that these kinds of projects are most appropriate for academic research and should become the target of agency- or foundation-funded research, with the longer time periods such grants allow.

As POE methods advance, and as the quality and reliability of data improve, it may be possible to move more assuredly into the area of predicting the impact of designs before they are executed. This kind of "preoccupancy evaluation" (called "ex ante assessment" by Marans, 1984) has, to date, usually been based on simulations of settings. Although simulations will no doubt become more sophisticated with the advent of high technology (e.g., computers and videodisks), the field may also approach the stage where a data base of findings linked to solid conceptual models and/or explanatory theories will allow extrapolations to as-yet-unbuilt settings. On a small scale, Archea's visual exposure–access theory (1980) provides such a potential.

For accumulated data and standardized research instruments to become available to evaluators, there needs to be a better system of access to resources. Although some instruments may become available through publishers (as is the MEAP), a centralized data and instrument clearinghouse may be necessary to encourage the sharing of data and instruments and to realize some of the potential of the increased understanding of environment–behavior relations that results from environmental evaluation.

ACKNOWLEDGMENT

I am grateful to the perceptive and very helpful comments of the editors—Erv Zube and Gary Moore—through the numerous drafts of this chapter. Virginia Kurshan and Anne-Claire Houacin also provided valuable services in preparing this manuscript, for which I am grateful.

REFERENCES

Anderson, J., & Butterfield, D. (1980). *Generalization and postoccupancy evaluation.* Housing Research and Development Program, Department of Architecture, University of Illinois–Urbana, Illinois.

Anderson, J., Lewis, W., Barker, G., Lund, R., & Roth, J. (1984). *Learning from the past to guide the future: Post occupancy evaluation of facilities of the Capital Development Board of the State of Illinois.* School of Architecture, University of Illinois, Champaign-Urbana, Illinois.

Archea, J. (1984). *Visual access and exposure: An architectural basis for interpersonal behavior.* unpublished Ph.D. dissertation, Pennsylvania State University, University Park, PA.

Bakos, M., Bozic, R., Chapin, D., & Neuman, S. (1980, October). Effects of environmental changes on elderly resident's behavior, *Hospital and Community Psychiatry, 31*(10), 677–682.

Barker, R., & Gump, P. (1964). *Big School, small school: High school size and student behavior.* Standford, CA: Stanford University Press.

Benedict, M., & Burnham, C. (1980). *Describing and perceiving the physical world: Research with isovists.* Presented at the 88th Annual Conventional of the American Psychological Association, Montreal, Canada.

Brill, M., with Margulis, S., Konar, E., & the Buffalo Organization for Social and Technological Innovation (BOSTI) in association with Westinghouse Furniture Systems (1984). *Office design to increase productivity,* Vols. 1, 2. Buffalo, NY: Workplace Design and Productivity, Inc.

Campbell, D. T., & Stanley, J. C. (1966). *Experimental and quasi-experimental designs for research.* Chicago: Rand McNally.

Canadian Hospital Design Evaluation Program. (1986). *Federal Provincial Ministry of Health.* Ottawa, Canada: Building Diagnostics, Inc.

Canter, D., & Kenny, S. (1977). *The need for systematic user evaluation in health buildings and the development of the User Survey Evaluation Package.* Guildford: Department of Psychiatry, University of Surrey.

Cook, T. D., & Campbell, D. T. (1979). *Quasi-experimentation: Design and analysis issues for field settings.* Boston: Houghton Mifflin.

Farbstein, J., & Wener, R. (1982). Evaluating correctional environments, *Environment and Behavior, 14,* 6.

Fitzhugh, J., & Anderson, J. (1979). *Personalization, control, security, and satisfaction: A study of causal relationships among four variables in multifamily housing.* Urbana-Champaign: Department of Architecture and Housing Research and Development Program, University of Illinois.

Francescato, G., Weidemann, S., Anderson, J., & Chenowith, R. (1979). *Research satisfaction and HUD assisted housing: design and management factors.* Washington, DC: U.S. Department of Housing and Urban Development.

Friedman, J., Zimring, C., & Zube, E. H. (1978). *Environmental design evaluation,* New York: Plenum.

Goldman, M., Zimring, C., & Dunne, D. (1987). Research to redesign to research to redesign: Improving California's prisons. In R. Shibley (ed.), *Proceedings of the Association of Collegiate Schools of Architecture Annual Research Conference.* Washington, DC: Association of Collegiate Schools of Architecture.

Gray, J., Watson, C., Daish, J., & Kernohan, D. (1985). Putting POE to work: A case study in which POE is combined with participatory planning. In S. Klein, R. Wener, & S. Lehman (Eds.), *Environmental change/social change.* Washington, DC: Environmental Design Research Association.

Kantrowitz, M., & Nordhaus, R. (1980). The impact of post-occupancy evaluation research: A case study. *Environment and Behavior, 12*(4), 508–519.

Kenny, C., & Canter, D. (1981). A facet structure for nurses' evaluations of ward designs. *Journal of Occupational Psychology, 54,* 93–108.

Keys, C., & Wener, R. (1980). Organization intervention Issues: A four-phased approach to post-occupancy evaluation. *Environment and Behavior, 12*(4), 533–540.

Knight, R. C., Weitzer, W., & Zimring, C. (1978). *Opportunity for control of the built environment: The E.L.M.E.R. Project.* Amherst: University of Massachusetts.

Lawton, M. P., Fulcomer, M., & Kleban, M. (1984). Architecture for the Mentally Impaired, *Environment and Behavior, 16,* 730–757.

Low, S. (1987) Developments in research design, data collection, and analysis: Qualitative methods. In E. H. Zube & G. T. Moore (Eds.), *Advances in environment, behavior, and design,* Vol. 1. New York: Plenum.

Marans, R. (1984). Evaluation research in architecture. In J. C. Snyder (Ed.), *Architectural research.* New York: Van Nostrand Rinehold.

Marans, R., & Ahrentzen, S. (1987). Developments in research design, data collection, and analysis: Quantitative methods. In E. H. Zube & G. T. Moore (Eds.), *Advances in environment, behavior, and design,* Vol. 1. New York: Plenum.

Margulis, S. (1981) *A methodology for evaluating housing in use: A case study approach.* Washington, DC: National Bureau of Standards, PB82-104969.

Moore, G. T. (Ed.). (1982). Applied architectural research: Post-occupancy evaluation of building. *Environment and Behavior, 14,* Whole No. 6.

Moore, G. T. (1987). Environment and behavior research in North America: History, developments, and unresolved issues. In D. Stokols and I. Altman (Eds.), *Handbook of Environmental Psychology,* Vol. 2. New York: Wiley.

Moore, G. T., Cohen, U., & McGinty, T. (1979). *Planning and design guidelines: Child care centers and outdoor play environments.* Milwaukee: Center for Architecture and Urban Planning Research. University of Wisconsin–Milwaukee.

Moos, R., & Lemke, S. (1984). *The Multiphasic Environmental Assessment Procedure: A method for comprehensively evaluating sheltered care settings.* Palo Alto, CA: Stanford University.

O'Reilly, J., Shettel-Neuber, J., & Vining, J. (1978). The use of post-occupancy evaluation in an aviary: Implication for a continuing assessment program in museums. In A.

Osterberg, C. Tierman, & R. Findlay (Eds.), *Design research interactions.* Washington, DC: Environmental Design Research Association.

Picasso, G. (1983). *Life after installation: Long term planning based on experimental research and post-occupancy evaluation.* Bedminster, NJ: AT&T.

Picasso, G. (1987) How to develop an in-house POE, *Facility Design and Management, 6*(10), 64–67.

Preiser, W. (1981). *Albuquerque High School: A post occupancy evaluation.* Albuquerque: University of New Mexico, Institute for Environmental Education.

Project for Public Spaces. (1978). *Exxon minipark: A re-design and management proposal.* New York: Project for Public Spaces.

Rossi, P., & Freeman, H. (1985). *Evaluation: A systematic approach.* Beverly Hills, CA: Sage.

Sanoff, H. (1988). Advances in facility programming. In E. H. Zube & G. T. Moore (Eds.), *Advances in environment, behavior, and design.* Vol. 2. New York: Plenum.

Schneekloth, L. (1987). Advances in practice in environment, behavior, and design. In E. H. Zube & G. T. Moore (Eds.), *Advances in environment, behavior, and design,* Vol. 1. New York: Plenum.

Shibley, R. (1985) Building evaluation in the main stream. *Environment and Behavior, 17,* 7–24.

Shibley, R., & Schneekloth, L. (1988). Risking collaboration: Professional dilemas in evaluation and design. *Journal of Architectural and Planning Research, 5*(4), 304–320.

Sinha, A. (1985). Continuity and branching in preschool playgrounds. In S. Klein, R. Wener, & S. Lehman (Eds.), *Environmental change/social change.* Washington, DC: Environmental Design Research Association.

Sommer, R. (1984). Action research is not business as usual. In D. Duerk & D. Campbell (Eds.), *The challenge of diversity.* Washington, DC: Environmental Design Research Association.

Stokols, D. (1984). Theoretical contributions of environmental psychology to the analysis of social behavior. *Representative Research in Social Psychology, 14,* 16–35.

Weidemann, S., & Anderson, J. (1980). *Using a multi-site evaluation of housing as the basis for post-occupancy evaluation.* Paper presented at the 86th Annual Convention of the American Psychological Association, Montreal.

Wener, R. (1988). Doing it right: Examples of successful uses of environmental evaluation. *Journal of Architectural and Planning Research, 5*(4), 284–303.

Wener, R., & Keys, C. (in press). The effects of changes in jail population on perceived crowding, spatial behavior, and sick call: Absolute and contrast effects. *Journal of Applied Social Psychology.*

Wener, R., & Olsen, R. (1980). Innovative correctional environments: A user assessment. *Environment and Behavior, 12,* 478–493.

Wener, R., Farbstein, J., & Frazier, B. (1985). Three generations of environmental evaluation and design. *Environment and Behavior, 17,* 71–95.

Wheeler, L. (1985). Behavior and design: A memoir. *Environment and Behavior, 17,* 133–144.

White, G. (1981). *The Senate Office Systems Research Project, Report to Senate Committee on Rules and Administration,* May.

Whyte, W. B. (1980). *The social life of small urban spaces,* Washington, DC: Conservation Foundation.

Wicker, A. (1987). Behavior settings reconsidered: Temporal stage responses, internal dynamics, context. In D. Stokols & I. Altman (Eds.), *Handbook of environmental psychology,* Vol. 1. New York: Wiley.

Winkel, G. (1983). The challenge of the case study for the environmental design researcher. In D. Amadeo, J. Griffen, & J. Potter (Eds.), *Proceedings of the Fourteenth*

International Conference of the Environmental Design Research Association. Washington, DC: Environmental Design Research Association.

Winkel, G. (1987). The implications of environmental context for validity assessments. In D. Stokols & I. Altman (Eds.), *Handbook of environmental psychology*, Vol. 1. New York: Wiley.

Yin, R. (1984). *Case study research: Design and methods*. Beverly Hills, CA: Sage.

Zimring, C., & Reizenstein, J. (1980). Postoccupancy evaluation. *Environment and Behavior, 12*, Whole No. 4.

Zimring, C., & Wener, R. (1985). Evaluating evaluations. *Environment and Behavior, 17*, 97–117.

Zube, E. H. (1980). *Environmental evaluation: Perception and public policy*. Cambridge, England: Cambridge University Press.

Zube, E., Crystal, J., & Palmer, J. (1976). *Visitor center design evaluation*. IME Report Number R-76-5, Institute for Man and Environment, University of Massachusetts, Amherst.

VI

ADVANCES IN RESEARCH UTILIZATION

10

The Policy Environment for Environment–Behavior Research

FRANCIS T. VENTRE

POLICY: A FORCE NEWLY RECOGNIZED

Environmental design is commonly defined as arranging the forces of nature to achieve the purposes of humans. The constructions manifesting that order, in turn, influence subsequent human activities, including the initiation of the next set of purposes. Thus, structural designers must take cognizance of the gravitational force, whereas lighting designers live intimately with the psychobiology of vision. These physical forces (mechanical, electromagnetic, and thermal, among others; the involuntary physiological reflexes; and the regularities of psychophysical responses) are universal, or nearly so, in their effect. To ignore them invites disaster, literally. But designers and managers of environments who operate in culturally complex societies have had to deal with an additional set of forces that, although frequently encountered, are neither as universal in their incidence nor as uniform in their efficacy as the physical, physiological, and psychophysical factors. These newer forces, no less real because they are analytically less tractable, are the institutions and policies that mediate human purposes and human action in

Francis T. Ventre • College of Architecture and Urban Studies, Virginia Polytechnic Institute and State University, Blacksburg, Virginia 24061.

the creation and operation of the built environment. Although academic analyses of these forces are comparatively recent emergents in policy studies, written building regulatory precedents may be traced as far as the Code of Hammurabi (ca. 1955–1913 B.C.E.). Present-day public policies are, as in ancient Babylon, instruments for achieving social purposes, on the one hand, and, on the other, boundaries limiting individual action.

Public policies are here viewed as being not only, or even chiefly, the actions of governments. Rather, public policies include, in addition to governmental activity, the sustained purpose-driven, formal programs—not isolated actions—of industrial firms and commercial enterprises in the private sector when those firms and enterprises cooperate in voluntary organizations like trade associations, professional societies, and similar institutions of voluntary self-governance. *Purpose-driven* and *formal* are important qualifiers here because this chapter leaves for others the vast, informal system of culturally regulated environmental design actions that occur relatively unself-consciously and that are communicated by general social mores and professional ethoses, on the one hand, or narrowly defined commercial transactions between consenting parties, on the other. This chapter reflects that medium-range definition—broader than governmental, narrower than cultural (including commerce)—for three reasons. First, governments—federal, state, and local—account for only just over one fifth of total economic activity in the United States, and frankly, four fifths of *anything* is too big a chunk to ignore. Second, the broader means of cultural regulation of technical activities has drawn the attention of a growing number of social scientists working in the anthropological (specifically, the ethnological) tradition (Nader & Nader, 1985). Moreover, the cultural regulation of environmental design and use has generated its own special-interest groups within the environment and behavior community (Carswell & Saile, 1986; Low, 1987). Last, although those "broader-than-government" public policies address widely shared and sometimes noble purposes, they are implemented by the use of often flawed and always prosaic instruments. It may be that the very mundaneness of these instruments has, until recently, repelled environmental design researchers, who seem to have been attracted to rather more salubrious, not to say sybaritic, topics like beauty in its many forms, or to more earnestly socially conscious ones like user involvement in the processes of planning and design.

Table 1 presents the three arenas of action that, together, comprise the policy environment for environmental design and therefore for environment and behavior research. This array of policy arenas will be differentiated to enhance our understanding of each element as it con-

TABLE 1. Differentiation in the Policy Environment

Row	1 Originating agent	2 Range of effect	3 Vehicle for promulgation	4 Mechanism for ensuring compliance	5 Speed of adoption
1	Public government	Broad	Direct (all of society: administrative regulation or code) Indirect (tax policy, jurisprudence)	Coercion by statute, constitutive authority	Deliberate, slow
2	Voluntary associations	Limited (members of consenting organizations	Convention (standard for materials, systems, and practices)	Consensus by peer pressure	Moderate
3	Private parties (individuals or firms)	Narrow (immediate signatories)	Contract specification (canon of commercial practice)	Cooperation by mutual advantage	Immediate, fast

tributes to our knowledge of the whole. The remainder of the chapter focuses on the more public two areas of action and is organized as follows: first, the terms listed in the headings of Table 1 are elaborated (where necessary) by definition; second, the terms in the body of Table 1 are illustrated by example; and third, opportunities for future environment–behavior and environmental-design research are identified.

POLICIES OF GOVERNMENTS

Vehicles for promulgating policies that bear on the design and operation of the built and induced environment are of two main types: direct and indirect (Table 1, Row 1, Column 3), where *direct* means that the stated purpose of the policy is to mediate in the physical environment to achieve an end, whereas *indirect* policies are those that, although they bear heavily on environmental design decisions, do so as a side effect of addressing other concerns. Some examples will clarify both types.

The "Mechanism for Assuring Compliance" (Column 4) and the "Speed of Adoption" (Column 5) are functionally very closely related. The power of the state in the modern world is awesome in its coercive finality, not only in totalitarian regimes but also in the countries whose governments are restrained by written (and, in the case of the United Kingdom, by unwritten) constitutions. Such great power must be administered scrupulously, lest the ruling elites tyrannize the ruled. To ensure this restraint, governments are enjoined from acting rashly by the interposition of mediating procedures—whether by the checks and balances of the tripartite division of power that we enjoy in the United States or by the federal systems operating in many other countries. But these mediating structures move slowly, deliberating at each step over what might be the long-term effects of any coercive action on the several affected constituencies. This measured pace brings the beneficial effect of allowing sufficient time for informing the affected parties of what's in the offing. Once apprised, these parties seek elimination or at least mitigation of the coercive rule's untoward effects. This is why constitutional governments guarantee that due process will be followed.

The inevitable price of due process is that new policies are adopted slowly. Take, for example, the extension of building regulations in the United States to include accessibility for the disabled: over 20 years elapsed between the emergence in 1959 of voluntary design standards, developed by the Easter Seal Society, to the amendments to both federal regulations and the state and local building codes mandating accessibility for nearly all places of public assembly and commerce (Steinfeld, 1986). To be sure, governments can and do move faster where an emergency threatens: "energy-efficient" building-code requirements where prepared and implemented within five years of the 1973 OPEC oil embargo. But it requires a "crisis" atmosphere to propel regulatory changes at other than a slow and deliberate pace.

In sharp contrast with the dilatoriness of coercive government action is the immediacy of commercial transaction contract, the form of cooperation by which private parties define their own narrow region of mutual restriction (Table 1, Row 3). Once the instrument is signed, and a "consideration" (usually money) is exchanged, the contract is in effect.

Lying somewhere between the expediency and immediate speed of a contract and the circumpection and studiedness of a regulation is the moderate amount of time required to achieve a consensus among peers to first fashion and then adhere to a convention or standard for expediting repetitive transactions (Ventre, 1975). Let us now consider the actions of governments and of voluntary associations.

Direct Policies

Among direct public policies are the legislative mandates, committing governmental resources—both tangible and intangible—to the achievement of specific ends. Toward the close of the "environmental decade," the Association for the Study of Man-Environment Relations (1978) amassed a 217-page compendium of federal statutes and executive orders that specifically addressed environmental objectives of the government of the United States. Governments make direct capital investments in the environment itself, although this has taken place at a sharply diminished rate since the near-completion of the interstate highway system in the 1970s and the curtailment of sewer and water and urban-development grants to state and local governments in the 1980s. Where governments wish to bring the benefits of improved environments to citizens but prefer not to become capital investors, they resort to transfer payments to, in effect, purchase environmental services on behalf of those not able to afford them. Ready examples are the housing allowance schemes whereby poor families are given vouchers or other forms of currency with which to make their own purchases of housing services, either as renters or buyers.

The last of the direct governmental public policies to be considered is regulation, the most pervasive influence of all, extending throughout all three sectors listed in Table 1: public, private, and voluntary. Physical development—at scales ranging from the room to the region—is regulated by governments to achieve such public purposes as health, safety, and welfare. The National Institute of Building Sciences (1981) identified these regulatory purviews. The U.S. Constitution reserved many regulatory powers to the states, most of which delegated them to their political subdivisions, the cities and towns. But as will become clear, the nationwide integration of the building and development industry has lead the federal government to recall (or usurp, according to strict constructionists of the doctrine of states' rights) those police powers initially reserved to the states. In the following sections, legislative mandates, capital investments, and regulation are more fully discussed, and connections are drawn with environment and behavior research.

Legislative Mandates. Legislative mandates express the will of the legislature and usually do so in the broadest of terms. Such breadth of language serves several purposes but leads to many problems and to intriguing research opportunities. A broadly stated legislative intent—whether of the U.S. Congress, a state assembly, or a city council—accomplishes several things.

By articulating the further vision of the people's representatives, the broadly stated mandate draws together a wider coalition of political supporters. If the language is vague enough, of course, each component of the coalition may interpret the mandate in terms of its own pursuits. But there's a sound managerial or administrative reason for expansive language: it guides the discretionary power necessarily granted to executive-branch agencies to implement the legislative will and at the same time provides to the judicial branch the legislative intent needed for informing subsequent legal interpretations (Galligan, 1986). It is no doubt true that broad and visionary language, like that found in legislative mandates, does lend itself to easily caricatured pomposity and the threat of uncompleted agendas and disappointed hopes, in the former instance (Edelman, 1977), and to protracted litigation, in the latter (Kester, 1984). But the hazard of using language that is narrowly precise has its own costs, too: recall the perennial difficulty that industry and government face in interpreting the Delaney Amendment, which prohibits the sale for human consumption of any carcinogen, no matter how epidemiologically or etiologically remote the cancer threat may be (Crandall & Lave, 1981). Or consider the recent debate over the likely inoperability of a constitutional amendment outlawing budget deficits. Even conservatives are hard put to define exactly what constitutes, in precise legal and accounting terms, a deficit situation. For instance, should accrual accounting or cash accounting methods be specified in the proposed amendment? Yes, even accountants differ: there is room for interpretation within the "Generally Accepted Accounting Principles" promulgated by the American Institute of Certified Public Accountants (Taylor & Turley, 1986).

The advantage of broad legislative mandates is that they leave the technicalities to the technicians and focus the public attention on what should be the long-run, most broadly supported pursuits of a legislative act. The challenge comes in operationalizing those pursuits. A vivid example is afforded by the Housing Act of 1949, envisioned by Truman Administration "Fair Dealers" and their allies in the 81st Congress to be the cornerstone of national housing policy. The act's most quoted phrases occur in its statement that the nation is committed to the goal of a "decent home and a suitable living environment for every American family." Definitional problems abound, and they are ripe for environment–behavior analysis. What, after all, constitutes a "home"? A warren of single-room occupancies? And a "suitable living environment"? Some of Washington's most exclusive (read, *expensive*) residential sections lie in an acoustically unsatisfactory situation—under the flight paths into and out of Washington National Airport. And an "American family"—What constitutes a family? Two gays living together? The one child in four being raised by a single parent (Current Population Reports, 1986)? Or the typical family of 1949: father at work and mother at home attending 2.*x* children?

Do government policies extend equally to such households? Should they? It took the Federal Trade Commission, in a celebrated dispute over truth in labeling, seven years to determine what constituted a frozen "cherry pie." If seven years of interpretation and litigation were required to define a cherry pie, then how long will it take to determine what constitutes a decent home in a suitable living environment for every American family? One easily imagines an infinite expansion of research opportunities with that single legislative mandate. And as the Association for the Study of Man-Environment Relations showed, scores of such exhortations lie (no pun) in the statute books. (By the way, the FTC ultimately defined a cherry pie as being 25% by weight, of cherries.)

But sometimes, the broadest, most generous and visionary legislative mandate goes neglected, even by the constituency instrumental to its realization. This surely was the case for the National Environmental Policy Act (NEPA) of 1969, 42 U.S.C. § 4332 (1970), the legislation that signaled the national commitment to a wholesome environment. Moreover, "little NEPAs" in 32 states and scores of cities made the commitment separately (C. P. Wolf, personal communication, 1987).

NEPA's Section 102 contained three subparagraphs, the first of which directed the agencies of the federal government to "utilize a systematic, interdisciplinary approach, which will ensure the integrated use of the natural and social sciences and the environment design arts" in any federally funded environmental intervention. Subparagraph B directed all federal agencies to "identify and develop methods . . . which will ensure that presently unquantifiable environmental amenities and values may be given appropriate consideration" in any future environmental decision-making. And Subparagraph C required that a statement be prepared describing the impact on the environment of any federally sponsored intervention. The legal profession in America created a virtual industry out of Subparagraph C and its by-now-notorious "environmental impact statements." But Subparagraphs A and B, which might have been the Magna Charta of enlightened environmental design and environment–behavior research in the United States, were neglected by the academic wing of the architecture profession in the United States, and sad to say, by, the environment–behavior research establishment. The landscape architects, on the other hand, seized the day, greatly enlarging their practices, enhancing their analytical methods, and even recasting their profession's self-definition (Simpson, 1985).

Government Investment. The governments of the United States—federal, state, and local—continue to invest directly in environmental improvements, although at a declining rate (Ventre, 1982). In 1964, for example, outlays for government buildings were equal to 1% of the gross national product (GNP). By 1984, however, that fraction had declined almost monotonically to 0.4% of the GNP. Governments are investing less in buildings and ancillary facilities for much the same reason as the private sector: greater productivity in the execution of government functions is promised from investment in equipment than from investment in buildings. Moreover, local politicians learned during the 1960s that municipal funds could be extended further, touching more individuals and families (read, *voters*), if paychecks were written to municipal workers performing long-running services, whether day care or playground maintenance, than if short-running payments were made to contractors and to the relatively well-paid building trades for new physical facilities. At about the same time, a few radical community organizers, some members of the mainstream liberal establishment, and most true-blue conservatives united in their opposition to the socially disruptive and market-defying programs of urban renewal.

This three-way alliance of strangers grew from three distinct perspectives on environmental intervention that deserves a second look because its effects are still being felt by the design and construction industries. Whether because of the displacement of poor families (which became the locus for radical community organizers), the destruction of culturally valued properties (which energized the liberal, middle-class urban intelligentsia), or the usurpation of the resource-allocating function of the urban land market (which engaged the conservatives), public construction became so controversial in so many cities that "bricks and mortar" became a pejorative term in electoral politics. What a twist! Municipal reformers and government activists from Baron Haussman to Nelson Rockefeller sought to use new environments as a means of social uplift, but by the end of the 1960s, that great tradition had been stood on its head, and what had once been slums had been transformed, Midas-like, into "urban villages," later, "new towns in town," and, later still, "Yuppievilles." These were effects felt largely at the local level and that, along with an increased—and increasingly vocal—demand for social programs, led to a curtailment of public construction in American cities and towns. Only legislation spawned by the current interest in refurbishing the frayed physical infrastructure is likely to reverse the situation.

For the federal government, the shift of investment out of structures and into productivity-enhancing equipment was intended to reduce the need for adding to the work force of highly paid civil servants. Another reason for curtailing building programs for the agencies was the shift to leasing rather than owning real property (Congressional Budget Office—CBO, 1983). Leasing space also reinforced the hope that a large federal bureaucracy might be only a temporary thing. But the main impetus for the investment shift is that federal workers are engaged not in direct service delivery but in transaction processing, a function increasingly automated during the 1970s. Although these electronically enhanced operations may need more expensively outfitted buildings to house them, the absolute amounts of new federally owned office space are likely to decrease (CBO, 1983), owing in great part to sharp reduction in General Services Administration (GSA) space standards: from 167 square feet per worker in 1981 to 135 square feet, to comply with the targets set in the Executive Order on Work Space Management Reform signed by President Reagan on March 29, 1983 ("GSA Commits to Cost Savings," 1987). Government office buildings do still matter, but more as energy-efficient shelters for increasingly automated information-processing equipment and less as testimonials to inspired civic vision.

Finally, as arguments over efficiency eclipsed aspirations of a benevolent civic vision, administrative efficiency itself gave way to demands for direct citizen participation. The symbolism embodied in public buildings changed accordingly. Goodsell (1984, 1988), for instance, documented and illustrated the changing architectural iconography of American municipal buildings, their legislative chambers particularly. Goodsell traced this evolution of council chambers through three distinct shifts in the "relatively clean and simple relationships between governors and the governed": the traditional chamber (1865–1919) representing "imposed authority"; the mid-century chamber (1920–1959) representing "confronted authority"; and the contemporary chamber (1960–present) representing "joined authority."

Regulatory Policy. It is in the sweep of the regulatory function as promulgated by governments (Table 1, Row 1, Column 3) and the standardization function as practiced by the design professions and the construction industry (Table 1, Row 2, Column 3) that environmental design researchers are likely to have a more direct influence (Margulis, 1986; Ventre, 1975). Regulations—broadly considered—are the means by which societies, using the coercive powers of government, mediate the private actions of individuals. Private actions know other limitations as well. Commercial transactions between informed individuals, for example, are limited by a mututally agreed-upon contract (Table 1, Row 3, Column 3). Contrast this with regulation: the reach of public policy is broad where commercial law is limited; public regulations are coercive where commercial contracts are subjects of mutual consent. Because they are universally and uniformly applied and coercively enforced, regulations must be carefully circumscribed either by statute, by legal precedent, or—most significantly for environment–behavior researchers—by technical knowledge. Statutes authorizing regulatory activities by public agencies are drafted by means of the same principles cited earlier, in the discussion of legislative mandates. Enforcing agencies must retain discretionary power in order to administer the regulations equitably and efficiently (Galligan, 1986). Legitimate use of this power requires a detailed knowledge of the social and physical technologies associated with the phenomena being regulated. Occupational safety and health agencies must be as knowledgeable about the etiology of industrial accidents and the epidemiology of industrial illness as they are about administrative procedure. The staff of a consumer protection agency must know the intricacies of commercial law as well as rudimentary ergonomics. The building official must be cognizant of current technology in design and construction as well as the rules of evidence and other forensic practices (Jones, 1985; Sanderson, 1969).

The private-sector firms and organizations being regulated expect and even demand this degree of familiarity with the technology under regulation. In the area of environmental design and development, for example, regulatory officials and the entities being regulated must, in a very specific sense, speak the same language. The technical provisions of regulations must be precisely stated so that all parties understand what requirements are to be satisfied by candidate materials, products, systems, or services. Test methods, by which candidate methods may be evaluated for regulatory approval, must reflect agreements about sampling and specimen selection as well as subsequent testing procedures required to determine the properties or performance of proposed designs (Heilstedt, 1986). Regulation is expedited, and so are professional design practices, by previously defined classifications or systematic arrangements or divisions of environments into groups based on characteristics relevant to the intent of the regulation. Similarly, the meanings of technical terms must be defined and explained to expedite communication. This is the self-assigned task of the voluntary system of standards generation and the incorporation, by reference, of those standards into regulation and professional practice. The workings of that system, and the role of environment–behavior researchers in it, are discussed in the "Nongovernmental Policies" section of this chapter.

Indirect Policies

Direct actions of governments, although ultimately coercive, are addressed to specific constituencies or clienteles within the economy. Of much broader reach, however, are indirect public policies, principally taxation and judicial doctrine.

Tax Policy. Tax policy and the responses it induces in the private sector of individuals and corporations, households, and businesses bear significantly on the nation's construction enterprise, and from both ends of the building process: first, in the decision to build or not build or to own or lease, and last, at the points of detailed design, down to the choice of furniture, fixtures, and equipment. But the influence of tax policy is so diffuse and its effects on design decisions are so roundabout that its exact impact may never be fully known. This elusiveness accounts for the vehemence of political discussion about changes in tax codes. But there is general agreement that the tax treatment of private investment favored (because it allowed faster depreciation of) office furnishings and equipment as opposed to investment in the permanent structure. This disparity of tax treatment had been, until the promulgation of the Tax Reform Act of 1986, an important contributor to the proliferation of "open office planning" in American businesses. Although environment–behavior researchers have identified and sometimes measured the advantages and disadvantages for organizational effectiveness of these interior design arrangements, the tax angle is likely to have been the business firm's dominant consideration, prompting greater investments in economically short-lived and tax-advantaged devices like moveable partitions and task (and sometimes ambient) lighting integrated with furnishings. Now that the tax advantage is eliminated (at least temporarily), environment–behavior researchers can evaluate a business firm's adoption, retention, or abandonment of the open office on its environmental advantages without having things muddled by the tax counsel's whispering in the owner's ear.

A recent U.S. Tax Court decision (*Shoney's Restaurants* v. *Commissioner of Internal Revenue,* 1984) with respect to the eligibility of certain lighting installations for investment-tax-credit consideration (an even greater advantage than rapid depreciation alone) seems to have turned on user perceptions about the contribution of that lighting not to the bare-boned operations of the building, but to the mood-enhancing ambience lent to the activity sheltered in the building. The Tax Court in the *Shoney's Restaurants* decision reasoned that, where lighting is a permanent structural element, integral to the *building's* operations, that lighting is *not eligible* for the investment tax credit; but that, where lighting is nonpermanent, "complementary," and accessory to the *business* operations, it *is*. The financial advantage brought by this ruling was worth Shoney's pursuit of it through 12 years of hearings, trials, and appeals! And the grounds for the opinion are shrouded in the mists of architectural metaphysics of the sort only an environmental scholar of Norberg-Schulzian capacities could fathom.

Although the federal investment tax credit was eliminated in the 1986 Tax Reform Act (a reform no doubt stirred by controversies like the one that the *Shoney's* case engendered), that Tax Court decision is worth recalling because it demonstrates just how far considerations first brought to the design profession's attention by the environmental psychologists have now informed the awareness of the broader community about the affective effects of environments. Like the "academic scribbler" of J. M. Keynes's anecdote on the power of ideas, the environmental psychologists who recorded and reported the effects of lighting on attitudes and performance deployed an influence where they never sought to: in the federal judiciary deciding questions on the tax status of building investments.

Tax expenditures is a term that was added to the political lexicon during the 1960s by the late Stanley Surrey, a Harvard Law School professor who served as an assistant secretary of the treasury in the Kennedy and Johnson administrations. Surrey (1985) argued, even before he got to Washington, that revenues not collected from taxpayers are as significant a governmental subvention as a direct cash subsidy and thus should be included in any calculation of the benefits and costs of government policies (Surrey & Daniel, 1985). One could argue, following Surrey's logic, that part of a homeowner's sense of satisfaction with her choice of residence may be the degree to which she considers it a tax-advantaged investment, bringing her a generous personal benefit. The Low Income Housing Information Service (B. Zigas, personal communication, 1987) calculates that, in fiscal year 1987, this tax expenditure to the homeowning middle class—including such deductions as those for interest paid on the mortgage loan, local property taxes, capital gains advantages, and energy conservation credits—amounted to $36.355 billion, whereas the dollar value of the U.S. Department of Housing and Urban Development's subsidies going to housing programs aimed at low-income families totalled $9.472 billion for the same period. Thus, the tax expenditure for middle-class housing was nearly four times as large as the congressionally mandated housing program itself.

Some tax expenditures are targeted to bring advantages to groups much smaller than the homeowning middle class. A provision of the same act that removed the investment tax credit allows businesses and professionals who serve the public to improve the accessibility of their facilities and to benefit from a yearly tax deduction of up to $35,000. Under Section 190 of the Internal Revenue Code, expenses for removing existing barriers, such as steps and narrow doors or those relating to inaccessible parking spaces, toilet facilities, or transportation vehicles, may be deducted (Paralyzed Veterans of America, 1987).

These examples illustrate that tax expenditures are themselves ethically neutral: they can be used either to widen or to narrow social inequities.

Judicial Doctrine. Judicial determinations have important effects on the design, construction, operation, and use of buildings and sites. Environment–behavior researchers have been effective in the judicial forum both as direct participants—say, expert witnesses—and indirectly, but more pervasively, as initiators of novel concepts of public welfare that have influenced judicial opinion. The careers of both the "social impact statement" and its progenitor, the NEPA-induced environmental impact statement, are instances of novel concepts of this sort.

The influence of environment–behavior knowledge (and specifically that of environmental psychology) became apparent almost immediately upon the promulgation of the NEPA in the early 1970s. Indeed, by 1975, this influence was extensive enough to require 19 pages of closely set type and 155 footnotes in the *Iowa Law Review* ("The Psychology of the Designed Environment," 1975) to summarize it. The *Review* concluded that, by 1975, the "courts have already recognized that the subject matter of this discipline—the relationship of man to the designed environment—is a relevant factor in the human environment" (p. 692). And that recognition of environment–behavior relationships need not be confined to the harmful or deleterious effects of environments, either. Numerous other courts have supported the positive, life-enhancing environmental attributes documented by environment–behavior researchers. A copious literature documenting the judicial endorsement of wilderness values and the beauty of unspoiled natural scenery appeared soon after the promulgation of the NEPA (Andersen, 1973), and judicial attention has since then been drawn to the cityscape. Brace (1980) reviewed nine such cases. The number has almost certainly grown by now, given the "discovery" of historic preservation and the emergence of gentrification in American cities.

The social impact statement (Finsterbusch & Wolf, 1981) is a more recent arrival. Freudenberg (1986) condensed and reviewed its career to the present. My own recollection of that career is that it was riddled with ironies. Participants in the annual meetings of the Environmental Design Research Associates, (EDRA) of the early 1970s will recall Wolf's early advocacy of social impact assessment (EDRA, 1974). The concept was embraced, initially, by proponents of "power to the people," particularly the people residing in urban areas about to be either "renewed" or transected with limited-access highways. (Remember the Battle of the Lower Manhattan Expressway that was to overrun Broome Street, in which Jane Jacobs and Charlie Wolf took on the transportation agencies and industries of New York, city and state?) But within 10 years, the social impact statement was embraced by a liberal Carter administration and a conservative Reagan one: the Carterites invoked a "family impact" test for federal domestic policies, including taxation (Giraldo,

1978), and the Reaganauts' defense of traditional values prompted a proposal that a "family fairness statement" be prepared by any legislation or regulations affecting families (Werner, 1986).

The social impact idea had gained sufficient currency so that, in 1986, a nationally influential Court of Appeals of the State of New York (that state's highest court) interpreted the term *environment* in New York City's Environmental Quality Review Act to include the quality of a neighborhood. City officials, ruled the court, must consider whether new construction will "accelerate the displacement of local low-income residences and businesses or alter the character of the community" ("Social Impact Added to Project Reviews," 1986; Schmalz, 1986). Until that ruling, environmental impacts had been required only for an intended project's effects on light, air pollution, and noise.

The courts not only legitimized novel conceptions of public welfare—which we may call the poetry of jurisprudential thought; they also established the evidentiary rules under which trials and appeals must operate—which we may call the prose of jurisprudence. Let us now consider the prose. The growth of knowledge and its incorporation into the social and physical technologies that undergird contemporary society have lead to a counterpart growth in the use of "authoritative evidence" in the resolution of disputes. Prosecutors, plaintiffs, and defendants may each employ (in two senses of the term) knowledgeable experts as witnesses. Moreover, the professional bodies and technical societies of the design and construction community often offer their general expertise not as witnesses for either the plaintiff, the prosecution, or the defense, but as providers of "friend-of-the-court" (*amicus curiae*) briefs, intended to place the dispute in question in a wider historical and intellectual context (Frankena, 1985; Schwartz & Schwartz, 1981). Providing expert testimony is a lucrative part-time activity for qualified environment–behavior researchers, as it is for specialists of all kinds ("Expert Witnesses," 1987). But that's not the principal reason for my including it here. Rather, it is that the initial testimony, counter-testimony, and subsequent cross-examination of expert witnesses provide what is, to my mind at least, a scrutiny of environment–behavior knowledge that is more exacting than that found in the annual meetings of that specialty's national and international bodies. This may be the reason that "mock trials" appear in the annual meeting programs of an increasing number of professional and technical organizations. (Readers who attended the seventh annual meeting of the Environmental Design Research Association (EDRA) in Vancouver, Canada, in 1976 will vividly recall the mock trial—presided over by a Canadian federal judge, no less). I, for one, welcome this educational innovation since adopted by many professional and technical groups.

NONGOVERNMENTAL POLICIES

This chapter's discussion of direct policies concluded with the issue of regulation and cited the regulators' reliance on a technical language shared with the regulated. But where do the specifications, test methods, classifications, and definitions that comprise this technological *lingua franca* originate, and why is it retained? To answer the second question first, the regulations themselves would be too voluminous if all the pertinent specifications, test methods, and so on, had to be defined by each regulatory agency confronting a novel technology (Heilstedt, 1986). Moreover, the millions of providers of environmental goods and services would be hard put to develop sets of specifications, test methods, classifications, and definitions for each individual product or service or for each regulatory agency. In fact, such terms would be needed by the industry itself even if it were not regulated, terms by which the industry might communicate internally and, ultimately, might govern itself. This has indeed happened, and it answers the first of our opening questions. The large industrial and professional aggregates in the Western economies are, in effect, self-governed by means of cooperatively written, consensually enforced conventions or standards that are the grammar and vocabulary of technical discourse. Not only do these industrial standards expedite intraindustry cooperation and efficiency, but they also serve as a bridge between the broad and coercive public regulations of governments and the limited commercial contracts that bind private actions.

The middle ground between regulation and contract is the consensual standard used in industrial and professional self-governance. Table 2, condensed from several publications of the American Society for Testing and Materials, identifies and defines the varieties of standards com-

TABLE 2. Varieties of Standards

Specifications—Precise statements of a set of requirements to be satisfied by a material, a product, a system, or a service.
Test methods—Cover sampling and specimen selection and describe the subsequent testing procedures used to determine the properties, composition, or performance of the materials, products, systems, or services that may be specified.
Classifications—Define a systematic arrangement of materials, products, systems, or services in groups, based on similar characteristics.
Definitions—The explanations of the meanings of terms as applied to materials, products, systems, services, and methods within the scope of standards-generating committees.
Practices—Procedures, guides, or services that may or may not be auxillary to a test method or a specification.

prising this middle ground. The varieties enumerated provide ample scope for environmental design researchers: "requirements," "determine . . . performance," "systematic arrangement," and "guides" are all terms that pepper EDRA Proceedings, *Environment and Behavior,* and the *Journal of Architectural and Planning Research,* and they're also the stuff of technical standards.

Voluntary Standards

The writing of these industry standards or technical conventions by which industries govern themselves internally are an important domain of inquiry for environment–behavior researchers. Ventre identified this opportunity in 1975, and the intervening decade has witnessed encouraging results. Several leading environmental researchers have enriched their academic pursuits by precisely describing, measuring, and evaluating such environmental attributes and concerns as habitability and health, accessibility, emergency egress, and way finding, and participation by these researchers on the national consensus committees has codified those subjects.

Special satisfactions come to those researchers involved in the consensus-based voluntary standards. Many of those standards, although initially intended for intraindustry communicaton, find their way into public policy in one of two ways. First, regulatory agencies, seeking more efficient ways to maintain technological currency, incorporate (by reference) those originally voluntary intra industry standards into regulations, imbuing them with the force of law.

Second, as government procurement agents seek the same degree of technical currency, they, too, incorporate (by reference) existing voluntary standards into purchase specifications. Governments buy vast and diverse quantities of goods and services, from highways to health care to part-occupancy evaluations (POEs) of completed buildings and other facilities, and where voluntary standards are not available, government purchasers or underwriters commission them. A conspicuous example is the work of the University of Wisconsin–Milwaukee's Center for Architecture and Urban Planning Research on design guide specifications for child-care centers and playgrounds prepared for the U.S. Army Corps of Engineers (Moore, Hill, Lane, Cohen, & McGinty, 1979). The influence of government as underwriter is illustrated by Steinfeld's work (1986) on a technical basis for accessibility standards. The U.S. Department of Housing and Urban Development—as cosecretariat to the 53-member (mostly trade, professional, and advocacy organizations) American National Standard A117.1 Committee—supported the effort whose results were adopted voluntarily by thousands of design firms and the building owners that employed them. By the middle of the 1980s, the American National Standards Institute ("CABO Assumes A117 Committee Secretariat," 1987) reported that 95% of local and state building-regulatory jurisdictions had adopted and were enforcing ANSI A117.1's technical provisions, the provisions also having been adopted by reference in the three major building codes and having been integrated into the Uniform Federal Accessibility Standard (UFAS) of the U.S. Government.

Third, the deliberations of those standards-generating organizations attract some of the nation's most knowledgeable experts on the given subject, and as a result, discussion is informed, exact, and exacting. Thus, environment–behavior researchers can test their results in the crucible of industry politics, and when successful, they find their research results bearing directly on the policy environment in which all designers—including the uninformed and the unwilling—must, by law, operate. Consider these examples:

Collin's multiyear effort (1983) to improve the technical basis for emergency signage design was undertaken for the U.S. Bureau of Mines, but the results are also being shared with the relevant professional communities (including a briefing for the Society of Environmental Graphic Designers). Beginning with a review of current practices, standards, and regulations, the effectiveness of two sets of symbols for 40 different safety messages was evaluated with the help of 267 miners from 10 sites at disparate geographic locations. The evaluation included an assessment of the effectiveness of six different "symbol-surround" shapes and colors (more representational symbols proved more effective than the highly abstract ones). What can lessons from mine environments teach designers of other work environments? For one thing, reduced visibility often accompanies fire emergencies, especially in the windowless areas of large buildings. Also, insights into orientation and way finding in labyrinthine mines may be instructive to designers and, especially, operators of labyrinthine multibuilding complexes such as are found at hospitals and universities, which have grown incrementally over many years.

Based on Collins's research and related efforts (Collins & Lerner, 1983), criteria for new designs were proposed and shared with the cognizant standards organizations. The fire safety community responded by adopting the National Bureau of Standards' Center for Building Technology (NBS/CBT) recommendations in the May 1986 revision of the National Fire Protection Association's NFPA Standard 171, "Fire Safety Symbols." This standard is incorporated into fire and safety regulations throughout the United States and in many foreign countries.

The empirical testing of candidate safety symbols is a practice pioneered in the late 1970s by Collins, Lerner, and Pierman (1982) of the National Bureau of Standards/Center for Building Technology and has itself now become the subject of an international standard. Previously, symbols had been chosen on the basis of committee consensus, unaided by rigorous evaluation. This practice is now superseded by formal International Standard Organization (ISO) test methods and criteria. Each symbol—or its variant—must now be recognized by at least 66% of those surveyed. The results have been "disturbing," according to Frascara (1987). The tests revealed poor recognition of 9 warnings out of 16.

Mention of orientation and way finding—whether in mines or in buildings—recalls Weisman's efforts (1985a,b) to respond to the regulatory challenges, identified by both Stahl, Crosson, and Margulis (1982) and by the Architectural and Transportation Barriers Compliance Board (1983, pp. 3–35), to, in the board's words, base "future [regulatory] provisions . . . on data and facts . . . rather than a majority opinion." Weisman, of course, is not the sole worker in this field, but he is virtually alone in addressing his effort to using the system of codes- and standards-generating organizations to test and disseminate his results. Indeed, a recent international conference on building safety and use (National Institute of Building Sciences, 1985) heard the reports of 16 orientation and way-finding research projects from several countries. Except for two reports performed under government contract, however, only Weisman's drew on the regulatory system as a propagator of new knowledge. All others addressed themselves to fellow academics, to enlightened designers, or to litigation-conscious owners and managers of buildings and sites. In so doing, these researchers attenuated rather than amplified the effects of their own best efforts on the design and construction community they ostensibly serve.

The wider use of voluntary consensus standards as a base for government procurement or regulation promises to enhance the engineering efficiency and the economic efficiency of regulation and of government operations. But what about the political equity of these regulations and operations? Equity emerges as an issue because that wider use of voluntary consensus standards carries with it a contradiction: Most standards are originally devised to expedite the transactions of contracting private parties, and standards used this way need satisfy only tests of cost and convenience. But once adopted by regulatory agencies and then imposed on all with the force of law or by government procurement officials (so that that they have great economic weight), these voluntary standards are invested with another public purpose, whose aim is not only, or even primarily, convenience or cost effectiveness but conviction: the belief that the wider public interest, in contrast with a narrower industry interest, is being addressed and advanced.

Not only does this shift in the uses to which technical standards are put raise political issues of equity and accountability, but it joins also the intellectual issue of epistemic warrant. The epistemic warrant for claims of technical knowledge—and technical standards ostensively embody such knowledge—is not established *a priori* but is made with reference to the specific domain of application for that technical knowledge (Dewey, 1929; Rescher, 1977). Thus, the epistemological burden that such environmental design research must bear is greater because it is imposed not only by scientific and professional peers but by politically accountable representatives of more diverse and potentially critical audiences (Ventre, 1979).

Effects on Design Freedom

Although standards, regulations, and regulatory policy may seduce into the public forum environmental researchers confident in their knowledge of the effects of environmental interventions, these topics are anathema to the design cohort of the design–research community. Many claim that regulations encumber design and consulting work and constrain the technical options that may be presented to a designer's client (the building industry case is argued in National Institute of Building Science, 1981, and in Field, 1980; for a general indictment, see "Special Report," 1983; the Reagan Administration's assay is made in Office Management and Budget's *Regulatory Program of the United States Government*, issued annually). In a National Science Foundation–supported study of engineers' access to and use of technical information, Schuchman (1980) reported that 92% of the respondents indicated that their work was affected by regulation, and 70% indicated that their professional work required them "specifically to react to" explicit or embodied regulatory provisions (5% indicated that their work was preparing regulations!). Manifestly, regulation is a large presence in design work and deserves more attention from the environmental-design research community.

But do regulations and standards stimulate or arrest and abort design creativity? Reconsider this chapter's opening in that light: Where would structural designers like Nervi, Salvadori, Candela, or Fazlur Khan be without gravity, or Alvar Aalto or Cyril Harris without the psychophysics of vision or of hearing? These pervasive physical, physiological, and psychophysical factors challenge rather than stifle design creativity. Can social factors induce a similar effect? Although some designers and managers may grumble about regulatory restraint, there is much evidence that others are responding creatively.

The trade press serving the building, facilities design, construction, and management industries publishes, with gratifying frequency, successful solutions to problems posed by especially challenging regulation. The July 1985 issue of *Progressive Architecture* (Fisher, 1985) devoted six pages to a discussion of accessibility regulation and concluded by dismissing the "almost dogma . . . that regulations inhibit initiative and creativity" (p. 124). On the contrary, wrote Fischer, the effect has been the opposite, and he substantiated his claim with descriptions and photographs of creative design responses to regulatory challenges.

Even the general press has reported the designs-recognition programs of the U.S. government that have cited research results that will aid the disabled. Nor are these obscure items: the President usually confers these honors personally in well-publicized White House ceremonies. And there will be more to come. Spreading the news of creative designs for accessible environments will be the Adaptive Environments Center in Boston, which undertook in 1986 the sponsorship through 1991 of an awards program set up to recognize designers and developers whose projects demonstrate "design excellence and thoughtful accessibility" (Dean, 1987).

IMPLICATIONS FOR FUTURE RESEARCH AND APPLICATION

This chapter has indicated the variety of public policies that constitute the environment for environment–behavior research. Table 1 organizes that variety into five attributes that all current policies manifest and that all future policies must accommodate or, at least, acknowledge. But what of the future?

The policies arrayed in Table 1 offer design challenges: First, they challenge creative designers to produce projects that satisfy the regulatory requirements imposed to maintain the public's environmental interests. Second, there is the challenge of designing the regulatory system itself. That is the ultimate design challenge and environmental-design application challenge: the artful, efficient, and scientifically informed design of the public policies themselves. Where would such a research program begin? I offer two suggestions.

My first suggestion regards the knowledge basis for current policy. We need a better understanding of the processes by which environment–behavior and environmental-design research informs the development of standards or of other public policies of the types described in this chapter and condensed in Table 1. Understanding is needed to ensure that the most reliable knowledge will form the basis for environmental design policy. I cite a specific challenge: What is the scientific basis for the Executive Order on Work Space Management Reform mentioned earlier?

Consider what happened here: a reduction, by one fifth, of the areal standard accorded federal office workers. The immediate effect is sweeping enough, applying to the real-estate inventory, valued at more than $9 billion and including 6,800 buildings, including offices, warehouses, and other space, under GSA management ("New PBS Commissioner Named," 1987), and affecting the work environments of 800,000 civilian workers, or over 230 million square feet (Congressional Budget Office, 1983, p. xi). But the likely further implications of that reduction are truly enormous and will affect many million more office workers because, as frequently happens with procurement and operating practices (the purchase of energy-efficient automobile fleets or the creation of smoke-free work environments, for instance) federal adoption leads to the diffusion of the novel practice to a much wider industry. The further effect of the GSA's reduced space standards on the nation's nonresidential building inventory, on the design and construction services industries that create that inventorv, and on the building maintenance industry that keeps it in repair is, of course, not immediately known. But it is calculable and is very likely to be large. Government buildings, it should be pointed out, are a large fraction of the work of the nation's architectural and engineering design firms; yet, attentive reading of the principal professional periodicals over the interval when this space reduction policy was being contemplated and after it was announced reveals no mention whatever of the policy's effect on either the functioning of federal buildings or the effectiveness of federal agencies or the productivity—not to mention the satisfaction—of federal workers. Nor, for that matter, did the periodicals or organizations serving the community of environment and behavior researchers provide notice to either that community or to the wider public.

This lapse, this "nodding off" of the professions and industries that design, build, and evaluate the physical environment, compares unfavorably with the derisive hoots and the fusillade of counterarguing research findings provoked by the Reagan administration's attempt to reclassify ketchup as a vegetable in the U.S. Department of Agriculture's school lunch program. One would be hard put to say whether the expert testimony of the nutritionists and food scientists or the cumulative effect of Johnny Carson's and David Letterman's jibes had greater effect, but the ketchup-for-tomato substitution policy was soon withdrawn. It may well be that the physical environment does not now enjoy the salience that school cafeteria menus do for stand-up comedians or the country's editorial writers. But that does not condone future lethargy within nor inattention by the environment and behavior research community. On the contrary, the promulgation of an environmental design policy or practice for which no scientific basis has been adduced provides an occasion for environment and behavior researchers to widen their constituency. Such a venture in public education could begin by initiating within the building community itself an informed discussion stating that work environments do matter and that their dimensions and other attributes are not to be specified without a consideration of the effects of those specifications on either the work force or the public that is ultimately to be served. A more fully informed cohort of design and construction professionals would stimulate a much wider public discussion of policy proposals. Participatory design could then extend "upstream," as it were, of the formation of specific project plans, to the more inclusive public policies that bring design projects in their train.

My second suggestion regards application: We need a counterpart of the *Iowa Law Review* ("Psychology of . . . ," 1975) article, cited earlier in this chapter, that would address the social impact statement and its legal effects. Similarly, counterparts to Brace's article (1980) on aesthetics are needed for other topics of environmental value besides amenity. At the very minimum, both these law review articles need to be updated. They need also to be extended to consider the same topics in an international reference. Why this wider address, and why at this time? International comparative studies on these topics would be instructive, particularly because design and construction services have been recognized as significant components of America's international trade (Office of Technology Assessment, 1987). A start is Prieser and Rohane's comparative analysis (1980) of visual and aesthetic controls, categorized by the issues regulated, in English-speaking countries. Environment and behavior researchers with foreign language skills, or those for whom English is a foreign language (and you are many), have a ready advantage that can be applied for the good of all countries.

"We shape our buildings, and afterwards our buildings shape us"—an epigram well known to, and repeated by, many environment and behavior researchers. Its source and the occasion of its utterance are also well known: Winston S. Churchill's October 28, 1941, speech to the House of Commons of the British Parliament arguing for the restoration of the blitz-damaged Commons chamber to its former condition, essentially unchanged (James, 1974). But Goodsell (1988) probed deeper and found Churchill's remarks rooted in the same deeper stratum that this chapter has attempted to prospect and then mine: the institutions of society.

Goodsell (1988) proposed that Churchill was very "likely echoing Montesquieu's words, 'At the birth of societies, the rulers of republics establish institutions; and afterwards the institutions mold the rulers'" (p. 209, fn. 7). This chapter has singled out for consideration the institutions that link the design of environments to the larger purposes and the wider aspirations of the societies doing the building. But in our case, the "rulers" who establish the institutions are neither benevolent princes nor tyrannical despots: our rulers are, on the contrary, drawn from among the ruled. The institutions that write the rules for environmental design are accessible to those equipped with two things: knowledge of environments, their origins and their effects, and the will to engage (and thereby test) that knowledge in the world of affairs. Rules will be written: they are, after all, the things that help keep societies intact and that have left us with the most stable political system in the modern world. All that remains to be decided is who will write those rules and to what ends, a decision that will never, ever, be final.

REFERENCES

Anderson, F. R. (1973). *NEPA in the Courts: a legal analysis of the National Environmental Policy Act.* Baltimore: Johns Hopkins University Press and Resources for the Future.

Architectural and Transportation Barriers Compliance Board. (1983, March). *Technical paper: Orientation and wayfinding.* Washington, DC: Author.

Association for the Study of Man-Environment Relations. (1978). *A review of federal mandates related to environmental design and the quality of the built environment, with implications for environmental design research* Bethesda, MD: Author. (Available from the National Technical Information Service, Springfield, VA 22161.)

Brace, P. (1980). Comment: urban aesthetics and the courts—A review of current judicial opinions on community appearance. *The Urban Lawyer: The (ABA) National Quarterly of Local Government Law,* 12:1, 151–156.

CABO assumes A117 committee secretariat. (1987). *ANSI Reporter* (May 8, 1987), p. 1.

Carswell, J. W., & Saile, D. G. (Eds.). (1986, November). *Proceedings of the 1986 International Conference on Built-Form and Culture Research: Purposes in Understanding Sociocultural Aspects of Built Form.* Lawrence: School of Architecture and Design, University of Kansas.

Collins, B. (1983). *Use of hazard pictorials/symbols in the minerals industry,* NBSIR 83-2732. Washington, DC: National Bureau of Standards/Center for Building Technology. (Available from NTIS, Springfield, VA 22161.)

Collins, B., & Lerner, N. (1983). *An evaluation of exit symbol visibility,* NBSIR 83-2675. Washington, DC: National Bureau of Standards/Center for Building Technology. (Available from NTIS, Springfield, VA 22161.)

Collins, B., Lerner, N., & Pierman, B., (1982). *Symbols for industrial safety,* NBSIR 82-2485.

Washington, DC: National Bureau of Standards/Center for Building Technology. (Available from NTIS, Springfield, VA 22161).

Congressional Budget Office. (1983). *The federal buildings program: Authorization and budget alternatives.* Washington, DC: Congress of the United States.

Crandall, R. W., & Lave, L. B. (Eds.). (1981). *The scientific basis of health and safety regulations.* Washington, DC: Brookings Institution.

Current Population Reports. (1986). *Marital status and living arrangements: March 1985,* Population Characteristics, Series P-20, No. 410. Washington, DC: Bureau of the Census.

Dean, A. O. (1987, January). The state of the art of design for accessibility. *Architecture, 76*(1), 58–61.

Dewey, J. (1929). *Quest for certainty: A study of the relation of knowledge to action* New York: Putnam.

Edelman, M. (1977). *Political language: words that succeed and policies that fail.* New York: Academic.

Environmental Design Research Association. (1974). C. P. Wolff (Ed)., *Man–environment interactions,* Vol. 2: *Social impact assessment.* Milwaukee: University of Wisconsin–Milwaukee

Expert witnesses: booming business for the specialists. *New York Times* (July 5, 1987), p. 1.

Field, C. G. (1980). *Building regulatory practices and the courts.* Report prepared for Federal Trade Commission, Office of Policy Planning. Washington, DC: Federal Trade Commission.

Finsterbusch, K., & Wolf, C. P. (Eds.). (1981). *Methodology of social impact assessment* (2nd ed.). Stroudsburg, PA: Hutchinson Ross.

A first time (sic) study of the efficiency of symbols. (1987). *ASTM Standardization News* (May, 1987), p. 13.

Fisher, T. (1985, July). Enabling the disabled. *Progressive Architecture, 66*(7), 119–124.

Frankena, F. (1985). *The social science practitioner as expert witness: A bibliography.* Monticello, IL: Vance Bibliographies.

Freudenberg, W. R. (1986). Social impact assessment. *Annual Review of Sociology, 12,* 451–478.

Galligan, D. J. (1986). *Discretionary powers: A legal study of official discretion* New York: Oxford University Press.

Giraldo, Z. I. (1978). *Tax policy and the dual-income family: The "marriage tax" and other inequities—A family impact statement,* Policy Paper #3. Durham, NC: Center for the Study of Family and the State, Duke University.

Goodsell, C. T. (1984, Winter). The city council chamber: From distance to intimacy. *The Public Interst, 74,* 116–131.

Goodsell, C. T. (1988). *The social meaning of civic space: studying political authority through architecture.* Lawrence: University Press of Kansas.

GSA commits to cost savings through systems furniture. (1987). *Building Economics,* 2(1), 1–2.

Heilstedt, P. K. (1986, June). Codes and standards. *ASTM Standardization News,* 36–39.

James, R. R. (Ed.). (1974). *Winston S. Churchill: His complete speeches, 1897–1963,* Vol. 7. New York: Chelsea House.

Jones, B. D. (1985). *Governing buildings and building government: New perspective on the old party.* University: University of Alabama Press.

Kester, J. G. (1984, February). Too many lawyers. *The Washingtonian, 19*(5), p. 114.

Low, S. M. (Ed.). (1987, March). *Cultural Aspects of Design Network Newsletter.* (Available

from Graduate School of Fine Arts, University of Pennsylvania, Philadelphia, PA, 19104-6311).
Margulis, S. T. (Ed.). (1986). Symposium on building regulations and systematic inquiry. In J. Wineman, R. Barnes, & C. Zimring (Eds.), *Proceedings of the seventeenth annual conference of the Environmental Design Research Association*. Washington, DC: EDRA.
Moore, G. T., Hill, A. B., Lane, C. G., Cohen, U., & McGinty, T. (1979). *Recommendations for child care centers*. Milwaukee: University of Wisconsin–Milwaukee, Center for Architecture and Urban Planning Research.
Nader, L., & Nader, C. (1985). A wide angle on regulation: An anthropological perspective. In R. G. Noll (Ed.), *Regulatory policy and the social sciences*. Berkeley: University of California Press.
National Institute of Building Sciences. (1981). *Federal regulatory programs that affect the building process*. Washington, DC: Author.
National Institute of Building Sciences. (1985). *Proceedings of the International Conference on Building Use and Safety Technology*. Washington, DC: Editor.
New PBS commissioner named. (1987). *Building Design and Construction* (June 1987), p. 5.
Office of Management and Budget. (1985, August). *Regulatory program of the United States Government*. Washington, DC: U.S. Government Printing Office.
Office of Technology Assessment. (1987). *International competition in services*. Washington, DC: Congress of the United States.
Paralyzed Veterans of America. (1987). *Tax deductions for improving accessibility*. (Seven-page brochure available from PVA, 801 Eighteenth St. NW, Washington, DC 20006.)
Preiser, W. F. E., & Rohane, K. P. (1980). *Survey of architectural control mechanisms in western industrialzed countries*. (Available from Center for Research and Development, School of Architecture and Planning, University of New Mexico, Albuquerque, NM 87131.)
The psychology of the designed environment: NEPA and public housing. (1975). *Iowa Law Review, 60*(3), 674–692.
Rescher, N. J. (1977). *Methodological pragmatism: A systems-theoretic approach to the theory of knowledge*. New York: New York University Press.
Sanderson, R. L. (1969). *Codes and code administration: An introduction to building regulation in the United States*. Chicago: Building Officials and Code Administration, International.
Schmalz, J. (1986). Appeals court bars Chinatown building till effect is studied. *The New York Times* (November 19, 1986), p. A1.
Schuchman, H. L. (1980, May). *Information transfer in engineering*. Glastonbury, CT: The Futures Group.
Schwartz, M., & Schwartz, N. F. (1981). *Engineering evidence*. Colorado Springs, CO: Shepard's/McGraw-Hill.
Shoney's South, Inc. and Shoney's Southwest, Inc. v. *Commissioner*. (1984). Docket No. 16361-80. U.S. Tax Court Memorandum 1984-413. (Filed August 6, 1984).
Simpson, J. W. (1985, January–February). Environmental planning in the backwash. *Landscape Architecture, 75*(1), 72–77.
Social impact added to project reviews. (1986, November). *ENR 27*, 16.
Special report: Deregulating America. *Business Week* (November 28, 1983), pp. 80–96.
Stahl, F., Crosson, J., & Margulis, S. (1982). *Time-based capabilities of occupants to escape fires in public buildings: A review of Code provisions and technical literature*, NBSIR 82-2480. (Available from NTIS, PB82-212887.)
Steinfeld, E. (1986). ANSI A117.1, a case study in the development of a research-based building accessibility standard. In Margulis, (1986).

Surrey, S. S., & Daniel, P. R. (1985). *Tax expenditures.* Cambridge: Harvard University Press.

Taylor, P., & Turley, S. (1986) *The regulation of accounting.* New York: Basil Blackwell.

Ventre, F. T. (1975). Transforming environmental research into regulatory policy. In B. Honikman (Ed.), *Responding to social change.* Stroudsburg, PA: Dowden, Hutchinson & Ross.

Ventre, F. T. (1979). How institutions shape a research style. *Journal of Architectural Education, 32*(4), 16–19.

Ventre, F. T. (1982). Building in eclipse, architecture in secession. *Progressive Architecture, 63*(12), 58–61.

Weisman, G. (1985a). Orientation, path finding, and architectural legibility: A review and theoretical integration. In *Proceedings of the international conference on building use and safety technology, 1985* Washington, DC: National Institute of Building Sciences.

Weisman, G. (1985b). Spatial orientation, way finding, and emergency egress behavior: Implications for architectural design and life safety standards. In *Proceedings of research and design 85: Architectural applications of design and technology research.* Washington, DC: American Institute of Architects Foundation.

Werner, L. M. (1986). U.S. Report Asserts Administration Halted Liberal "Anti-Family Agenda." *The New York Times.*

Contents of Previous Volume

Volume 1

Index